PUBLIC TECHNOLOGY PROCUREMENT AND INNOVATION

Economics of Science, Technology and Innovation

VOLUME 16

The titles published in this series are listed at the end of this volume.

PUBLIC TECHNOLOGY PROCUREMENT AND INNOVATION

Edited by:

Charles Edquist, Leif Hommen and Lena Tsipouri

KLUWER ACADEMIC PUBLISHERS
Boston/Dordrecth/London

Distributors for North, Central and South America:
Kluwer Academic Publishers
101 Philip Drive
Assinippi Park
Norwell, Massachusetts 02061 USA
Telephone (781) 871-6600
Fax (781) 871-6528
E-Mail <kluwer@wkap.com>

Distributors for all other countries:
Kluwer Academic Publishers Group
Distribution Centre
Post Office Box 322
3300 AH Dordrecht, THE NETHERLANDS
Telephone 31 78 6392 392
Fax 31 78 6546 474
E-Mail <services@wkap.nl>

Electronic Services <http://www.wkap.nl>

Library of Congress Cataloging-in-Publication Data

Public technology procurement and innovation / edited by Charles Edquist, Lief Hommen, and Lena Tsipouri.
p.cm.--(Economics of science, technology, and innovation ; v. 16)
ISBN 0-7923-8685-X (alk. paper)
1.Technological innovations--Government policy--Sweden.2.Technological innovations--Government policy--European Union countries.3.Government purchasing--Sweden.4.Government purchasing--European Union countries. I.Edquist, Charles, 1947-II.Hommen, Leif.III.Tsipouri,Lena J., 1953-IV. Series.

HC380.T4 P83 1999
338'.0644'09485--dc21 99-046682

Printed on acid-free paper.
Printed in the United States of America

CONTENTS

3. Procuring Products and Power: Developing International Competitiveness in Swedish Electrotechnology and Electric Power – *M. Fridlund*

4. A Case Study of the Swedish Public Technology Procurement Project "The Computer in the School" (COMPIS), 1981-1988 – *T. Kaiserfeld*

5. Switching Relations and Trajectories: The Development Procurement of the AXE Swedish Switching Technology – *M. Fridlund*

10. On Implementing the Austrian Computerised Digital Switching System (OES) – *M. Husz*

11. Analysis, Findings and Conclusions – *C. Edquist, L. Hommen and L. Tsipouri*

12. Policy Implications – *C. Edquist, L. Hommen and L. Tsipouri*

List of Contributors

Charles Edquist — Professor, Department of Technology and Social Change, University of Linköping, Sweden.

Mats Fridlund — Researcher, Department of History of Technology, Royal Institute of Technology (KTH), Stockholm, Sweden.

Per Hammarqvist — Former research assistant, Department of Technology and Social Change, University of Linköping, Sweden. Currently a consultant with Cap Gemini, Stockholm, Sweden.

Leif Hommen — Researcher, Department of Technology and Social Change, University of Linköping, Sweden.

Martin Husz — Former researcher, Department of Technology Studies, Austrian Research Centre, Seibersdorf, Vienna, Austria. Currently an independent entrepreneur in Vienna, Austria.

Thomas Kaiserfeld — Researcher, Department of History of Technology, Royal Institute of Technology (KTH), Stockholm, Sweden.

Patrick Llerena — Professor, Bureau d'Économie Theorique et Appliquée (BETA), Louis Pasteur University, Strasbourg, France.

Mireille Matt — Researcher, Bureau d'Économie Theorique et Appliquée (BETA), Louis Pasteur University, Strasbourg, France.

Christopher Palmberg — Researcher, Group for Technology Studies, Technical Research Centre of Finland (VTT), Helsinki, Finland.

Stefania Trenti — Former researcher, Bureau d'Économie Theorique et Appliquée (BETA), Louis Pasteur University, Strasbourg, France. Currently a doctoral candidate in the Institute of Political Economy, Bocconi University, Milan, Italy.

Lena Tsipouri — Professor, Centre for Financial Studies, University of Athens, Greece.

PREFACE

This book studies public technology procurement as an instrument of innovation policy. In the past few years, public technology procurement has been a relatively neglected topic in the theoretical and research literature on the economics of innovation. Similarly, a preoccupation with 'supply-side' measures has led policy-makers to avoid making very extensive use of this important 'demand-side' instrument. These trends have been especially pronounced in the European Union. There, (as will be argued in this book) existing legislation governing public procurement presents obstacles to the use of public technology procurement as a means of stimulating and supporting technological innovation. Recently, however, there has been a gradual re-awakening of practical interest in such measures among policy-makers in the EU and elsewhere. For these and other, related reasons, this volume is meant to contribute to a serious reconsideration of public technology procurement from the complementary standpoints of innovation theory and innovation policy.

The book originates from studies initially planned and carried out as a sub-project of the Innovation Systems and European Integration (ISE) research project, which ran from 1996 to 1998. ISE was supported by the European Commission (EC), through funding for Targeted Socio-Economic Research (TSER) provided under the EC's Fourth Framework Programme (Contract no. SOE1-CT95-1004 DG XII SOLS). Charles Edquist was coordinator of both ISE and its empirical sub-project on Public Technology Procurement as an Innovation Policy Instrument. ISE involved nine research groups in nine European countries. It had the overall purpose of elaborating the Systems of Innovation approach with respect to: 1) evaluating this new understanding of the development of science, technology, innovation and the implications thereof for economic growth and employment; and 2) developing new policy options and implications on this basis. The empirical sub-project on Public Technology Procurement was particularly successful in fulfilling these purposes. This volume reflects the sub-project's achievements in responding constructively to difficult theoretical issues and policy problems with conclusions based on well-researched empirical evidence.

As editors of this book, we would like to acknowledge and thank a number of people. We are, of course, indebted to the European Commission for the funding which made our research possible. The authors who have contributed to this book, and who were also participants in the ISE sub-project on Public Technology Procurement, deserve special mention and thanks for their co-operation in what was truly a collective research effort. In addition, we owe a debt of gratitude to all of our other colleagues in ISE, from whose collegial advice and assistance we were able to benefit. We are also grateful to Simon Ivarsson, François Texier, and other members of the Department of Technology and Social Change at the University of Linköping, who lent us valuable assistance in producing this book.

Charles Edquist and Leif Hommen
Linköping, Sweden, 1999

Lena Tsipouri
Athens, Greece, 1999

INTRODUCTION

C. Edquist, L. Hommen and L. Tsipouri

In this volume we discuss theory, evidence, and policy perspectives concerning the use of public technology procurement as an instrument of innovation policy. Public technology procurement (as defined in Chapter 1, part 1), occurs when a public agency places an order for a product or system that does not yet exist, requiring technological innovation for the order to be met.

The most extensive part (Part II) of this volume presents empirical material, in the form of nine case studies. These studies document important instances of success and failure in public technology procurement in various countries. The remainder of the book (Parts I and III) considers theoretical, analytical, and policy questions.

The opening part (Part I) of the book provides a review of theoretical and policy issues related to public technology procurement. The initial chapter devotes considerable attention to definitional questions. It also relates a broad theoretical and conceptual framework to problems that are of specific relevance to policy development.

The central part (Part II) of the book contains our collection of empirical case studies. An Overview of the case studies is provided at the beginning of Part II. In addition to briefly summarising the cases, the Overview discusses our rationale for the selection of cases and addresses the matter of making valid analytical generalisations from a 'purposive sample' of cases.

The closing part (Part III) of the book returns to questions raised at the outset. First, it presents findings and conclusions drawn from a comparative analysis of the empirical case studies. It then outlines policy implications developed on this basis. In what follows, we will briefly foreshadow the policy issues that will be addressed.

1. SYSTEMS OF INNOVATION: A DEMAND-SIDE APPROACH

Our work builds upon a particular theoretical perspective – the so-called 'systems of innovation approach' (Edquist, 1997). The authors represented in this book have all used variants of this approach in their contributions to the description, analysis, and understanding of public technology procurement. Our work is also motivated by eminently practical concerns, which are emphasised in systems approaches to the study of innovation. In particular, we hope to redress an imbalance in current innovation policy-making.

Recently, there has been an over-riding concern with the development of 'supply-side' policies and a general neglect of the 'demand-side'. The demand-side, we contend, is of vital importance to the development of innovations and innovative economies (Edquist & Hommen, 1999). The examination and evaluation of public technology procurement – one of the main demand-side instruments available to innovation policy – provides a natural focus for developing this argument.

Although we proceed from a common theoretical standpoint, the findings and conclusions that we will present at a later point, together with the policy implications that flow from them, depend upon the analysis of empirical data. This takes the form of the aforementioned case studies of public technology procurement, which are presented in Part II of this book. Our analytical framework and strategy are discussed in Part III, at the beginning of Chapter 11.

In the remainder of our introductory remarks we will outline some of the ideas and debates addressed in this book. We will relate the study and practice of public technology procurement to general theoretical, conceptual, and policy issues that are elaborated more fully in Chapter 1.

2. THEORETICAL, CONCEPTUAL, AND POLICY ISSUES

Our analysis of the case studies (in Chapter 11) involves the concepts of 'developmental' and 'adaptive' public technology procurement (see Chapter 1, section 2.2). 'Developmental' public technology procurement means that (completely) new products or systems are created.[1] In 'adaptive' public technology procurement, the product or system procured is not new to the world but still new to the country of procurement. It must still be adapted to specific 'local' conditions, and this involves innovation.[2]

The purpose of these and other terms used in our work is to identify elements and modes of interaction related to innovation in the public technology procurement process (not applicable in regular, non-innovative, public procurement, i.e., the public purchasing of standard products "off the shelf"). These categories are used to suggest new patterns of policy making based on concepts that go beyond auction theory. As we explain more fully in Chapter 1 (section 1.4), auction theory is the 'conventional wisdom' drawn from standard economic theory that normally informs the regulation of public procurement processes.

2.1 Direct and Indirect Innovation Policy

Two modes of innovation policy action become relevant for our purposes: indirect and direct policies.[3] Indirect policies are those "which are closely related to the public in-

---

[1] This could also be called 'creation-oriented' public technology procurement.

[2] This could also be called 'diffusion-oriented' or 'absorption-oriented' public technology procurement.

[3] *Innovation policy* includes all public activities which influence innovation. These involve elements of R&D policy, education policy, infrastructure policy, regional policy and industrial policy.

fluence on ... framework conditions", including the institutional context, of innovation processes (Edquist et al., 1998: 43). Direct policies are initiatives or interventions in the creation or development of new technology "where a public organisation is directly involved in the interaction" (Edquist et al., 1998: 43). In public technology procurement these two kinds of policy action should ideally complement and be co-ordinated with one another. In practice, however, they might become opposed to one another. For example, a public agency might take the direct measure of procuring a particular technology and succeed in meeting its immediate goals with respect to satisfying social needs. At the same time, however, such an initiative might fail to develop national capacities for innovation in this technology, thereby forfeiting the competitive advantage that might have been gained through the procurement. This might, for instance, occur where the public procurement of new technology is not combined and co-ordinated with more 'indirect' policies that would aid the formation of innovation networks and the growth of competent firms by "supporting the knowledge-generating process in public research organisations and universities" (Edquist et al., 1998: 44).

In this book, we focus on an emergent conflict between the direct public technology procurement practices of individual countries and indirect policies affecting the institutional framework within which such procurement is carried out, i.e., the creation of public procurement rules by the EU. In the case of the European Union, the latter kind of policy has been framed and introduced at a supra-national level, although it has subsequently been implemented at the national level of individual member states. In the EU procurement regime – the rules governing public procurement in general – public technology procurement in the member states has been associated, at least implicitly, with protectionist industrial policies. This is the basis of the potential conflict to which we refer.

The EU policy perspective has concentrated almost exclusively on indirect measures that will create more *competition*. In contrast, we focus on public technology procurement as a means of building or enhancing capacities for *innovation*, thereby making firms and nations more *competitive*. Thus, we contradict the current orthodoxy which identifies any attempt to use public procurement as an ipso facto leverage for protectionism affecting optimal allocation of resources. Within the EU such attempts are considered as threatening the Single Market.

2.2 Competition Rules

Public procurement has been widely used as an industrial policy instrument in the past. This practice found some of its earliest justifications in the 'infant industry' arguments of List in the 19th century in Germany (List, 1904). The practice of public *technology* procurement was developed substantially in the 20th century. Like 'regular' public procurement, it was particularly relied upon when combined with infant industry arguments. It was also especially prevalent where the public sector was the sole or lead user and when substantial growth in international demand or spillovers were anticipated. This has led M. Best (Best & Forrant, 1996) to consider the public technology procurement of the US Department of Defence as the basis of American industrial policy.

In contrast, most industrial policies in the EU in the 1980s involved a protection in public procurement of, for example, the electronics industries. But all these policies of EU member states during the neo-protectionist period of the oil crises were dictated by the need to create national champions. Technological considerations were indirect and secondary, and this policy often proved disastrous for the countries and companies it was trying to protect (Pitelis, 1994).

This kind of industrial policy, widely attacked in the Cecchini Report (WS Atkins Management Consultants & Associates, 1988), triggered a negative perception of policies for public procurement. This, in turn, has led to suspicion (if not outright rejection) of any public procurement conducted on a 'non-auction' basis, regardless of the target and the nature of the procurement.

2.3 Deciding under Uncertainty: A Case for New Rules

Our purpose in this volume is to reposition the idea of public technology procurement in relation to innovation policy. We seek to separate public technology procurement policies from earlier industrial policy connotations. We also view the element of competition in a dynamic rather than static perspective. Competition rules are clear for existing products and may also be clear when asymmetric information is concerned. But here we deal with *procurement under uncertainty*.

In theory, awards of public technology procurement contracts involve fairly clear and definite agreements about costs and functional specifications. In practice, however, the final price and quality parameters of new and innovative technologies are unknown at the moment the contract is awarded. This changes the rules of the game.

REFERENCES

Best, M. H., & Forrant, R. (1996). Creating industrial capacity: Pentagon-led versus production-led industrial policies. In J. Michie & J. G. Smith (Eds.), *Creating industrial capacity*. Oxford: Oxford University Press.

Edquist, C. (1997). Systems of Innovation approaches: Their emergence and characteristics. In C. Edquist (Ed.), *Systems of innovation: Technologies, organisations and institutions*. London: Pinter Publishers/ Casell Academic.

Edquist, C., & Hommen, L. (1999). Systems of innovation: Theory and policy for the demand side. *Technology In Society, 21*, 63 - 79.

Edquist, C., Hommen, L., Johnson, B., Lemola, T., Malerba, F., Reiss, T., & Smith, K. (1998). *The ISE policy statement: The innovation policy implications of the 'Innovation Systems and European Integration' research project*. Linköping, Sweden: Unitryck (University of Linköping Press).

List, F. (1904). *The national system of political economy*. New York: Longmans Green.

Pitelis, C. (1994). Industrial strategy: For Britain, in Europe and the world. *Journal of Economic Studies, 21*(5).

WS Atkins Management Consultants, & Associates. (1988). *The "cost of non-Europe" in public sector procurement*. (Vol. 5). Luxembourg: Office for Official Publications of the European Communities.

1. PUBLIC TECHNOLOGY PROCUREMENT AND INNOVATION THEORY

C. Edquist and L. Hommen

1. INTRODUCTION

Public technology procurement (PTP) occurs when a public agency places an order for a product or system which does not exist at the time, but which could (probably) be developed within a reasonable period. Additional or new technological development work is required to fulfil the demands of the buyer. This is the 'ideal type' of public technology procurement.

In contrast to 'public technology procurement', regular 'public procurement' occurs when public agencies buy ready made 'simple' products such as pens and paper – where no R&D is involved. Only price and performance of the (existing) product is taken into consideration when the supplier is selected.

Public technology procurement can also be contrasted with private technology procurement. Both processes involve a buyer's purchase from a supplier of a not-yet-existing product or system whose design and production will require further, if not completely novel, technological development work. The main difference between the two kinds of technology procurement is that in private technology procurement, the buyer is a private business organisation, not a public agency.[1]

In capitalist economic systems, where markets are effective mechanisms for articulating and satisfying most economic needs or demands, the point of departure in the application of public technology procurement must be the satisfaction of *genuine* social needs – in other words, specific societal needs unlikely to be met by the market. The products and systems that are developed – and the technical change that enables their provision – as the result of public technology procurement must therefore be targeted to solve specific problems.

On the basis of the socio-economic needs identified, the procuring agency specifies the 'functional demands' of the product or system required to satisfy these needs.

[1] We will deal extensively with private technology procurement in part 3 of this chapter, below.

This demands a highly developed competence on the part of the procuring organisation. Close collaboration between procurers and suppliers is also a necessary element in effective technology procurement.

Public technology procurement has been used as an instrument of innovation policy in the defence material sector in many countries. It has also proven to be an extremely potent means of influencing the speed and direction of innovation on the civilian side in some countries – usually in traditional infrastructure development. In addition, it has served to enhance the competitiveness of those firms successfully meeting the demands and thereby winning the contracts. Such procurement was, for example, highly instrumental in the consolidation of some firms of a Swedish origin which are currently large and international. Examples are Ericsson and ABB.[2]

In spite of its great practical value as an innovation policy instrument, public technology procurement has not received a commensurate level of serious attention in the research and theoretical literature on innovation. In economic policy making, moreover, there now appears to be a corresponding disregard for public technology procurement. Recent policy developments in the European Union, in particular, indicate that this situation is not simply the consequence of oversight, but is to a large extent the product of systematic neglect.[3] For policy makers and public authorities in the European Union, the long-term result of such neglect could be a seriously diminished capability for initiating processes of economic growth based on innovation.

The project of economic integration, central to the policy agenda of the European Union for the past decade, has focused on the creation of a common framework for economic activity. Both the Single European Act and the Maastricht Treaty have sought to lay key foundations for a convergence in real incomes and living standards across the countries and regions of the Union. This was to be achieved through means such as completion of the internal market, common regulatory frameworks, and ultimately a common currency. In this policy context, stimulating innovation has figured less importantly than achieving economies of scale, and only 'regular public procurement', in the sense discussed above, has been the focus of debate and legislation. Public technology procurement has largely been 'exempted' from the policy arena.

In this discussion, we will present the case for a serious reconsideration of public technology procurement as an instrument of innovation policy, making special reference to the current situation in the European Union and its member states. We will begin, in this first part, by outlining the main elements of past and present debates about public technology procurement, the current EU policies regarding public procurememfi, and their underlying theoretical rationale. In subsequent parts of the discussion we will develop an alternative perspective, based on theory and research concerning the economics of technological innovation. We will proceed from developing a typology of different kinds of public technology procurement and their relation to other kinds of innovation policy to considering the nature of (non-public) technology procurement as

---

[2] Public technology procurement has been dealt with previously elsewhere (Edquist, 1994; Edquist, 1996).

[3] This situation may be changing. Very recently, DG XVII of the European Commission has begun to discuss the use of public technology procurement as a policy instrument.

an economic phenomenon. Returning to our main topic, we will elaborate a general theoretical perspective on public technology procurement – a perspective that can serve as the basis for its utilisation as an instrument of innovation policy.

1.1 Debating Public Technology Procurement

The 'exemption' of public technology procurement from policy making and policy debates in the European Union is clearly indicated by the absence, in the recent EU "Green Paper" (European Commission, 1997) on public procurement policy, of any reference to the role of governments and public agencies in stimulating innovation through public technology procurement.[4] Instead, the "Green Paper" deals mainly with regular public procurement from the standpoint of achieving savings in public expenditure and introducing competitive conditions into the award of public supply contracts. The regulatory framework already created by the EU for these purposes is simply reaffirmed.

The current system of EU procurement rules and their enforcement by the EU and member states were developed in the early 1990s, as a result of both the establishment of a single European market and the need to rationalise and enforce effectively previously existing EU procurement regulations. The objectives were to eliminate artificial barriers to trade and reduce unnecessary differences in regulations. The EU policy regarding public technology procurement has been contained within this broader initiative, which has been almost exclusively concerned with the *regulatory* aspect of policy – i.e., the rules governing public procurement, including public technology procurement. However it is possible to identify another policy dimension that has been largely ignored. It concerns the *strategic* aspect – i.e., the use of public technology procurement as an instrument of innovation policy by the EU or by national government agencies.

When it is carried out by public agencies, procurement is not simply an economic but also a political phenomenon.[5] It is not surprising, therefore, that the development of EU policy on public procurement has been ideologically charged to some extent. Generally speaking, two ideologies have been counterposed to one another: a 'free market' orientation, which "emphasises the need to exclusively apply commercial criteria when awarding the contract" and an 'interventionist' orientation, which "regards public procurement as an instrument to realise social and economic objectives wider than mere efficiency in the use of public money" (Martin 1996: 41).

Of these two approaches to procurement, the first, or 'free market', orientation has clearly been the dominant influence in the development of the EU rules regarding public procurement. It finds a theoretical basis in classical 'economic efficiency' principles

---

[4] The only exception to this rule is a discussion of the potential role of public procurement as a vehicle for the achievement of environmental policy objectives. Even in this context, however, the main emphasis is on regulatory standards. Technological innovation, as such, is not mentioned. (European Commission, 1997: chap. 5 - VI)

[5] The political and institutional dimensions of procurement are clearly indicated by the existence of various national models of public technology procurement, as discussed below in section 1.2 of this chapter.

that regard the maximisation of competition as the chief means of reducing costs, increasing public savings, and guaranteeing the 'fair and equitable' use of taxpayers' money (Ponssard and Pouvourville 1982: 6 - 7). This position has been buttressed by a large body of economic research and analysis indicating that broader competition in tendering or bidding on public contracts in the European Union will not only result in public savings but also promote macro-economic growth through restructuring and adjustment (Baldwin, 1987; Finsinger, 1988; McLachlan, 1985; Parker, 1990; Tovias, 1990; Utterly & Harper, 1993).

The second, or 'interventionist', orientation in public procurement has been severely limited by the EU procurement rules, though some allowances are still made for it. Although alien to the spirit of the EU rules, it is highly consistent with the past practices of most national governments of EU member states. Many of them, especially under the influence of Keynesian doctrines during the postwar period (Roll, 1982), have made extensive use of public procurement, including PTP (Public Technology Procurement), as an instrument of socio-economic policy, including industrial policy (Geroski, 1990).

It is possible to distinguish at least five instrumental uses of this general approach to the use of public procurement as a tool of policy (Jeanrenaud, 1984). First, procurement (though not necessarily PTP) has been used as a means of increasing overall demand and stimulating economic activity, thereby creating employment (Keyzer, 1968). Second, it has been used to protect national industry against foreign competition (Goodman & Saunders, 1985; McLachlan, 1985). Third, procurement has been used to improve the competitiveness of certain industrial sectors, by linking secure access to public supply contracts to commitments on the part of national champions to invest in R&D (Jeanrenaud & Meyer, 1984). Fourth, it has been used to remedy regional disparities, so as to attain redistributory objectives (Jeanrenaud, 1984). A fifth use is that of using procurement to create jobs for marginal sections of the labour force (McCrudden, 1994).

Of these five instrumental uses of public procurement, nearly all have in effect been rejected in principle, if not entirely in practice, by the EU regime governing public procurement. From an integration policy perspective, "the Commission has followed a straightforward economic efficiency approach to the regulation of public procurement, thereby neglecting other perspectives applied at national level" (Martin 1996: 89). However, some recognition has been afforded to the third use, that of enhancing "competitiveness". This may be largely due to the fact that this use of procurement is most consistent with the general principles and aims of the economic efficiency approach, which ultimately seeks to promote economic growth through improving the competitiveness of firms and nations.

A concern with the competitiveness of European industry and firms based in EU member states can be said to be fundamental to the EU regime for public procurement. The Cecchini Report, which ushered in the current EU procurement regime, gave considerable attention to the importance of improving "the rate of innovation, investment and growth in the sectors enjoying the benefits of restructuring, with positive effects on their international competitiveness" (WS Atkins Management Consultants and Associates 1988: 7). The report stressed viewing competition policy on an EU-wide basis, permitting both national and transnational mergers so that world-league EU firms could be built (Business International Ltd. 1991: 6 - 7). In this respect, the EU can be said to remain committed to "competitiveness" as an instrumental use of procurement, particularly in sectors where "European champions" may now be required to replace national ones (Hartley, 1987).

In addition to its 'competitiveness' ethic and ultimate concern with fostering economic growth, the EU procurement regime has, in practice, been required to make some concessions to national-level public interest. In case law, it has had to achieve compromises between the forces of economic cohesion and those of market integration (Snyder, 1990). And, more generally, it has had to recognise that in certain contexts, "overwhelming importance granted to the pursuit of efficiency is regarded as detrimental to wider administrative and political aims" (Parsons, 1988).

Where the 'wider administrative aims' in question are those of enhancing competitiveness in the form of innovation, this may involve the limitation or even suspension of 'competition'. The EU procurement regime is therefore largely permissive, rather than restrictive, in such cases – providing that the argument for competitiveness can be made. Even though EU policy may have altered some of the basic 'rules of the game', public technology procurement remains an area in which national governments and public agencies continue to enjoy considerable autonomy. Thus, the continued existence of various national models of PTP should not be considered moribund or irrelevant in the context of European integration.

The EU has neglected the *strategic* dimension of public technology procurement, which has been dealt with only by relegating it to the national level. Even at this level, moreover, the EU has not directly addressed the 'interventionist' aspects of public technology procurement, except in a negative way. The EU procurement rules have, it is true, allowed for the continuation of public technology procurement in public supply contracts through certain special tendering procedures, permissible exemptions from the regular procurement rules, and a flexible regime of enforcement (Westling, 1996: Appendix 5).[6] However, this has been done without an explicit policy rationale – only the implicit understanding that these are necessary accommodations of national and sectoral interests.

Concerning the *regulatory* dimension, EU legislation has paid considerable attention to the establishment of competitive market conditions affecting the technology procurements of public agencies. Competition has been strengthened through the reduction of protectionism, the imposition of requirements for more transparent tendering procedures in formerly 'excluded' sectors (water, energy, transport and telecommunications), and complementary initiatives aimed at the liberalisation of publicly owned 'natural monopolies' (Business International Ltd., 1991). These measures have diminished the number of sectors in which public technology procurement can be readily used as an instrument of innovation policy, and it has discouraged this from being done as a matter of course, even in those sectors where it remains possible. The present EU legislation implicitly regards such intervention as an aberration – a deviation from the legalised norm of anonymous 'market' relationships.

---

[6] See the further discussion in section 1.4 of this chapter, below.

Given these circumstances, it is not immediately possible to open a dialogue with current EU policy on public technology procurement. Several preliminary steps are required. First, it is necessary to refer to practical and theoretical arguments concerning the interventionist uses of public technology procurement which have hitherto been excluded from debates concerning the development of the EU regime governing public procurement. Second, it is necessary to investigate the theoretical rationale which underlies the current EU policy and motivates its neglect of innovation. This rationale can then be counterposed to an alternative perspective.

Later parts of this discussion are devoted to an examination of the theoretical and research literature concerning technology procurement (including technology procurement in the private sector), and the arguments that are offered for and against the use of public technology procurement as an instrument of innovation policy. The last section (1.4) of this part of the chapter is largely concerned with the theoretical rationale underlying the present EU procurement regime. Before proceeding to take up that topic, it will be appropriate to refer, at least briefly, to arguments about the actual practice of public technology procurement in different national contexts. These arguments, reviewed in the next two sections (1.2 and 1.3) of this chapter, foreshadow the alternative perspective that will be developed in subsequent parts of the chapter.

1.2 Practices of Public Technology Procurement

Implicitly, at least, the EU regime governing public procurement regards public technology procurement as a practice at variance with sound economics, and for this reason merely tolerates its practice at the national level. Accordingly, it can be fairly stated that EU policy does not directly address public technology procurement. To understand the reasons for this conscious 'oversight' it is useful to consider some of the main features of the practice of public technology procurement in EU member states and in other countries from an historical and comparative perspective.

In many Western European countries, there has been a relative lack of reference to technological innovation in public procurement policies. Where this condition has been present – as, for example, in countries such as Sweden – much of the national effort with respect to public technology procurement has often been concentrated in defence and other strategically important public sector domains. However, there have been relatively weak attempts to encourage the commercialisation of results, by such means as requiring contractors to diffuse research results and promoting the development of spin-off companies capable of commercially exploiting the new technologies developed through PTP.

Instead, the main focus has been on developing close relationships between public agencies and large firms deemed capable of becoming 'national champions'. As we will illustrate below, this policy orientation has not always worked to the benefit of the national economies involved. It is not surprising, therefore, that current EU policies regarding public procurement express little interest or confidence in public technology procurement as a potential engine of economic growth.

With respect to the problematic economic behaviour of 'national champions', the experience of Sweden is typical of the more general pattern in Western European countries that are now EU member states. In Sweden, public technology procurement has been characterised by informed public purchasing and the exercise of a fairly high level of technological capacity (or 'user competence') on the part of clients during the strategic stages of technical specification and product development. These features of PTP have, for example, been credited with playing an important role in the successful development of the telecommunications industry in the Nordic countries (Granstrand & Sigurdson, 1985). The fairly high success rate of public technology procurement in promoting industrial development in Sweden has meant that PTP has been widely accepted by the general public, by a large segment of industry, by trade unions and by most political parties.

The Swedish practice of PTP has featured the common occurrence of close relationships between public agencies and major firms: "This has, to a large extent, taken the form of long-term intimate collaboration on joint development between large manufacturers and large customers – a type of industrial relationship for which the ... term 'development pair' has been coined" (Fridlund 1993: 4). This is perhaps an extreme case of a more "generally accepted" historical tendency in Europe for government selection, through the advantage conferred to innovators by the concentration of purchases by public agencies, "to favour the emergence of national champions" (Dalpé 1994: 73). However, national economies have not always been well served by national champions. This is also increasingly the case in Sweden – a country which has in the past excelled in 'growing' large, multi-national corporations but has not succeeded in developing a very large population of technology-based small and medium-sized enterprises. Trends towards internationalisation have made the multi-national corporations increasingly independent of Sweden, but Sweden remains as dependent as ever upon them (Edquist & Lundvall, 1993: 291 - 292). During the later half of the 1990s this problem has become more severe. Several large Swedish companies have merged with firms in other countries, with the consequence that the head offices have left Sweden. Examples are Pharmacia, Stora, Nordbanken and Astra.

However, the past and current problems that Sweden and other member states have experienced with capturing the benefits of PTP should not be regarded by EU policy makers simply as a justification for disregarding the developmental potential of public technology procurement. There are various 'national' policy traditions and approaches to public technology procurement from which it is possible to draw both positive and negative lessons that might contribute to a 'strategic' perspective on PTP for the European Union.[7] Of particular relevance to the present discussion is the US approach to public technology procurement, which emerged in the developmental context of the United States' postwar technology policy.

---

[7] For example, Japan minimises direct public purchasing (Freeman 1988: 331) but relies on public agencies to play an important role in the organisation of industrial 'clusters' (Carlsson and Stankiewicz 1995: 43). Japan's policy offers a 'network' model (Allen 1981; Vogel 1980) that may be of some interest to EU policy makers.

Certain features of the US experience are well known – in particular, the central role of the US defence program. The extent to which defence-related spending and defence-related procurement historically dominated the US innovation system has been widely recognised (Mowery & Rosenberg, 1993). Similarly, the degree to which defence contracting was responsible for the origin and subsequent development of several major commercial industries in the US – for example, aeronautics (Mowery & Rosenberg, 1982), semiconductors (Levin, 1982), and computers (Flamm, 1987) – has been documented in numerous studies. The protectionism of US procurement policies under the "Buy American Act" is also a widely known feature of the US model of PTP (Goodman & Saunders, 1985).

There are, however, other equally important though less frequently noted elements of the US model – including the influence of anti-trust policy and the commercial orientation of much US procurement for defence purposes. It has been observed, for example, that in public technology procurement supporting the development of computers and semi-conductors, "the US military and intelligence communities pursued a very different path from that followed by Western European governments" (Mowery 1995: 7). In particular, they awarded contracts not only to established firms but also to new ones, and supported projects undertaken by firms that were mainly interested in commercial markets (Flamm 1988: 134).

These agencies also sought to disseminate technical information on both computers and semi-conductors across a broad industrial community, rather than confining it to a 'military enclave' (Levin, 1982). And, they made sure of having more than one supplier for a given procurement through a policy of 'second sourcing', which "served both to diffuse knowledge of advanced technology amongst a range of firms, and to facilitate the entry of new firms to the market" (Geroski 1990: 188). These practices led to substantial "spillovers" from military procurement of defence technology into civilian applications. They also fostered the development of competition in procurement contracts. Such practices "appear to have been motivated by policymakers' belief that in order to create a viable military supplier community, a larger commercial industry was needed" (Mowery 1995: 7).

Of course, the various identifiable national models of PTP are not static. The US model has been in transition since the 1980s, and the Swedish model is also changing. Both have been affected by intensified international competition in sectors that were formerly national 'strongholds', the rise of liberal or neo-liberal orientations in economic policy and, more recently, the end of the cold war. In the US, in particular, this has led to an ambiguous "new direction" in technology policy (Ham & Mowery, 1995), motivated in part by extensive controversy on the effectiveness of defence procurement practices (Kanz, 1993).

In the EU, which now includes Sweden, there has been since 1985 a "revolution in ... public procurement policy", marked by the subsequent passage of legislation and other measures related to the consolidation in 1992 of a single European "Internal Market" (Martin 1996: 16). However, this revolution has occurred with little if any reference to public *technology* procurement and there has been no apparent attempt to learn from the experience of the US and other countries in the area of PTP. This is unfortunate, since in many other jurisdictions the debate on public procurement, including technology procurement, has focused attention on failures and successes in public technology procurement and thereby helped to clarify the justifications for its continued use as an instrument of public innovation policies.

1.3 Failures and Successes in Public Technology Procurement

In the US, at least, debate on public technology procurement has been fuelled by the publication of works documenting notable failures in PTP (Cohen & Noll, 1991). To name but one fairly recent US example, the attempt to develop civilian uses for an 'on the shelf' military technology for the generation of nuclear power was spectacularly unsuccessful. Due to premature commitment to a specific commercial application, the liquid metal fast breeder reactor (LMFBR) project at Clinch River, Tennessee, was terminated in 1983 in view of the complete lack of commercial consequences after the expenditure of more than five billion US dollars (Nelson, 1984).[8]

Much of the recent debate on public procurement, including but certainly not limited to the discussion of failures in PTP, has drawn attention to the distorting effects on rational economic decision making of what may be broadly termed 'political considerations' (Burton, 1983). Risk avoidance on the part of elected officials, their tendency to favour short-term projects with fairly immediate pay-offs, and their preference for decisions that have no clear distributional impact (and hence, no electoral consequences) are three such characteristics of public sector demand that have been alleged to have adverse consequences on procurement decisions in particular (Cohen & Noll, 1991). On these grounds, the case can be made that "elected officials tend to be reluctant to make procurement decisions, which, according to innovation theory, could favour innovation" (Dalpé 1994: 74).

These problematic aspects of behaviour on the part of public procurers can be exacerbated by the effects of public procurement contracts on the behaviour of potentially innovative firms. Public buyers tend to maintain the same suppliers over long periods of time, often on a sole-source basis, thus inducing them to specialise in public supply contracts (Ponssard & Pouvourville, 1982). Firms with guaranteed access to public supply contracts may tend to neglect acquiring new markets and commercial applications for their products (Cohendet & Lebeau, 1987; Zysman, 1977). Both the buyer and the supplier may tend to maintain the same technology, in order to avoid the risks of innovation and the costs of adjusting to new relationships (Williams & Smellie, 1985). And, as recent evidence from the US suggests, when suppliers limited to a single client do innovate, the potential for transfer to other applications may be minimal, even though this is an intended goal of a procurement programme (Barfield, 1994).

---

[8] Europe, too, has had its failures. In Sweden, an often cited procurement failure is the Marviken nuclear power reactor contracted by Statens Vattenfallsverk (Granstrand and Sigurdson 1985: 10). The British-French Concorde programme is widely acknowledged to have resulted in a kind of technological success — but not a commercial one (Eads & Nelson, 1971).

There are thus both occasional failures and enduring problems in public technology procurement. Governments and public agencies may fail, just like markets. However, there is also a strong case to be made for PTP from the standpoint of the economics of innovation. Arguments for public technology procurement can be derived from more general arguments for public intervention in technological development. Following Blume (1981: 9 - 10), these can be summarised as private sector problems of underinvestment, knowledge imperfections, underdevelopment, high risk, undercapacity in terms of necessary economies of scale, and disinterest in strategic issues such as national security (c.f. Granstrand and Sigurdson 1985: 7 - 8).

Since public technology procurement is a demand side policy instrument, the arguments in its favour are more specifically related to the nature of demand, rather than that of supply or investment. A general argument for specifically *public* technology procurement is that public agencies are better able to make demands for technological solutions to problems that private actors are either unable or insufficiently motivated to address. The case for PTP can thus be made on the grounds that strong social needs or demands often correspond to normally weak rates of short term private return on investments in innovation (Mansfield & Rapoport, 1977). However, the most frequently cited arguments in favour of PTP are those which refer primarily to certain special characteristics of demand: strategic importance, largeness of scale, high risks, and high costs (Rothwell & Zegveld, 1982). With respect to these problems, there is a fairly large body of empirical research demonstrating that public demand can be a positive, though not exclusive, factor in the promotion of technological development.[9] Further, it has been demonstrated convincingly that under certain market conditions, in sectors of significant technological content, the high concentration of public demand early in the product cycle acts as a potential catalyst for innovative activity (Faucher & Fitzgibbons, 1993).

To illustrate arguments in favour of public technology procurement under certain conditions, Rothwell (1994) cites several prominent examples of economic and technological success in PTP. One is drawn from a comparison of the regulatory and public procurement policies pursued by the UK and Norway in their exploration and exploitation of North Sea oil resources (Cook & Surrey, 1982). Both countries used 'buy national' policies to make domestic firms the main beneficiaries of economic opportunities in the development of the oil resource, and in both countries state-owned corporations were established to lead this activity.

In the UK, however, the most technologically advanced and profitable segments of the supply industry came to be dominated by subsidiaries of foreign (mainly US) firms using imported technology. There was little effort on the part of either regulators or buyers, including the British National Oil Company, to enhance the technological capabilities of British suppliers or to move them into more sophisticated areas of production.

Norway, in contrast, pursued a policy of 'Norwegianization'. Thereby, the state-owned oil company, Statoil, had a mandate to build up the technological capabilities of domestic supply firms to a point where they could compete internationally. Statoil did this through direct public technology procurement, which it combined with other

[9] For a review, see Dalpé (1987).

measures, such as the promotion of international joint ventures and support for industrial R&D based in Norway. Other oil companies were required by regulators to follow similar practices. As a result, Norwegian firms not only gained a growing proportion of supply contracts in Norway's offshore oil industry. They also acquired competencies needed to compete successfully in other countries.

Another notable example of successful PTP was the development of long-face coal mining in the UK, involving supplier companies, the National Coal Board's R&D laboratories and mining companies (Townsend, 1976). In this context, the National Coal Board played a leading role as an equipment user with superior technological capabilities. During the 1960s and 1970s, the UK coal mining machinery industry became a leading exporter within its sector as a consequence of this innovation process.

To summarise, there are arguments and bodies of evidence both 'for' and 'against' the use of public technology procurement as an instrument of innovation policy. These considerations have informed theoretical and practical debate about public purchasing as a means of stimulating innovation. However, this is a debate which has not been addressed in European Union policy making and in the development of the EU rules governing public procurement. For the most part, both the debate and the formation of a new regulatory regime within the European Union have been more concerned with public procurement in general than with the use of public technology procurement as an innovation policy instrument. In addition to the historical reasons suggested in section 1.2 of this chapter, this may be due in large part to the fact that the 'economics of innovation' does not enter directly into the underlying theoretical rationale of the EU procurement rules. Let us now consider that rationale and its implications for innovation.

1.4 Auction Theory and the Regulation of Public Procurement [10]

We have already noted that the EU regime regulating public procurement is characterised by a strong 'free market' orientation firmly rooted in standard economic theory. For its theoretical rationale, then, the EU procurement regime draws upon a large literature on public procurement produced by neo-classical economists. The model of public purchasing usually employed in such work is that of a monopsonistic buyer (the public agency) confronting an oligopoly of potential suppliers (producer firms) who possess information that is not available to, or known by, the buyer. The problem facing the buyer is to select the best (i.e., least expensive) supplier. The solutions available to the buyer are conceived as 'auctions'.

In this situation, the asymmetry of information becomes a crucial problem, one that can not be resolved by the conventional market mechanism (Akerlof, 1970). Potential suppliers have more information about the item being procured than does the buyer. The suppliers can be expected to hide information relevant to costs and quality, thus

[10] We acknowledge here our colleague Martin Husz (Department of Technology Studies, Austrian Research Centre Seibersdorf). This discussion refers to points made in Husz's more extensive work on *Auctions as Procurement Mechanisms*, included as Appendix B of Edquist and Hommen (1998).

obtaining information rents, or they can establish formal or informal cartels and co-operate against the procurer. The buyer is therefore compelled to design a pricing mechanism that will require potential suppliers to produce the item at a reasonable cost. The buyer has this possibility as a monopsonist.

The alternative designs available to the procurer can be described as auction models. Following the work of Arrow (Arrow, 1985) on adverse selection mechanisms, auctions can be conceived as mechanisms whose inputs are bids and whose outputs are decisions on winners and payments (McAfee & McMillan, 1987). There are a number of different design criteria that may be used in auction models: public *versus* secret bidding; acceptance of highest bid or second highest bid (lowest or second lowest offers, respectively, in the case of procurement); different stages of bidding, i.e., one-step *versus* recurrent bidding (as in ascending auctions). Some basic assumptions must also be made in designing auctions, as to whether bidders are: risk-neutral or risk-averse; have independent or correlated valuations of cost; and tend to bid symmetrically or asymmetrically (depending on cost structures and beliefs).

Four basic types of auction models have been identified (Vickrey, 1961). (1) The *English* auction is an open-bid, first price, ascending-bid auction in which payment is equal to the winner's bid. (2) The *Dutch* auction is an open-bid, first-price, descending bid auction, in which payment is equal to the auctioneer's bid that is first accepted. (3) The *First-Price, Sealed-Bid* auction is a one-step auction in which each bidder submits a secret bid, the highest bid wins, and payment is equal to the bid. (4) The *Vickrey* auction is a second-price, sealed-bid, one-step auction in which each bidder submits a secret bid, the highest bid wins, and payment is equal to the amount determined by the second-highest bid.

These various auction models may be analysed by applying to them different sets of the basic assumptions briefly referred to above. Assuming independent valuation, payment as a function of bids, symmetric, risk-neutral bidders and non-collusion, the English auction and the Vickrey auction yield equivalent dominant equilibria (Vickrey, 1961). On average, though, the outcome of all four auctions is equal according to the revenue equivalence theorem (Milgrom, 1989). Choice of an optimal mechanism can be facilitated by the revelation principle, which dictates that there should be direct reports of bidders' valuations and that they should have sufficient incentive to reveal their true preferences (Harris & Townsend, 1985).

Under standard assumptions, any traditional auction can be chosen as long as a reserve price is set that will maximise competition (Riley & Samuelson, 1981). If the further assumption is made that the bidders are risk-averse, then the auction should be a First-Price, Sealed-Bid auction (ibid.). Under conditions of asymmetric bidding in procurement, if domestic contractors have a lower productivity than foreign ones, they should be favoured to increase competitive pressure on the latter but they (any of the former) should not be given high odds of winning (McAfee & McMillan, 1987).

If it is assumed that the beliefs of bidders are correlated, then the procurer should provide as much information as possible to make beliefs more realistic, and any other auction will be superior to the First-Price Sealed-Bid auction (Milgrom, 1989; Milgrom & Weber, 1982). The latter can be improved by paying for the procurement in the form of royalties (McAfee & McMillan, 1986; McAfee & McMillan, 1987). In cases of collusive bidding or cartels, reserve prices should be increased and sealed or secret bidding should be practised (McAfee & McMillan, 1987). Alternatively, negotiations should be closed (Stigler, 1964). If in later procurements quantity and quality can be varied depending on price, the auctioneer should do this to increase competition and total surplus while minimising procurement costs (Hansen, 1988).

The EU procurement rules, which govern public, not private, procurement, appear to follow the recommendations of 'auction theory' fairly closely. The rules prescribe three basic procedures of source selection: the open, restricted and negotiated procedures. The preferred procedure is the 'open' one. Corresponding to the 'First-Price, Sealed-Bid' auction, this procedure is one of unrestricted tendering, according to established rules of advertisement, closed bidding and contracting (Business International Ltd. 1991: 27). It applies primarily to the purchase of existing, standardised supplies, works and services. Contracts can be awarded on the basis of 'the most economically advantageous offer' – a criterion that takes into account not only cost but also technical merit, the reliability of the tenderer, and what might be termed transaction costs.

The EU procurement rules make the open procedure – a First-Price, Sealed-Bid auction – a normal practice for both 'regular procurement' and 'technology procurement'. This accommodates assumptions of risk-averse bidding and also allows the already high odds of winning enjoyed by domestic firms to be counteracted to some extent by foreign competitors. In the latter respect, the rules comply with 'auction theory's' advice concerning asymmetrical bidding.

In cases of technology procurement, the rules make allowance for other types of auctions: the so-called restricted and negotiated procedures (the latter being a negotiation rather than a true auction). Development projects are more likely to be dealt with under the restricted procedure. This allows for invited tendering and provides more latitude for procurers (particularly utilities) in processes of pre-qualifying prospective suppliers and advertising contracts. It is justified by "a need to maintain a balance between contract value and procedural costs, and by the specific nature of the goods to be procured" (Business International Ltd. 1991: 27). The negotiated procedure is highly exclusive and is most likely to be pursued in highly innovative development projects. Innovation is, in fact, one of its primary justifications. The negotiated procedure can also be used for technical or artistic reasons, or to protect exclusive rights, when the goods can only be provided by a particular supplier. Additionally, its use can be justified in cases of extreme urgency, or where a change of suppliers would not be feasible for technical reasons. (Business International Ltd. 1991: 28)

The two alternative procedures just mentioned respond to auction theory's advice concerning cases of correlated beliefs. In both, the auctioneer is obligated to make all relevant information available to bidders. The restricted procedure allows for reserve prices to be set and secret bidding to be practised. The negotiated procedure, as its name suggests, allows for closed negotiations. Thus, the rules enable procurers to follow anti-collusive strategies. There is no prohibition on variation of quantity and quality in successive auctions to increase competition and minimise costs. In these respects, the EU procurement rules might be considered to approximate 'best practice' according to 'auction theory'.

There are, however, some inherent problems involved in applying 'auction theory' to public *technology* procurement, as distinct from regular public procurement. While auction theory may be appropriate to instances of regular public procurement, some of its main assumptions are highly questionable in the case of public technology procurement.

Applied to the analysis of public technology procurement, auction theory tends to assume that the asymmetry of information (mentioned at the outset of this section of the chapter) stems from the indeterminate character of innovative (i.e., previously unknown) products and the possibilities of strategic use of private information or collusion on the part of an oligopoly of suppliers. In other words, the potential suppliers of an innovative product are assumed to know more about it than the buyer, and they are expected to take advantage of their superior knowledge. These conditions dictate that in public technology procurement the procurer, who has the advantage of the first move, must design a fixed mechanism for pricing that will lead potential suppliers to develop and deliver the product or system being procured at a competitive cost. Thus, well designed procurement rules are supposed to minimise the burden of public technology procurement on taxpayers.

For these reasons, the EU procurement directives pressure public agencies to conform to a norm of non-interactive, 'market' relations in public procurement and other dealings with their suppliers. As the EU directives on public procurement are introduced into formerly excluded sectors (see below), they make 'open' tendering – i.e., First-Price, Sealed-Bid auctions – the normal procedure to be followed by the responsible public agencies in awarding contracts to suppliers. 'Restricted' and 'negotiated' tendering – respectively, the second and third of the procedures described above – can only be used under special circumstances, which must be justified to the appropriate regulatory authorities and, ultimately, to the European Commission.[11] The use of these procedures is thus closely circumscribed. In particular, the negotiated procedure – i.e., the one most appropriate to highly innovative development projects – is applicable only under exceptional circumstances (explained above).

The EU directives on public procurement are also highly restrictive with respect to the scope of action that they allow public agencies in the use of public technology procurement as a means of initiating innovative development projects. Over the past decade, EU legislation in this area has progressively widened the application of the directives on public procurement to cover many formerly 'excluded' sectors.[12] These sec-

[11] Until 1992 the EC's main mechanism for ensuring compliance with the procurement rules was the administration of remedies imposed by the European Court of Justice. This approach proved to be unwieldy and ineffective. In the 1990s, a more decentralised approach has been adopted, in which Member States administer Community Law. The rationale was to avoid "apoplexy at the centre and paralysis at the extremities" by achieving closer cooperation of Member States through the harmonization of national procedures for examining complaints (European Commission, 1992).

[12] The current EU procurement rules were developed in the early 1990s, as part of a larger effort to consolidate a single 'internal market' for public purchasing in Europe. It was argued in this connection that the opening up of hitherto nationally protected internal markets in areas of public sector ownership such as railways, power generation, and telecommunications equipment would reduce over-capacity, and facilitate rationalisation in the form of mergers and investments in new technology (WS Atkins Management Consultants and Associates 1988: 7).

tors have included energy, transport, and telecommunications. These are infrastructural sectors that have, in the past, tended to constitute 'natural monopolies', usually subject to management by public agencies and closely regulated by national authorities.

'Natural monopolies' of this kind are, of course, precisely those sectors in which specifically *public* technology procurement has played an important role in the past and can be expected to do so in the future. They have also tended to be highly protected markets, in which there has often been nationalistic purchasing for strategic reasons. The EU's unambiguous purpose in 'opening up' the excluded sectors and bringing them into the 'Single Market' has been to break down protectionist barriers around these sectors. The rationale has been to provide for international competition for supply contracts that will lead, over time, to the restructuring of the affected supplier industries on a 'European' scale. This approach to creating economies of scale in the supplier industries depends on restricting the scope of action with respect to strategic purchasing by public agencies operating at the national level.

Thus, in compliance with the advice of auction theory, the general trend in the public procurement activities of public agencies in member states of the EU is towards international competition (within the bounds of the EU), and away from (national) protectionism. The trend is also away from close relationships with specific suppliers and towards arm's-length dealings with a broader range of suppliers in an increasing number of public sector domains. The framework nature of EU legislation in this area does not completely prevent the development of national strategies to encourage technologically-oriented procurement actions. The existence of the three different forms of tendering procedures introduces a certain degree of flexibility in the enforcement, and it does not *a priori* discourage the use of public tenders for the stimulation of sophisticated technological demand with clearly defined socio-economic, environmental, or military objectives. Nevertheless, the EU legislation has definitely restricted both the scope and character of innovative public technology procurement.

In subsequent parts of this discussion, we consider innovation theory as an alternative to auction theory. We explicitly consider innovation theory and research as the most appropriate basis for developing policies and regulations concerning public technology procurement. We further consider that innovation theory and research also provide strong grounds for a critical perspective on current EU policy perspectives regarding public technology procurement.

The theoretical and research literature reviewed below indicates that successful examples of public procurement aimed at achieving important technological innovations involve close collaboration and interactive learning between users and producers – phenomena that are not fully considered in auction theory and the EU procurement rules. The literature also indicates that, contrary to auction theory's assumptions, buyers must often be more knowledgeable than suppliers about innovative (i.e., previously unknown) products. Before proceeding to consider these and other aspects of an alternative to the dominant EU perspective on public technology procurement, however, we will first consider, in the next part of this discussion, the varieties of public technology procurement and their relation to other instruments of innovation policy.

2. DIFFERENT KINDS OF PUBLIC TECHNOLOGY PROCUREMENT

We will now proceed to discuss the relations between public technology procurement and other innovation policy instruments. Policy can influence both the supply and demand sides of innovation. Both the supply of knowledge, competence, etc., and qualified demand are important determinants of innovation, and hence can be related to innovation policy.

2.1 Types of Innovation Policies and Policy Instruments

As just indicated, policy issues and instruments can be divided into the supply side and the demand side. Examples of supply side policy relate to R&D, education, firm competences, the establishment of institutions which influence communication between firms and between firms and universities, etc. Examples of demand side policy include public technology procurement, laws, regulations, standards and other institutions which may influence the development and diffusion of technologies.[13] The supply and demand side innovation policy instruments are illustrated in the four-field table shown in Figure 1.

The most common supply side policies often have to do with generation of knowledge, or science and technology. Public funding of research and subsidies to private R&D efforts are the most common examples of such policy. However, there are other options such as emphasising the diffusion and use of existing knowledge by encouraging co-operation and trust (Foray, 1997). One method of doing this is the Swedish system of technological attachés abroad, who report home to Sweden on the existence and use of new technologies in technologically advanced countries. Both of these policy instruments are shown in Figure 1.

Another method of diffusion is developing co-operative arrangements for technology transfers between universities and companies. Building such longer term relationships between firms and universities can involve both benefits and risks. The benefits have to do with transferring information about new technological opportunities and new user wants. The risks have to do with the potential for conflict between the culture of industry and that of the university. For example, the introduction of material incentives and secrecy into the academic community might eventually erode its ethical standards, decrease the speed of scientific progress, impair its contribution to economic welfare, and frustrate its broader social purposes. This is one example of how public policy can stimulate product innovations through new instruments or through new use of existing instruments – and the problems that may be involved in doing so.

One important demand side innovation policy instrument, which has historically enhanced product development very efficiently, is public technology procurement. In Figure 1 we make a distinction between 'developmental public technology procurement' and 'adaptive technology public procurement'. We will now discuss that distinction.

[13] When referring to 'institutions', we employ the following definition: "Institutions are sets of common habits, routines, established practices, rules or laws that regulate the relations and interactions between individuals and groups" (Edquist & Johnson, 1997: 46).

	Technology Policy Instruments Operating on:	
	The Supply Side	**The Demand Side**
Technology Development	**(1) R&D (Research and Development) Policy** (Public Funding of Research)	**(4) Developing Technology** (Developmental Public Technology Procurement)
Technology Diffusion	**(2) Accessing Technology** (Technical Attachés)	**(3) Adapting Technology** (Adaptive Public Technology Procurement of Goods and Services)

Figure 1. Technology Policy Instruments

(Adapted from Edquist 1994: Figure 1)

2.2 'Developmental' and 'Adaptive' Technology Procurement

The distinction made here means that we do not include under the heading of public technology procurement only 'developmental' public technology procurement, i.e., cases where completely new products, processes or systems are created. Cases of 'adaptive' public technology procurement are also included. In such cases, the product or system procured is not new to the world but still new to the country of procurement. Emphasis is put on the fact that the system needs adaptation to local circumstances – not simply that it is new to the country. In such cases a formal contract between the procurer and the supplier should also have been written.

Together, 'developmental' and 'adaptive' public technology procurement constitute what we call an 'extended ideal type' of public technology procurement. The majority of cases studied in the public technology procurement literature are of the first kind, although the second category is just as interesting and important. Much innovation is incremental and is here represented by the 'adaptive' category. Including such procurement increases the scope for a public technology procurement policy.

'Adaptive' cases should also include R&D or at least (incremental) innovation on the producer side. At a minimum, the products, processes or systems are adapted to local circumstances, or to the buyer – which includes some amount of R&D or technical change. This is, for example, the case for software compatibility. The capability of making such adaptations is taken into account when the supplier is selected. Both kinds of procurement should therefore be called public technology procurement (PTP).

'Public procurement', then, is a term reserved for cases where public authorities buy ready made 'simple' products such as pens and papers – where no R&D is involved. Only price and performance of the (existing) product is taken into consideration when the supplier is selected.

Hence, we have three kinds of public procurement, two of which we call public technology procurement. In order to distinguish these, we will call the 'developmental' kind 'developmental public technology procurement'. The 'adaptive' sort will be labelled 'adaptive public technology procurement'.

2.3 Procurers as End-Users versus Procurers as Catalysts

We will also introduce a distinction between cases where the procuring agency is also the end-user of the product or system and when it is not. In the 'classical' cases, the buying agency, e.g., the electricity authority, the railway company or the PTT (public telegraph and telephone company) will use the product procured itself. It simply uses its own demand to influence innovation.

Alternatively, the agency may serve as a catalyst, co-ordinator and technical resource to the benefit of the end-users. This latter kind of procurement has been practised by NUTEK (The Swedish Board for Industrial and Technical Development) with energy saving objectives. This 'catalysing' approach is quite different from the more conventional kind of PTP that NUTEK, in its earlier incarnation as STU, practised in the Swedish attempt to develop a school computer.[14]

An example of the 'catalysing' approach is NUTEK's activity in energy-saving, involving the procurement of new refrigerators in the early 1990s.[15] The requirement was that much less freon – which damages the atmosphere's ozone layer – should be used in production and that the refrigerator's energy use should be considerably lower than with earlier designs. A bidding contest was announced where the prize was an order of at least 500 refrigerators – which went to the company which could satisfy the demands. A design which could meet the demands was presented by the Electrolux company within a relatively short time. This example illustrates clearly that innovation policy through technology procurement can have other objectives besides economic ones; in this case the goal was environmental.

In the case discussed it was not a question of using the public agency's own demand as it was in the cases of trains and telephone exchanges. NUTEK was not the end user of the products that were developed. Instead support was given to the buyers – builders and administrators of apartment buildings – so that they could influence the suppliers. Here it concerned products for mass markets. As well as serving as a catalyst and technical resource, NUTEK also served the role of 'organising' the demand into a customer group.

The 'catalysing' model of public technology procurement exemplified above is a way of "empowering the users". This can be generalised by giving resources directly to the users. These resources could then be employed by the users to specify their own

---

[14] The school computer case is discussed in a case study by Thomas Kaiserfeld (Chapter 4, this volume).

[15] This case is discussed in relation to a broader theoretical framework and policy context by Hans Nilsson (1996).

requirements and then to procure products meeting these requirements from potential suppliers.

A central role of both kinds of public technology procurement is to articulate demand. Hence the state agencies serve as sophisticated customers. They take societal problems and needs as their point of departure. However, scientific and technological progress is necessary to solve the problems and satisfy the needs being focused upon.

2.4 Procurement, Demand Articulation and Policy Co-ordination

Since public technology procurement takes societal problems and needs as its point of departure and, accordingly, is a demand side policy instrument, it is likely to be effective in identifying and articulating new demands. Therefore this instrument might be a particularly powerful one in an overall policy of supporting the development of *new products*.[16]

We have shown in other work that product innovation – particularly in R&D intensive goods and services meeting new or unsatisfied consumer needs – is an essential means of creating employment (Edquist, Hommen, & McKelvey, 1998). Public technology procurement could thus be a powerful instrument not only for solving specific societal problems but also for addressing the more general societal problem of high unemployment. Powerful instruments are certainly needed – in a European situation of the coexistence of massive unemployment with large unsatisfied needs.

It is also important to relate developmental public technology procurement and adaptive public technology procurement to other kinds of innovation policy instruments mentioned in Figure 1. For example, public technology procurement can sometimes in a natural way be combined with R&D subsidies as part of a demand-induced process of technical change. The organisation of research projects might be closely related to a procurement case. For example, it could be helpful in preparing for procurement deals.

Such research projects could focus on special 'needs and technology areas' as candidates for public technology procurement, e.g., 'infrastructure for information technology', 'urban traffic systems', long-distance passenger transportation systems in Europe, 'the clean car', etc. The research projects would normally have an interdisciplinary composition and could include behavioural scientists, economists, engineers, natural scientists and innovation researchers – depending on what the character of the projects demands. Drawing on this competence, such research projects should deal with questions about what is *motivated from the needs point of view* and what is *socially desirable*. The research projects it should also investigate whether it is *technically possible* to satisfy the needs by technical development within reasonable *economic limits*. In this way, results should be reached which could be a basis for the formulation of functional demands. It is a matter of getting the relationship between needs and production to work better in the long term.

---

[16] Other demand side innovation policy instruments might also strongly influence product innovation.

With respect to policy co-ordination in public technology procurement, it is useful to distinguish between 'direct' and 'indirect' forms of innovation policy. *Innovation policy* involves all public action "influencing technical development" in its speed, direction, or diversity "so that socio-economic problems could be solved and societal needs could be met" (Edquist, 1996: 152). *Indirect* innovation policies "are closely related to the public influence on ... framework conditions" (Edquist et al., 1998: 43). In public technology procurement, such conditions would include, for example, the institutional context, or rules and procedures, governing procurement processes. *Direct* innovation policies are those interventions in innovation processes "where a public organisation is directly involved in the interaction" (Edquist et al., 1998: 43). Public technology procurement is itself an example of this kind of policy action.

Public technology procurement, as a direct innovation policy, should ideally be carried out in conjunction with complementary forms of indirect policy. In practice, however, direct and indirect forms of innovation policy, can sometimes be opposed to one another, or work at cross-purposes. Under circumstances where indirect policies, such as rules and procedures, do not constrain public agencies to pursue innovative goals in technology procurement, public technology procurement might be implemented in a manner that is counter-productive from the standpoint of innovation. For example: "If orders are given to a single company with whom [it] has a long term relationship, there is a risk that the procuring [agency] becomes 'locked' into a collaborative arrangement which leads to the development of inferior products and firms" (Edquist, 1996: 156). This risk is especially high when there is a restriction in public technology procurement to dealing solely with domestic suppliers and there is little or no priority given to meeting international standards in order to be competitive in other countries.

2.5 Procurement and Standard Setting[17]

The relation between the innovation policy instruments of standard setting and public technology procurement is also a very important and interesting aspect of policy coordination. Standards can be regarded as 'institutions' in the sense of 'things that pattern behaviour' of organisations when they are involved in innovation processes.[18] One example is the formulation of the NMT 450 standard for mobile telephony by the Nordic PTT's, which defined a standard very early compared to other regions and countries. The technical standard was very specific and gave the supplying firms something to work against in their development work. This is one important reason why mobile telephony took off early and rapidly in the Nordic countries, that the current density of mobile phones is very high in these countries, and that two of the three large global

[17] We are indebted to our colleague Christopher Palmberg (Group for Technology Studies, Technical Research Centre of Finland) for many of the ideas discussed in this section of the chapter, as well as for its inspiration.

[18] The relations between institutions, organisations and innovation processes have been discussed in Edquist and Johnson (1997).

players (Ericsson and Nokia) in the mobile telephony equipment industry developed in these countries.[19] Because of the close relation between public technology procurement and standard setting in some cases we will below briefly touch upon the large literature on theory and praxis of standards.

To begin with, it is useful to distinguish between cases where standard setting occurs in close connection with public technology procurement, and those where standard setting is simply a matter of regulating products and product markets. For example, there may be standards set by public agencies for products which are not purchased by any public agencies. In this type of case, standard setting may be regarded as a purely regulatory activity with no relation to procurement. However, there is often some degree of overlap between these two activities. Even in regular public procurement of existing goods and services (i.e., those which can be purchased off the shelf) public agencies as 'buyers' will be constrained to observe whatever standards have been set by public authorities.

Where an emerging or evolving technology is involved, either or both purchasing and standard setting by public agencies may – or may not – decisively influence the technology's further development.[20] There may also be cases in which these two instruments are used together, so that the setting of a new standard serves to define the technical specifications of a technologically 'new' product or system that is being purchased by a public agency. Where this close co-ordination occurs, it may be legitimate to refer to standard setting as an aspect of public technology procurement (PTP).[21]

Much of the discussion in the literature concerns intermediate kinds of standard setting – i.e., those which are neither 'purely' regulation nor 'effectively' PTP. Of these, standardisation affecting the development of emerging or evolving technologies has attracted the most attention and is most pertinent to the present discussion. For the most part, recent debate has concerned technologies for which the pubic sector is a large and influential purchaser, but for which there is also a (potentially, at least) large market in the private sector. In the case of information technology (IT), therefore, the policy issue addressed has been how to manage public sector purchasing of IT, through standard setting and other means, in such a way as to "receive the most benefit" in other (private) sectors of the economy "without impeding the desired efficiency gains in the government operations themselves" (Cowan 1995: 201).

With regard to standard setting, at least, one response to this issue has been to insist that the role of governments and public agencies in the standardisation process

---

[19] Note that this standard-setting was of importance very early in the development of the technology and defining the standard took ten years. Later standard-setting has gone more rapidly, has had larger numbers involved, and has led to more open standards, which allow for greater flexibility of design.

[20] Rankine (1995), for example, cites a number of instances where government attempts to set standards for the development of information technology have either failed or proven to be ineffectual — in some cases because they were at odds with government buying practices, in others because the standards set did not adequately represent the needs of users.

[21] Case studies on telecommunications in the Nordic countries indicate that this use of standards as functional specifications in PTP was what occurred in the formulation of the NMT 450 standard for mobile telephony by the Nordic PTTs (Fridlund, Chapter 5, this volume; Palmberg, Chapter 6, this volume).

should be minimal. This *first position* is against standard setting by governments and public agencies. Governments and public agencies should only "reflect their needs as users, rather than ... speak for the needs of non-governmental users" – because the latter function "can often be distorted by political objectives, as well as by the abuse of government procurement powers" (Rankine 1995: 193 - 194).

There is a *second position*, which accepts standard setting by governments and public agencies. In this perspective, it is recognised that even if public agencies were to refrain from setting standards *de jure* the market weight of the public sector would still lead to *de facto* standard setting. The public sector's status as a large user means that its purchasing decisions can often have major consequences in terms of the establishment of one technological standard or another. Therefore, rather than abdicating their responsibilities, governments and public agencies should make "careful" choices leading to adoption of "good" standards (Cowan 1995: 214).

The *second position*, which accepts standard setting by governments and public agencies, finds considerable support in recent literature on the economics of innovation dealing with technological 'choices' (or competitions). There is wide consensus that markets tend to 'lock in' on one of several options in technology, due to positive feedback effects of initially influential choices (Arthur, 1988; Foray, 1989). 'Bandwagons' are formed – and they may be bandwagons for either superior or inferior technologies. Since the public sector is often a very large user, its purchasing decisions can have decisive consequences for steering this process towards the adoption of one technology standard or another. It is not a matter of setting standards or not. Public agencies as buyers of new technology therefore face an implicit challenge to set standards that are 'right' in a qualitative sense – i.e., that codify a correct anticipation of user needs. Standards can be set informally, through the simple exercise of buying power, or the process can be formalised (Schmidt, 1992). The second strategy is favoured, for example, by the EU (European Commission, 1997: 37).

The *first position*, against standard setting by public authorities, emphasises the usual ineffectiveness of the 'informal' strategy – i.e., that of influencing standards through the weight of uniform public purchasing, without also establishing 'formal' standards – on the grounds that the normal tendency of public agencies (especially large ones) is to act in a fragmented way (Rankine, 1995). It cites examples of how governmental attempts at achieving central co-ordination can fail. In addition, this position is sceptical about the abilities of public agencies to set 'good' standards. Rather than compensating for the tendency of markets to under-supply experimentation in the early life of a technology, a premature standardisation initiative by a public agency can exacerbate this problem. However, this argument does not address those cases where standard setting and other regulatory activity by governments and public agencies can have a beneficial effect on industrial innovation. This tends to occur "when standards are set that challenge current practices and provide a clearer pointer to the types of user needs for which firms should attempt to cater" (Geroski 1990: 183).

The caution against premature standard setting is, nevertheless, a valid one within limits. There are other reasons, as well, for governments and public agencies to take a non-interventionist approach to standard-setting. Economic 'globalisation' has created both opportunities and incentives for governments and public agencies to take a more passive approach to standards – that is, simply to adopt world standards and require domestic (or other) suppliers to observe them (OECD, 1991). Costs – including development costs – will be lowered and multiple suppliers of a given technology will be forced to compete on price. And, setting a standard other than the world standard (if one exists) may disadvantage domestic producers.[22]

Of course, passivity by governments and public agencies with regard to standards may do nothing to advantage domestic producers either. One way of balancing these priorities is to adopt national standards that do not conflict with world standards but rather encourage compatibility between different technological systems. Such standards permit, and may promote, the development of so-called 'gateway' technologies (Cohendet and Llerena 1997: 236 - 238).[23] This solution allows for the importation of foreign technologies and the export of complementary or equivalent domestic technologies, thus resolving a potential trade dilemma.

Finally, and notwithstanding the considerations raised by non-interventionists, there may be compelling reasons for governments and public agencies to attempt 'early' standardisation in certain, strategically important cases. For example, the early establishment of a 'good' national standard for an emerging technology can be crucial to developing a large domestic manufacturing base and capturing a large share of world markets for that technology (OECD 1991: 95n). Governments and public agencies will have the strongest justifications for doing this where technological specialisations peculiar to their own national systems of innovation are involved. In such cases, there will also be good grounds for the use of public technology procurement (PTP) as an instrument for the further development of these technological specialisations.

There are parallels to be drawn between issues raised in the discussion of standards and questions addressed in section 1.3 of this chapter, which discussed 'failures and problems' in PTP. Both discussions suggest that there are sound reasons to consider PTP in relation to private technology procurement, i.e., where two private organisations are involved in the development process. In the next two sections of this chapter, therefore, we will deal, among other things, with similarities and differences between public technology procurement and private technology procurement.

3. TECHNOLOGY PROCUREMENT IN CONTEXT: ASPECTS OF THE 'DEMAND SIDE'

Recognising that public technology procurement is one viable instrument of innovation policy among others and indicating the circumstances under which PTP can prove to be particularly useful and effective does not mean either apologising for public technology procurement or conducting missionary activity on its behalf. Nevertheless, there is some danger that even the most objective and disinterested discussion of public

---

[22] This argument, however, tends to overlook the fact that *someone* must first establish a world standard. It remains necessary in some technologies — e.g., telecommunications — to set world standards for reasons of compatibility. If private actors are not able to do this, standard setting by public agencies may be called for.

technology procurement, and demand-side innovation policy more generally, may be viewed in this light. This risk stems in large part from a predisposition on the part of both the attentive public and policy-makers to view demand-side interventions by public authorities with great suspicion. There is thus a tendency to neglect "that government regulation often has a beneficial effect on industrial innovation, and is not, as popular discussion would have us believe, wholly inimical to progress" (Geroski 1990: 183). It is matched by similar neglect that "procurement policy is, in general, a far more efficient instrument to use in stimulating innovation than any of a wide range of frequently used R&D subsidies" (ibid.: 183).

Demand side innovation policy and public technology procurement often appear, then, to be ideologically charged topics. Public technology procurement, especially, is often associated, fairly or unfairly, with protectionism, monopoly and even corruption. Apparently, what is often most objectionable or provocative about the term 'public technology procurement' is the word 'public', which is associated with 'government'. We will devote much of this part of the discussion, therefore, to an examination of 'technology procurement' – that is, generic technology procurement in which 'government', in the form of one or another public agency, is not the buyer – and how it is affected by various economic and technological factors. In other words, we will devote all of part 3 to a discussion of *private technology procurement*. The main factors considered include demand structures, circumstances affecting the choice of procurement procedures or mechanisms, and considerations related to the technology life cycle. We will then relate key points made in this discussion to a policy perspective on the 'demand-side' that favours neither protectionism nor monopoly. Subsequently, in part 4 of this chapter, we will return to arguments concerning public technology procurement. There, we will address the same issues in the various sections and subsections of that part of this chapter, relying strongly on the basis of the discussion of 'private technology procurement' in part 3 of this chapter.

3.1 Technology Procurement in the Private Sector

While technology procurement does occur within the private sector, it is not the dominant practice in many markets. In consumer markets, especially, demand-pull is seldom a force for innovation, since individual producers tend to have greater (more concentrated) control over these markets and more information about them than do individual users. In such markets, products may be improved or modified by producers in response to changing patterns of demand, but the role of users in exercising 'demand-pull' tends to be reactive rather than pro-active.[24] In some consumer market contexts,

---

[23] "Gateway technologies are technological solutions of compatibility used when irreversible processes of adopting different technologies have made the interactive learning mechanism inefficient for allowing natural compatibility. Setting up norms within the network of different local systems of innovation during the stage in which the technology is being created seems to be the key to the diffusion process." (Ibid.).

[24] Producers, on the other hand, may actively engage in market research.

though, users may take a more active role, demanding changes in products through organised 'customer initiatives'. The market for paper products is one well-known example of this sort of development, and the market for detergents is another.[25]

Demand pull more often induces innovation in markets for investment products, even though it is difficult to exercise and "requires an articulation of customer will" (Nilsson, 1994) The most concentrated and forceful expressions of customer will are found in industrial markets, which are associated with high levels of 'buying power'. In such settings, effective demand-pull will often take the form of technology procurement.

From a 'systems' perspective on innovation, technology procurement is an important form of user-producer interaction (Lundvall, 1988). Building upon insights central to the 'chain-linked' model of innovation (Kline & Rosenberg, 1986), a systems-oriented view of innovation accords great importance to the 'demand side', rather than concentrating primarily, if not exclusively, on the supply side, as in the overly simplistic 'linear' model (Edquist, 1997: 20 - 22).

A systemic view of the innovation process explicitly recognises the potentially complex interdependencies and possibilities for multiple kinds of interactions between the various elements (or 'stages') of the innovation process (Edquist, 1997a: 20 - 22).[26] This perspective is consistent with a view of technology procurement as a matter of interaction between users and producers, leading to innovation based on 'interactive learning' (Lundvall, 1992: 8 - 10). Such interaction is not only a matter of price signals and quantities bought and sold. It contains other kinds of information and also knowledge, and it can initiate learning processes (Edquist & Hommen, 1999).[27]

3.1.1 Defining and modelling private technology procurement

Although it is seldom recognised as such, technology procurement is widely practised between private organisations in industrial markets. The practice can be defined as "the procurement by a buyer of products, services or systems, which at the time being are

[25] Thus, Nilsson (1994: 2) comments, "during the last few years we have experienced demand-pull as a factor of growing importance when working with goods having impact on the environment". He notes that in Sweden from 1990 to 1994 there was a build-up of market shares for environmentally adapted detergents from less than 1% to approximately 70%, and that the market for environmentally adapted paper held, as of 1994, a market share of approximately 60%.

[26] The chain linked model as originally developed referred to stages in a process, rather than elements or components of a system. Thus, while certain types of organisations or organisational sub-units could be associated with certain stages — for example universities or other research organisations with the 'research' stage, and the marketing or distribution departments of firms with the 'distribute and market' stage — the model did not refer explicitly to interactions among organisations as elements of a system. Instead, it referred only to feedback paths connecting stages. In this sense, the chain-linked model fell short of a 'systems of innovation' (SI) approach, which focuses on interdependencies among organisations.

[27] See Edquist and Hommen (1999) for a more detailed account of how 'interactive learning' theory, together with closely related ideas such as the 'chain-linked model' (described above), has contributed to the development of a broader 'systems of innovation' perspective on the role of the 'demand side' in processes of innovation.

not available on the market and for which some element of technical development is needed" (Granstrand 1984: 9).[28]

Procurement contracting may be accomplished through several different buying procedures, including closed (or sealed bid) tendering, negotiated tendering, and direct contracting with no tenders. There are a number of identifiable steps in the procurement process, though they may not always follow the same order. There are also numerous forms of technology procurement. Variation in procedures and contracts is affected by factors such as volume and frequency of orders, complexity and pace of technological development, the number and relations of buyers and sellers, and their respective competences.

There are some well established general models of buying behaviour that are applicable to the practice of technology procurement (Choffray & Lilien, 1978; Robinson, Faris, & Wind, 1967; Sheth, 1973; Webster & Wind, 1972). Technology procurement departs from these models mainly by adding some further stages and sequences of contingent decision making. It is more similar in some ways, therefore, to the interactions between buyers and sellers that have been modelled for systems and plant selling (Bergström, 1980; Mattson, 1978).

Granstrand (1984) uses a 'transaction costs' framework to identify technology procurement as a special form of buyer-seller interaction. His analysis begins with two ideal-typical cases of involvement that are polar opposites – a 'customer-active' paradigm in which the buyer is fully integrated "backwards" into R&D and production for internal use (full vertical integration) and a 'manufacturer-active' paradigm in which this form of integration does not exist (a 'normal' market situation). These two polar cases can also be called, respectively, the 'fully integrated' form of interaction and the 'market' form of interaction. Between them, it is possible to identify a third, intermediate form corresponding to the relations involved in technology procurement. Namely, this is "a (buyer/seller) co-operative paradigm or, alternatively expressed, a quasi-integrated ... form" (Granstrand 1984: 19).[29] This third form is then compared with the first and second.

The first analytic comparison of the *co-operative* paradigm of interaction (private technology procurement) is with ordinary purchase contracting in a *market* form of interaction. Here, Granstrand finds that "the attractive feature of technology procurement is its potential to smooth peaks in the perception of risk, essentially by shifting some part of the risks from seller to buyer by affecting uncertainties and sharing the cash flows" (Granstrand 1984: 22). Normally, perfect markets make innovation too risky for producers. But commercial uncertainties are greatly reduced for producers if buyers' specifications can be met. Technical uncertainties are reduced for the buyer by realism in technical specifications. While the buyer's economic uncertainties are actu-

---

[28] This definition parallels our own definition of public technology procurement (part 1) but does not include the word 'public'.

[29] Other, less formalised models of buyer-seller cooperation or quasi-integration for innovation and new product development are also possible within this framework (Granstrand and Jacobsson 1983; Williamson 1975: 205 - 207).

ally increased to some extent, they are compensated for by an overall decrease in the level of total uncertainty and risk resulting in reduced transaction costs.

The second analytic comparison of the *co-operative* paradigm of interaction (private technology procurement) is with procurement within a *fully* vertically *integrated* organisation. In this instance, Granstrand finds it more difficult to conduct an analysis of uncertainty reduction but accomplishes this by 'splitting' the vertically integrated organisation into buyer and seller components. This results in a reduction of total risk-bearing ability without any necessary improvement in terms of efficiency. The implication of this analysis is that "the disadvantages of a large organisation (internal procurement bias, organisational persistence and loss of entrepreneurial incentives a. o.) are substantial in innovative work, thus calling for some degree of disintegration or use of market mechanisms instead of administrative ones" (Granstrand 1984: 25).

The results of the second analytic comparison (just above) are consistent with other work in the 'transaction costs' tradition pointing to the disadvantages of large organisations with respect to innovation and the need for the introduction of greater flexibility into vertically integrated organisations, particularly at later stages in the development of an industry (Phillips, 1980). The results of the first analytic comparison (further above) indicate that "perfect" markets are normally unable to generate product innovations. This is also a centrally important point of 'interactive learning' theory (Lundvall 1985). This approach differs from transaction costs analysis primarily in that it refers less to the problems of (un)certainty and risk and focuses more on the qualitative aspects of interactive learning, communication flows, and knowledge creation.

The interactive learning approach opposes standard economic theory's perspective on innovation in order to relate innovation as a learning process to the dynamics of 'organised markets'. The main point of departure is a critique of the orthodox view of innovation as a process of "learning-by-doing" that (ideally) takes place within firms situated in perfectly competitive markets where firm behaviour is adjusted according to signals in the form of prices and quantities (Arrow, 1962). In such a world, cost-reducing process innovation (e.g., learning-by-doing) induced by price signals is possible and can even take place simultaneously within all producer units.

However, "it should be obvious that perfect competition does not induce product innovation" (Lundvall 1985: 17). This is because in an anonymous market, where all communication takes the form of price signals, producers can not acquire information about user needs not already served by the market. Conversely, users have no means of assessing new products, especially complex ones. In the 'real world', however, product innovation is ubiquitous, comprising the majority of important innovations and indicating the existence of (non-anonymous) user-producer relations (Pavitt, 1984). This necessitates an explanation of product innovation as a phenomenon associated with 'imperfect markets'. Moreover, "a user-producer perspective introduces the need for qualitative information about new use-values used as inputs and about the needs of users" (Lundvall 1985: 69).

Up to this point, the argument is essentially the presentation of a strong case for vertical integration. If perfect competition militates against product innovation by restricting access to information about user needs, the logical solution will be for a producer to integrate with a user. The producer will then be able to acquire the necessary information about new technical opportunities and the user will become more competitive in relation to other users. This dynamic undermines the anonymous market and "gives rise to concentration both on the producer and the user side of the market" (Lundvall 1985: 20).

However, the product innovation made possible by a partial integration of users and producers and the concomitant creation of a market characterised by 'small numbers' would result in unacceptably high levels of uncertainty and potential for 'opportunistic behaviour', since innovating units would have much better information about new products than potential users. Trends towards vertical integration would be further reinforced, and complete vertical integration in the long run would reproduce one of the main effects of perfect competition: "All innovations should be in-house process innovations" (Lundvall 1985: 68).[30] For product innovation to occur, there must, therefore, be limits to vertical integration. These arise from the competitive need for both users and producers to maintain broad access to information about product capabilities and user needs, respectively. Vertical integration restricts such access, particularly for users. The main flaw of perfect competition is over-compensated for, resulting in an equivalent problem.

The answer posed to this apparent paradox is the solution of "organised markets". This involves a regulation, by mutually agreed to (i.e., consensual) "codes of conduct", of economic exchanges between users and producers. It diminishes opportunities for "cheating" and rewards "trustworthiness" while increasing possibilities for the flow of information between users and producers. The quality of such information, moreover, is improved through co-ordinating mechanisms such as user associations.

In the organised market, then, trustworthy producers are rewarded by users with reliable information, and relatively stable relationships develop between users and producers. This theoretical solution presents a world in which product innovation is made possible by rendering the market 'non-anonymous' while at the same time avoiding the pitfalls of complete vertical integration: "This vertical 'semi-integration' ... is a more easily reversed relationship that will not have as strong a negative impact on the flow of information as full integration" (Lundvall 1985: 28). Thus, organised markets allow for the existence of elements of hierarchy, but demand their coexistence with countervailing elements of co-operation – i.e., "mutual trust and responsibility" (Lundvall 1985: 29).

Broader implications of this theoretical formulation only become apparent when the 'learning' dimension of organised markets is brought into clearer focus (Lundvall, 1988). In markets where products are simple and relatively inexpensive, the theory does not expect either market organisation or learning to be pronounced, but it anticipates that both will be important in markets where products are complex and their use-value characteristics change rapidly. In such cases, the requirements for direct co-operation and exchange of qualitative information between users and producers will lead to the

[30] This would only be true for industrial markets. There would still be atomised consumer markets, for which there could be product innovation.

establishment, through such interactions, of specialised channels and codes of information (Arrow, 1974). The strong association between complex technologies and differentiated communications systems has, in fact, long been well established in organisational research on innovation (Burns & Stalker, 1961; Tushman, 1977).

3.1.2 Exemplifying private technology procurement

Although there are few systematic empirical comparisons of technology procurement with other forms of buyer-seller interaction, there is a large literature on alternative modes of contracting and subcontracting between buyers and suppliers and their relation to competitive strategies within various industries – for example, the automotive industry (Holmes, 1986). Some of these studies are particularly relevant to technology procurement (Grant & Gadde, 1984). Here, we draw upon one such study to exemplify and illustrate the practice.

Hellman (1993) describes the three-year co-operation between Toyota (as customer) and Nippon Steel (as supplier) which led to the latter's production in 1983 of a new type of corrosion-resistant steel sheet. The introduction of this innovation into its production process allowed Toyota to make significant improvements in anti-corrosive protection for autobodies. It not only benefited Toyota but also strengthened the competitive position of Nippon Steel.

Nippon Steel is the world's largest steel maker with a 30% domestic share in Japan's crude steel production. During the 1970s and 1980s, however, the company faced poor market conditions and undertook a major rationalisation, as well as an aggressive product diversification initiative. Central to the company's strategy for product innovation was a policy and organisational structure supporting long-term co-operation agreements with key customers. Toyota was during this same period not only the number one automotive manufacturer in Japan but also a main customer of Nippon Steel and a pioneer in the use of coated steel sheet. Toyota had acquired considerable expertise and competence in this area, complementary to competences and proprietary knowledge that had been developed within Nippon steel.

Discussions between the two companies, each of which was in search of a partner for technology development and both of which had collaborated in the past, led to a proposal from Nippon Steel for co-operation with Toyota. This led, a year afterwards, to the establishment of a joint R&D group. In the ensuing year, both companies participated in refining functional specifications and conducting laboratory tests and product trials, though they followed a division of labour that reflected each firm's special expertise. Co-ordination between the two firms made it possible for Nippon Steel to retool its production facilities prior to Toyota's formal decision to use the new material in automotive manufacture. This resulted in "a very rapid change, compared to the time it usually takes to introduce a new material in the automotive industry" (Hellman 1993: 10).

Although only a very simple contract governed the initial collaboration between the two companies, a more comprehensive agreement concerning exploitation of research results was concluded before large-scale use of the new product commenced.

This agreement met Toyota's demands that Nippon Steel license the patent rights to other Japanese steelmakers, allowing Toyota to continue its policy of maintaining multiple suppliers of all components and materials.[31] In return, the contract allowed Nippon Steel to be Toyota's sole supplier for a period of 18 months. The agreement gave Toyota, over the long term, the benefit of competitive sourcing. It also benefited Nippon Steel by stimulating the growth of the Japanese steel industry as a whole via patent licensing. "For Nippon Steel, it is better to get a smaller share of a rapidly growing market than having a monopoly position in a small and slowly growing market" (Hellman 1993: 11).

One 'lesson' to be drawn from this case is that close, long-term co-operation between a supplier and a customer in the development of a technologically new product need not preclude 'competitive' sourcing on the part of the customer and competition with other producers on the part of the supplier. This type of supplier-customer relationship is not uncommon in many industrial markets – including the market for steel (Håkansson, 1987). For example, the history of the Swedish steel industry – and, more generally, the European – also features long-term development co-operation and stable relationships between suppliers and large customers.[32] Conversely, countries whose steel industries have lacked these kinds of relationships have tended to lag behind in more advanced production and product technologies. For example, the MIT study, *Made in America*, attributed much of the US automotive industry's poor performance in adopting advanced coating technologies to traditionally tenuous links between American steelmakers and auto producers (Dertoutzos, Lester, and Solow 1989: 103). Thus, too strong an orientation towards "perfect" competition may, at times, undermine important sources of national competitive advantage.

3.2 Technology Procurement and Demand Structures[33]

We have already noted that private technology procurement does not occur in all markets, but only in some types of 'buyer markets', which are characterised by certain structures of demand. Here we will consider the relations between different kinds of demand structures and private technology procurement. Attention to demand structures, their determinants and effects on innovation processes has been emphasised in many systems-oriented approaches to the study of innovation. In the 1980s, for example, the 'chain-linked model' suggested that reforms of the financing and management of innovation in the US were a necessary response to "the very high costs for the develop-

---

[31] This policy is similar to the 'second sourcing' policy of the US Department of Defence, as discussed earlier in section 1.2 and later in subsection 4.2.1 of this chapter.

[32] However, these relationships have tended to be less extensive and more informal in character than the rather unique combination of legal contracts and formalised project organisations, such as joint R&D committees, found in the case of Nippon Steel and Toyota.

[33] We are indebted to our colleague Christopher Palmberg (Group for Technology Studies, Technical Research Centre of Finland) for many of the ideas discussed in this section of the chapter, as well as for its inspiration.

ment of new products, the shortening product life-cycle times, and the forces tending to squeeze out independent entrepreneurs in some heavy industrial sectors" (Kline and Rosenberg 1986: 303 - 304). The demand structures of these sectors as 'buyer markets' for innovative investment products were in danger of being dominated by large firms with a propensity for making "small, cumulative, evolutionary changes that reduce costs and bring better fit of the[ir] product to various market niches"; conversely, these demand structures were also in danger of eliminating the complementary influence of smaller, more entrepreneurial organisations that were "very effective at high-risk, radical innovation" (ibid.: 304).

Rothwell and Zegveld (1982: 93 - 99) distinguish three basic types of demand structures within which technology procurement might occur.[34] These 'market types', which refer to different configurations of demand (rather than supply), are: monopsony, oligopsony, and polypsony. According to standard definitions, monopsony refers to "the situation when there is only a single buyer in a market", oligopsony to "a type of [buyer] market in which there is a fairly high degree of concentration" (i.e., a few buyers), and polypsony to a very diffuse market in which there are many buyers with no large shares (Bannock, Baxter, & Rees, 1986).[35]

The use of concepts and models of demand structure in mainstream economic analysis is often essentially static in character. A 'systems' perspective, however, is quite different from that of standard economic theory. The latter has tended to neglect both product innovation and the structural character of interfirm relationships (Andersen 1992: section 4.5.1). Its primary focus has been on process improvements achieved through "learning by doing" in competitive markets where there are no fundamental differences among firms (Arrow, 1962). Relationships between buyers and sellers are here assumed to be standardised to the extent that price signals are sufficient to convey all of the relevant information required by the parties to any transaction (Hayek, 1948). In contrast to this orthodox view of an "extreme degree of flexibility" in the relationships of the economic system, 'systems' approaches to the study of innovation devote much greater attention to issues of "stability" and "'linkage' structure" (Andersen 1992: 82).

The relevance of demand structures for a systems-oriented analysis of technology procurement is that they indicate different kinds of relatively stable linkage structure. A 'systems' view of demand structures makes possible an analysis of dynamic processes of knowledge creation and learning occurring within differentiated communication systems possessing different structural or 'relational' characteristics. Such an analysis goes far beyond mainstream economic theory's use of concepts of demand structure, which is mainly concerned with how these structures may affect the way that 'buyers', using only information about prices, select among 'suppliers'.

[34] These distinctions are made by Rothwell and Zeveld in the course of discussing technology procurement as an instrument of innovation-oriented public policies — or, what we have defined as public technology procurement. However, the basic concepts employed in the discussion are also fully applicable to cases of technology procurement 'without government'.

[35] Polypsony — 'many buyers' — is one, though certainly not the only, condition of perfect competition.

3.2.1 Monopsony

Monopsonistic markets are, of course, those in which "demand pull" is most highly concentrated – a necessary, though not sufficient, condition of effective technology procurement (Mowery & Rosenberg, 1979; Parkinson, 1982). However "pure" monopsony seldom occurs in the real world, except in conjunction with certain "natural monopolies" – which have all, or nearly all, been publicly controlled in most countries. Thus, there is much evidence and analysis available for a discussion of technology procurement under monopsonistic conditions in public sector "natural monopolies" (Dalpé, 1994; Faucher & Fitzgibbons, 1993). However, relatively little is known about technology procurement under monopsonistic conditions in the private sector, due to its comparatively rare occurrence.

The strong association of monopsonistic demand structures with 'natural monopolies' indicates that monopsonists in industrial markets are often monopolists on other markets – for example, consumer markets. In the telecommunications industry, for instance, service providers that operate as monopolists vis-à-vis the users of telecommunications services also act as monopsonists vis-à-vis the suppliers of telecommunications equipment.

'Natural monopolies' are sometimes owned privately, and the firms that control them can exert a major influence over their technological development. Monopoly can also come into existence temporarily in the private sector on the basis of innovation. Systems perspectives on innovation recognise the linkage between monopolistic and monopsonistic tendencies in the private sector and dwell upon their implications for innovation. Their usual point of departure is the Schumpeterian concept of an entrepreneur as the *sine qua non* of successful innovation processes. In order to innovate, the entrepreneur – which is often "a major business firm" in the case of complex technologies requiring extensive resource mobilisation – "has to perceive the [future] need, identify the necessary ingredients, secure the resources that may be missing initially, and communicate ... vision to the relevant agents – capitalists, suppliers of raw materials, people with the required skills, etc." (Carlsson and Stankiewicz 1995: 40 - 41)

The systems-oriented theory of "development blocks" (Andersen, 1992; Dahmén, 1970; Edquist & Lundvall, 1993; Stenberg, 1987) suggests that monopsonistic entrepreneurs are required for the evolution of networks formed in relation to innovative products.[36] The theory recognises that in such processes there is a basic developmental problem of 'critical mass' to be overcome, the solution of which requires vertical integration and co-ordination of information flows between users and producers through the centralising agency of an exceptionally capable 'entrepreneur'. The same point is also made in 'network analysis', which is a complementary systems-oriented approach (Gelsing, 1992; Håkansson, 1987; Håkansson, 1989; Håkansson, 1990; Håkansson & Lundgren, 1993).

In the analysis of the formation of networks around new (or 'novel') capital goods, it has been argued that there is often no means of centralised communication between

[36] A similar theoretical conception was the later theory of "growth poles" (Perroux, 1969a; Perroux, 1969b).

producers and a broad range of users (Teubal, Yinnon, and Zuscovitch 1991: 382). The absence of an agency capable of these functions not only slows the pace of product innovation dependent upon user-producer interaction – i.e., learning. It also limits producers' capability to match product types to new users – i.e., market creation (Teubal, Yinnon, and Zuscovitch 1991: 386). To overcome these barriers, the entrepreneur requires extraordinary capabilities with respect to the co-ordination and scope of R&D in a 'new market' environment where the expansion of sales opportunities through the gathering and diffusion of relevant information can be extremely difficult.

A special emphasis of 'development block' theory is that the need for linkage and co-ordination arises in the context of structural economic tensions emanating from 'gaps' in technological development (Dahmén 1988). These are indicated, not only by price and cost 'signals' but also by other forms of communication among actors in economic networks. "In both cases, the challenge is in 'gap filling' which tends to eliminate structural tensions but may also lead to new tensions by overshooting, as technical and other solutions sometimes run ahead of the immediate goal" (Dahmén 1988: 6). From the standpoint of innovation, then, a continuing drive towards monopsonistic market power vis-à-vis producers on the part of a 'focal entrepreneur' representing diverse user needs is justified, perhaps demanded, so long as the development potential of the core technology has not been exhausted. However, when "development blocks have reached a state of saturation and, perhaps, overaccumulation" monopsonistic entrepreneurs become candidates for "forced adaptation or creative destruction" (Andersen, 1992: 89 - 90).

Empirical instances of monopsony or near-monopsony in "high technology" industrial markets located in the private sector tend to bear out this perspective. A well known and well studied historical example is the domination of the US telecommunications sector by the very large private telecommunications operator, AT&T. AT&T's monopoly over the provision of long-distance telecommunications made it a monopsonist with respect to the demand for certain types of telecommunications equipment. Porter (1990) credits AT&T with having made "large scale investments" and providing (through activities such as technology procurement) "stimulus and investment" contributing to the development of American infrastructure as "arguably the most advanced in the world" (Porter 1990: 297). AT&T was in this respect an exemplar of the lead role played in technology development by privately owned (and often monopolistic) companies in "communications, power generation and transportation" (Porter 1990: 297).

However, a prolonged lack of domestic rivalry eventually caused AT&T to suffer a loss of dynamism and innovation. This not only resulted in "the provision of less innovative products or services to domestic customer industries" but was also "reflected in less advanced and sophisticated needs for inputs" (Porter 1990: 664). Thus, monopsony began to stall innovation. This situation contrasted dramatically with more recent developments: "Telecommunications services are a hotbed of improvement and innovation in America today after the breakup of AT&T, for example, despite some misplaced early concerns" (Porter 1990: 664).

This judgement is shared by D. Mowery (1995). The latter author draws attention to the "the adoption of new technologies by mature industries, a phenomenon that has sparked productivity growth and even the appearance of new products ... in these industries" (Mowery 1995: 3). He attributes much of the current "renaissance" of US competitiveness to "the ability of the US market to sustain a position as the home of the largest mass of 'lead users' in information technologies in particular" – and in this connection he makes special note of recent "rapid deregulation of the domestic telecommunications system" (Mowery 1995: 19 - 20).

International evidence lends further support to this interpretation of the US experience. Other countries have followed the US example in deregulating and in some cases privatising telecommunications, with positive effects on the rate of innovation among telecommunications operators and their suppliers (OECD, 1996a; OECD/ICCP, 1996). Although the continued existence of public telecommunications operating monopolies in many countries has slowed the pace of innovation and its diffusion, "countries such as Japan and the United Kingdom experienced a twofold increase in employment in the mobile cellular phone service market after moving from a duopoly to a competitive market and Australia experienced a three-fold increase when moving from a monopoly to a competitive market" (OECD 1996b: 108).[37] By similarly transforming formerly monopsonistic structures of demand for telecommunications equipment, the restructuring of public telecommunications operators (PTOs) has also led to increased technology procurement. For example, "specialist expertise needed for core telecommunication operations (e.g., software development) is also outsourced ... creating new jobs in and beyond the telecommunications sector, but reducing direct employment by PTOs" (OECD 1996b: 116). Supplier industries have, as a result of such trends, experienced considerable growth in employment (Glassman, 1993).

On the whole then, the more recent evidence from the US and other telecommunications sectors indicates that, despite the advantage conferred by highly concentrated demand-pull, or 'buying power', (fully developed) monopsony is not the ideal demand structure for effective technology procurement.[38] One of the main disadvantages associated with this type of demand structure is that in later or mature stages of development, the interactions of monopsonistic entrepreneurs with their suppliers tend to take on a highly routine character, mirroring the entrepreneurs' monopolistic interaction with their clientele. Product innovation is abandoned in favour of standardisation and price reduction.

3.2.2 Polypsony

Polypsony is much less favourable to the 'articulation of demand' than monopsony. Here, there are many buyers and none have very high concentrations of buying power. Technology procurement, as we understand the term (see subsection 3.1.1, this chapter), is by definition very unlikely to occur within this type of demand structure. This is not to say that it is impossible. The 'many buyers' might conceivably organise them-

[37] What is described by this source as a shift to 'competitive markets' is really a shift to oligopoly.

[38] This finding agrees with the account given in subsection 3.1.1 in this chapter of the disadvantages for technology procurement under conditions of complete vertical integration.

selves into one or several large buyers. This is done, e.g., in buyers' co-operatives. The NUTEK case of 'catalysing' public technology procurement mentioned earlier, in section 2.3 of this chapter, involved this kind of organisation of concentrated buying power on the part of normally diffuse end-users within a mass market.

As explained in the introductory section of part 3 of this chapter, the lack of concentrated buying power in polypsony means that producers normally tend to have more control over the market than do users. The obvious example is markets for consumer goods. Occasionally there are 'customer initiatives' organised around demands for improvements to products – the demand for 'environmentally friendly' household goods, for example – but it is normally producers, not users, who attempt to tailor product designs to perceived changes in consumer demand.[39] This is product innovation based on 'market research', not 'technology procurement' (Rothwell 1994: 635 - 636).

The conventional pattern in supply and demand interactions underpinning the development of manufactured goods for such markets is thus one of "supply-push", not "demand-pull". Manufacturers will often attempt to assess customer response through various kinds of market research. Customers may also try to seek out better products, but in many cases their search will not coincide with that of the manufacturer. Both parties may be aware that the technology could be improved, but doing this on a competitive, mass-production basis will depend on the direct coincidence of manufacturer and customer awareness. The problem is one of 'timing'. Usually, markets will eventually act to resolve this kind of problem, but in some important cases, such as energy efficiency, situations can arise where "optima will not be sought, nor maintained, under normal market conditions" (Nilsson, 1994: 7n).

Another problem with polypsonistic demand is that of user competence. Much of the existing evidence suggests that users corresponding to those who would typify buyers in a truly polypsonistic demand structure generally possess low levels of technical competence. As a rule, they have a severely limited capability to make enlightened choices about functional specifications. Thus, even those opponents of representation of users by public authorities, who champion instead the technical wisdom of free markets, tend to hold a rather low opinion of the individual consumer as a user. One expert, for example, says of these "level 3 users" of information technology that they "have little or no interest in the technology involved apart from considerations of safety, ease of use and other ergonomic factors" (Rankine 1995: 180).

'Professional' users tend to be more competent and more capable of "controlling" their suppliers than individual consumers (Lundvall, 1991). Therefore, one solution to the problem of user competence in polypsonistic settings is for professional users to procure new technology on behalf of individual consumers. In this connection, an ensuing process of "market transformation" is also needed. Following the introduction of new products with superior performance characteristics, it is also necessary to ensure "penetration to the market", by securing or testing market uptake and response,

[39] See, however, section 2.3 of this chapter, where we discuss the 'catalysing kind' of PTP as a policy response to problems generated by this type of situation.

and to ensure that "partners in dissemination are participating in, rather than slowing down the process" (Nilsson, 1996: 7).

'Systems' approaches to the study of innovation have, for the most part, dealt with relations among organisations, and they have addressed 'polypsony' primarily in the form of so-called 'networks of innovators' (Freeman, 1991). In particular, Håkansson's (1989) work on industrial networks has been cited as an important influence (Carlsson, 1995; Gelsing, 1992). Due to its interest in studying patterns of linkage structure among firms and other organisations within polypsonistic settings, a major theme in this school's exploration of industrial networks devoted to collaborative development of new technologies has been the investigation of 'vertical' and 'horizontal' relations within networks.

In this approach, it is recognised that all user-producer (or customer-supplier) relations constituting inter-firm *networks* must, by definition, involve some degree of vertical integration (although, by definition, vertical integration can never be 'complete' in the case of an *inter-firm* network). Therefore, the term 'horizontal' is used to describe networks based on other kinds of relations – including, for example, the "co-operation between rivals" or "informal know-how trading" discussed by von Hippel (1988). This usage has generated a basic distinction between 'trade' networks with a strong vertical aspect and 'knowledge' networks with a strong horizontal aspect (Gelsing 1992: 120).[40]

Network analysis has demonstrated that product innovation in polypsonistic markets is associated with a drive towards the consolidation of 'buying power' and 'demand articulation' on the part of users, leading to tendencies towards vertical integration. Håkansson's (1989) research has suggested that "horizontal" development relationships formed primarily for the purpose of product innovation, are intended to develop both new markets and new 'vertical' development relationships (user-producer relations) (Håkansson 1990: 378). Market creation, in the sense of developing "organised markets", is thus a major concern of networks formed around technological innovations within polypsonistic settings.

Teubal, Yinnon and Zuscovitch's (1991) work on network analysis addressed the relationship between networks and market creation, with particular reference to "the user-producer relationships characterising ... many emerging capital-goods markets" (Teubal, Yinnon, and Zuscovitch 1991: 381). In their argument, an 'entrepreneur', in the form of a focal organisation, was considered essential to market creation. While 'collective learning' often depends on networks, networks without focal organisations are likely to remain in "a low level equilibrium trap" (Teubal, Yinnon, and Zuscovitch 1991: 382). An organised articulation of demand is required to resolve the co-ordination problems involved in market creation. Concentrated 'demand' exercised by an entrepreneurial 'focal organisation' is equivalent to the formation of a core of innova-

---

[40] "The specification to be made here refers to the distinction between trade networks and knowledge networks. The trade network consists of relations between users and producers and the flow of information is connected to flows of commodities with a certain price. The knowledge network focuses on the flow of information irrespective of its connection to the flow of goods. The knowledge network must pay equal attention to information exchange between users and producers on one hand and competitors on the other." (Gelsing 1992: 120)

tive users sufficiently large to dissolve the "trap". This argument refers back to our earlier discussion of 'entrepreneurship' under the heading of 'monopsony'. It also points to the relevance of the 'catalysing' kind of technology procurement (see section 2.3, this chapter).

To conclude, technology procurement is difficult, though not impossible, under polypsony. Users have to be "co-ordinated", and their competence must increase. A successful example of entrepreneurial activity directed to such ends is NUTEK's role as a 'focal organisation' in the 'catalysing' type of technology procurement, first discussed in section 2.3 of this chapter. (However, this is an example of public, not private, technology procurement.)

3.2.3 Oligopsony

Given that technology procurement is very difficult in the context of a truly polypsonistic demand structure, and given also that monopsonistic demand structures, despite their high concentration of "demand-pull", ultimately tend to inhibit product innovation, oligopsony might be the 'ideal' demand structure for effective technology procurement.

In oligopsony, there are several large buyers. Though none of these large buyers completely controls the market, all have fairly high levels of buying power. Private technology procurement is a relatively common practice within such settings and much of the available evidence suggests that oligopsonistic demand structures are highly favourable to product innovation through technology procurement. In subsection 3.2.1 of this chapter we discussed the recent transition from monopsony to oligopsony in markets for telecommunications equipment, resulting from the transition from monopoly to oligopoly in telecommunications service provision. We pointed out that this restructuring has increased the need for more advanced and sophisticated technological inputs into telecommunications operating systems, thereby stimulating product innovation.

Oligopsony appears to offer incentives to firms to engage in competition based on innovation. We have illustrated this point with the example of private technology procurement given in subsection 3.1.2 of this chapter. There, it was shown that a supplier and a customer firm who had realised great short term benefits from close collaboration in product innovation through technology procurement both had strong incentives to maintain a less exclusive relationship between buyer and seller over the longer term. This arrangement allowed the buyer to benefit from efficiencies created through competitive sourcing, as well as avoiding technological 'lock in'. It also enabled the supplier firm to benefit from the longer term growth of its industry, both indirectly and directly. The firm was able, for example, to license its patents to competitors.

Oligopsonistic demand structures allow one oligopsonist to function as a 'quality leader' capable of *de facto* standard setting that will influence the decisions of other oligopsonists. Thus, the procurement activities of a single oligopsonist may have decisive effects on others and also for the suppliers to these firms, in terms of the direction and rate of innovation in technological development (Arthur, 1988; Foray, 1989).

This point (and its policy implications) was discussed at length in section 2.5 of this chapter. Oligopsony also favours the formation of 'co-operative' relationships between firms, in particular those forms of 'quasi-integration' that are most conducive to product innovation (Granstrand, 1984). (See subsection 3.1.1, this chapter.)

In theory and research contributing to the development of 'systems' approaches to the study of innovation, the dynamics of technology procurement occurring within the context of oligopsony are best captured by von Hippel's (1988) empirically grounded theory of variation in the functional sources of innovation. This theory challenged the long-standing assumption that "product innovations are typically developed by product manufacturers" and replaced it with a model of a "distributed innovation process" in which product innovations could originate from any one (or combinations) of at least three distinct (functional) sources: suppliers, producers and users (von Hippel 1988: 3 - 5). The theory held that the "source" of innovation would vary according to the share of temporary benefits, or "economic rents", that firms in each category could expect from a potential innovation (von Hippel 1988: 6, chap. 4).

In exploring strategic implications of the effects of incentive structures on "distributed innovation processes", the theory identified the "lead user concept" as one of the most viable means of "shifting the sources of innovation" in those product markets where the locus of innovation naturally resides with users, rather than producers. This strategy meant altering the normal patterns of interaction and information exchange among users and producers within a given product market so that the modified incentive structure resulting from these changes would induce more rapid and successful product innovation (von Hippel 1988: chap. 8).

The main policy example given in this connection was that of the US semiconductor industry, which lost much of its market share in the production of 'leading edge' computer memory chips to Japanese competitors during the 1980s. Among US policy-makers, it was alleged, the common wisdom was that US semiconductor process equipment firms "should somehow be strengthened and helped to innovate so that US semiconductor process equipment users (makers of semiconductors) will not also fall behind" (von Hippel 1988: 9). However, the finding that users, rather than producers, were the developers of most innovations in semiconductor process equipment (von Hippel 1988: Table 1-1), indicated that the problem had a reverse causality and suggested that it could best be remedied by assisting US users to resume innovation at the 'leading edge' (von Hippel 1988: 122).

This was a strategy which relied upon the dynamics of international competition among the relatively few large buyers of 'leading-edge' computer chips to supply the primary incentive for product innovation, in the form of innovation-based "rents" accruing to buyers. The main point was to shift the geographical locus of innovation at the leading edge by increasing the probability of achieving such rents for certain oligopsonistic 'users' of chips – namely, those based in the US. This, in turn, would advantage the oligopoly of producers based in the US. Thus, the overall result of the proposed reorganisation of the existing incentive structure affecting the demand side would be not only to encourage the emergence (or re-emergence) of "lead users" of advanced computer chips in the US, but also to create (or re-create) a specific kind of relationship, conducive to further innovation, between the users and producers of semiconductor process equipment. That is, one where (as previously) "US equipment manufacturing firms ..., owing to geographic proximity to the innovating US users, were typically the first equipment builders to gain access to the innovation" (von Hippel 1988: 121).

3.3 Technology Procurement and Technology Life Cycles[41]

The practice of technology procurement illustrates the very close connection between two aspects of innovation that are often treated separately. On the one hand, there is the development of a given technology, in which the producers can be viewed historically as progressing through the various stages of a life-cycle. On the other hand, there is the adoption of a technology, in which the buying decisions of users can be viewed as following stages of diffusion over time. Models of the technology life cycle tend to represent these two processes as mirror images of one another. They also tend to suggest that the only significant procurement decisions in the development of a technology are those made by first or early adopters, in the initial stages of the technology life cycle. However, a closer look at recent theory on technological evolution indicates that significant procurement decisions can also be made in the later stages of a technology's diffusion. There is thus in private technology procurement a distinction to be made which is equivalent to that we made in section 2.2 of this chapter between 'developmental' and 'adaptive' public technology procurement.

3.3.1 The conventional model

The conventional model of the technology life-cycle, as developed by Utterback and Abernathy (Utterback & Abernathy, 1975) is well known. It describes how, for a given industry, the type of innovation (product vs. process), the location of innovation, and the barriers to innovation change over time. It also relates these changes to an account of changing relations between and among sellers and buyers (or producers and users).

The model begins with a 'fluid' phase of technology-competition, in which there is a high level of product innovation and a low level of process innovation. A fairly large population of producing firms within the industry, each with a distinctive product concept, is in competition to capture sufficiently large shares of a new market. In addition to technological uncertainties, therefore, there are also market uncertainties, constituting a major problem for product developers. Thus, on one hand, "[t]he main source, or stimulating factor, of innovation in this phase is insight into user needs" (Hidefjäll 1997: 20). On the other hand, "many potential customers will not be able to perceive the relevance and value of the innovation" (ibid.: 20).

---

[41] We are indebted to our colleague Lena Tsipouri (Centre of Financial Studies, University of Athens) for many of the ideas discussed in this section of the chapter, as well as for its inspiration.

In a subsequent, 'transitory' phase of the technology life cycle, the technology-competition is resolved in favour of one or another 'dominant design', which is then allowed to mature as a product-concept (Utterback 1994: ch. 2). In effect, this is accomplished through the procurement decisions of influential early adopters. Subsequently, market uncertainty decreases for the (one or a few) winner(s) of the technology-competition and greater attention can be given to investment in product development and production technology. The profile of process innovation is raised dramatically, while that of product innovation enters into a decline, becoming increasingly minor in character. Eventually, there is a shift to a 'specific' phase in which the industry structure becomes highly concentrated and both product and process innovation decline. Here, "the primary stimulus of innovation comes from the need to reduce costs and improve quality in production"; there is "little room for major innovations, which may disrupt the smooth and efficient production machinery or make given investments appear as sunk costs" (Hidefjäll 1997: 21).

In broad outline, the technology-life cycle model presents an account of technology development in which being 'first' matters – especially at the point where a dominant design emerges. It is the producing firms first associated with the product-concept winning the technology-competition and the influential early adopters whose procurement decisions first lead to the establishment of a dominant design who gain the greatest advantages. For the original producers of dominant designs, the market "takes off" and they, rather than imitators, are in the best position to profit from market growth by making the transition to standardised, high-volume production (Utterback & Suarez, 1993). The early adopters are the first to master the ideas and approaches that govern subsequent problem solving with the technology (Clark 1985: 245); they can therefore realise greater benefits from the technology than the laggards for whom they have set the agenda. This account has been discussed as a 'unidirectional' Schumpeterian perspective in which the dominant economic factors and the key actors may change over time, but not the technology (Silverberg, 1990a).

3.3.2 Further theoretical development

There are, however, important variations on this basic story-line. For example, the idea that technology life-cycles can be reversed and restarted, through the occurrence of either 'competence-enhancing' or 'competence-destroying' "technological discontinuities" has been introduced (Tushman & Anderson, 1986). The first type of discontinuity tends to consolidate the dominant position of established firms, while the second type poses a threat to them from new entrants to their industry.

The idea that technology, though strongly influenced by 'demand-pull', retains an evolutionary dynamic somewhat autonomous from the influence of market forces has been elaborated in the related concepts of internal 'inducement mechanisms' and 'focusing devices' (Rosenberg, 1976) and 'regimes' and 'natural trajectories' in the development of technologies (Nelson and Winter 1977: 57). These elements have been brought together in Dosi's (1982) influential model of technological paradigms and technological trajectories. In this model, the 'supply-side' is initially responsible for creating a paradigm – "the 'universe' of possible modalities" – but the demand-side subsequently makes progressive gains in importance, by acting (through means such as technology procurement) to select developmental trajectories within the paradigm, according to their "economic significance" (Dosi 1982: 155 - 156).

Dosi's (1982) formulation makes possible a more discerning perspective on the relations among and between sellers and buyers in the historical development of a given technology. According to the classic Schumpeterian narrative, both sellers and buyers of a new technology are motivated by the desire to capture temporary monopoly gains from innovation. The first adopter(s), whose buying decision to procure an innovative technology leads to the establishment of one or another seller's product-concept as a dominant design, will enjoy this same status within the buyer industry. The decisive influence of the demand-side, however, is exercised only at this time. After the dominant design has been established, an oligopoly of sellers comes into being that effectively controls the technology's further development.

The Dosi model agrees with this in part, describing a shift in the source of oligopolistic gains from (at first) Schumpeterian 'rents' on innovation to (later) the benefits of static entry barriers (Dosi 1982: 157 - 158). However, it also points to the continuing and growing influence of the demand-side in acting to select alternative developmental 'trajectories'. This was already reflected in earlier observations of how the technical evolution of an innovation is bound up with and related to its pattern of diffusion (Rosenberg, 1972; Sahal, 1981). On this basis, it is possible to conceptualise, in a systematic way, how a technology can undergo significant changes, not only because of the technology procurement decisions made by early adopters, but also by later ones.

These implications have been brought out more clearly in later applications of the Dosian perspective to the development of modelling frameworks for the analysis of technical change (Silverberg, 1987; Silverberg, 1990b; Silverberg, Dosi, & Orsenigo, 1988). In this work, the observation that technical change is an ongoing and uneven process leads to an evolutionary treatment of entrepreneurial technology replacement decisions based on the assumption that these require collective learning about 'paybacks' (Silverberg, 1987). This approach seeks to explain, among other things, why there are systematic differences in the 'payback periods' calculated by different entrepreneurs. To model the 'feedbacks' presumed to generate these patterns, it relies on the concept of technological trajectory (Dosi, 1982; Nelson & Winter, 1982; Sahal, 1985). On this basis, the so-called 'self-organisation' approach develops the idea that improvements or modifications to a technology that make it competitive within one or another market segment "could be represented as 'branchings' along an evolutionary 'tree' and allow for classification, taxonomy and rigorous definition of technologies in question" (Grübler, 1989). Thus, new 'bandwagons' can be identified within later stages of technology diffusion.

Within this framework, entrepreneurs face dilemmas regarding the choice between earlier or later adoption (Rosenberg, 1976). Early adopters will be, essentially, the procurers of new technology. In order to do this successfully, they must develop 'absorptive capacity' (Cohen & Levinthal, 1990). The competency or 'learning' levels of such firms must be exceptional, since they must be able to anticipate a sufficient payback to justify the adoption of a new technology (Silverberg et al., 1988). At the highest levels of competency and learning, 'first movers' will make the greatest gains from the introduction of new technology. At intermediate levels, "a second mover or imitator may be able to capture more of the benefits by letting early innovators bear an excessive share of the development costs" (Silverberg 1990a: 185). Or, at low levels of competency and learning, diffusion may fail altogether.

Thus, a sufficiently high level of competence and learning is an essential precondition for any diffusion to occur, or any benefits to be realised. In order to meet this requirement, firms may be compelled to pool what would normally be regarded by them as proprietary knowledge in a process of "collective invention" – i.e., "the free exchange of information about new techniques and plant designs among firms in an industry" through which "fruitful lines of technical advance (can be) identified and pursued" (Allen 1983: 2). Such efforts can be reproduced or amplified through certain forms of public policy such as the successful Japanese system of "pre-competitive co-operation and co-ordination" (Silverberg 1990a: 189).

3.3.3 Implications: 'adaptive' procurement

The theoretical revisions to the conventional model of the technology life cycle and the analytic framework reviewed above make it possible to speak legitimately of 'adaptive' as well as 'developmental' technology procurement. It is to be emphasised that the former is not simply a matter of diffusing an unchanged technology but, rather, a form of innovation in its own right – and one that has important economic consequences. The critical importance of high levels of competence and learning required for potential adopters to become 'early' procurers of a technology in the later stages of its development and diffusion reinforces this point. In particular, learning is required to gauge correctly the timing that will yield a sufficiently high payback to justify the adoption of a new technology. Once determined, the relationship between timing and benefit can provide the procurer with the basis for a strong negotiating position with suppliers and an accurate assessment of the competitive advantages to be gained from (relatively) early adoption.

3.4 Summing up the Demand Side: Porter

Porter (1990) has elaborated one of the most comprehensive views of the strategic importance of the demand side for innovation and technological development. For this reason, his perspective provides a useful framework for relating private technology procurement to other strategically important aspects of the demand side as a whole. It is used in this way below.

In addressing public policy, Porter (1990) is fundamentally concerned with private sector demand, including demand taking the form of private technology procurement.

His work therefore outlines a number of 'levers' that public authorities can use to improve the quality of domestic demand and so improve national competitiveness (Porter 1990: 647 - 653). With one main exception (i.e., the consideration of 'foreign aid and political ties') these levers correspond roughly to those aspects of the demand side that have been considered thus far. Porter's levers include: 'regulation of products and processes', 'buyer industry structure', 'stimulating early or sophisticated demand' and 'technical standards' (ibid.: 647 - 653). We will review his main points under these headings, relating them to preceding sections of this chapter.

3.4.1 Regulation of products and processes

With regard to the regulation of products and processes, Porter draws a fundamental distinction between regulation relating to standards and that relating to competition (Porter 1990: 647). A very positive view is offered of standards, which can "pressure firms to improve quality, upgrade technology, and provide features in important areas of customer (and social) concern" (ibid.: 647).[42] The regulation of competition is approved of only if it encourages domestic rivalry and new firms (ibid.: 647).

With respect to the regulation of competition, there is encouragement for policies allowing firms to outsource less productive activities but discouraging the export of highly productive domestic activities (ibid.: 647). Firms should be under competitive pressure promoting "innovation at home". In this context it is stressed that "Developing domestic suppliers produces more sustainable advantage than relying solely on foreign suppliers" (Porter 1990: 658). One means of accomplishing these purposes might be the institutionalisation of procurement mechanisms or auctions of the kind discussed in section 1.4 of this chapter – particularly, those appropriate to conditions of asymmetric bidding. According to 'auction theory', if domestic contractors have a lower productivity than foreign ones, they should be favoured to increase competitive pressure on the latter, but they should not be given high odds of winning.

There is also support for policies ensuring "vigorous domestic rivalry". This means strong anti-trust legislation, avoidance of policies supporting the emergence of 'national champions', and a strong policy against "horizontal mergers, alliances, and collusive behaviour" (Porter 1990: 663). However, antitrust law should not "be a barrier to vertical collaboration between suppliers and buyers that is so integral to the innovation process" (ibid.: 663). Such advice agrees well with the discussion of demand structure effects on innovation in section 3.2 of this chapter. It also agrees with 'auction theory's' advice (in section 1.4, this chapter) concerning the need to institutionalise anti-collusive procurement mechanisms. Porter's recommendations about the importance of non-exclusive forms of vertical collaboration are in keeping with the discussion in section 3.1 in this chapter of technology procurement as a form of 'quasi-integration' or partial vertical integration.

[42] We will return to this theme in subsection 3.4.4 of this chapter.

3.4.2 Buyer industry structure

Regarding buyer industry structure in fields such as health care, electric power, and telecommunications, Porter (1990: 650) stresses that "regulation or state ownership is beneficial to supplier industries if it encourages an industry to act as a more sophisticated buyer with more stringent and advanced needs". He considers, however, that this is seldom the case with state-owned monopolies, and that "In most instances, private ownership and exposure to competitive pressure create the best environment leading industries to play the role of demanding and anticipatory buyers" (ibid.: 650). The high innovation profile of the privately owned US health care sector is cited as a leading example here– although elsewhere Denmark's socialised health care system is credited with having stimulated the innovativeness and competitiveness of certain supplier industries (ibid.: 651). This perspective agrees well with some of the main points made in the preceding discussion of demand structure in section 3.2 of this chapter – in particular, the point that oligopsony generally has a more positive effect than monopsony on technology procurement leading to product innovation in supplier industries.

Porter's case against regulation supporting monopsony is that "it works against local suppliers if it retards innovation and introduces conservatism into purchasing decisions" (Porter 1990: 650). We have arrived at similar conclusions in our discussion of technology procurement as a form of economic relations in subsection 3.1.1 of this chapter. In subsection 3.1.2 of this chapter we discussed the empirical example of Nippon Steel and Toyota to illustrate the institutionalisation of barriers to vertical integration between suppliers and buyers in technology procurement. The effect was to create an incentive structure that counteracted tendencies towards producer inertia resulting from exclusive supply arrangements with monopsonistic buyers. Insistence by the latter on multiple sources of supply acts as a stimulus for further innovation, introducing competitive dynamics into the supply side that may ultimately lead to a transition from monopsony to oligopsony on the demand side.

3.4.3 Stimulating early or sophisticated demand

In view of the critically important role played by "early or sophisticated demand" in the development of technologically innovative products, Porter (1990: 651) places considerable emphasis on the need to "reduce the risk perceived by (producer) firms that demand will fail to materialise". Assurances of future demand are required to encourage a sufficient investment in R&D and production. Technology procurement, as we have pointed out in our discussion in subsection 3.1.1 of this chapter, is a form of economic arrangement that has "potential to smooth peaks in the perception of risk, essentially by shifting some part of the risks from seller to buyer " (Granstrand 1984: 22).

Of course, shifting risks from sellers to buyers still leaves the problem of risks to be borne by buyers. Porter therefore urges a "policy of providing incentives to buyers to be early purchasers of sophisticated products" – which, he considers, "is often more beneficial to innovation and to competitive advantage than directly subsidising firms" (op. cit.: 651). Subsidising buyers instead of producers subjects the latter to the discipline of having to meet the formers' demands, thus stimulating competitive rivalry among producer firms. Buyers are also helped to improve, and this "encourages the mutually reinforcing process in which a nation's buyers become more sophisticated and in turn stimulate local producers" (ibid.: 651). In our discussion of 'oligopsony' in subsection 3.2.3 of this chapter we drew upon von Hippel's (1988) theoretical and empirical work on innovation as a 'distributed process' to give an example of how the same basic principles were applied in an analysis of how to regain the 'leading edge' in the US semiconductor industry. What Porter adds to this particular perspective on the sources of innovation is that innovative competition among oligopsonistic users is closely linked to, and can be used to encourage, similar competition among oligopolistic producers.

Porter's perspective on early demand is also in essential agreement with our observations in section 3.3 of this chapter on the technology life cycle and its implications for technology procurement. According to the conventional model of the technology life cycle, those who derive the greatest benefits from technological innovation are the firms who first win 'technology competitions' as suppliers, and the influential early adopters, whose influential procurement decisions as buyers initiate the 'bandwagon' effects leading to establishment of a dominant design. Subsequent imitators among both buyers and sellers gain less and less benefit from the technology. The technology does not evolve; instead, it is eventually exhausted as an resource capable of generating significant economic 'rents'. This is the 'classical' Schumpeterian account of technological evolution, and it provides the basic rationale for encouraging 'early demand' in the form of 'developmental' technology procurement.

However, we have in subsection 3.3.2 of this chapter reviewed recent theoretical and analytical work that points to the continuing influence of the demand side in later stages of a technology's development, making possible further new branchings along a trajectory. In such contexts, the buyer's role in technological innovation continues to be decisive. Those buyers who become innovative procurers require particularly high levels of competence and learning. They must be able to specify the functional and technical requirements of the product. They also need to determine the correct timing and to negotiate with suppliers from a position of strength. If, however, they can (possibly through co-operative action) accomplish these things successfully, they can resolve the associated problems of risk and reap the post-Schumpeterian benefits. There is a case to be made, therefore, for industrial policy to encourage 'early demand' not only in 'developmental' technology procurement, but also in 'adaptive' technology procurement.

3.4.4 Technical standards

As noted earlier, in subsection 3.4.1 of this chapter, Porter (1990: 647) advances a positive view of regulation related to standards. He does not agree with the position that standards regulation is a counter-productive intrusion into competition. His case has two complementary 'pillars'. One is the argument that there are many standards related to social (e.g., environmental or health) concerns that firms can not be expected to set independently. However, they can become important sources of national competitive advantage since "selling poorly performing, unsafe, or environmentally damaging products is not a real route to competitiveness in sophisticated industries and industry segments" (ibid.: 648). Tough standards in these and other areas will require buyer firms to be demanding customers in technology procurement.

The other argument is that those standards which are most beneficial will not be laggard or anachronistic. Rather, what is to be desired are "stringent regulations that anticipate standards that will spread internationally". Such regulations "give a nation's firms a head start in developing products and services that will be valued elsewhere" (ibid.: 648). To have positive effects on technology procurement, public policy in this area needs to have a strong focus on improving demand quality "by providing accurate and complete information to buyers" (Porter 1990: 652). Standard setting also affects the rate of supplier innovation and buyer improvement in industry, since "when basic standards are set, firms then turn their attention to rapidly developing and improving products and processes to meet them" (ibid.: 652).

This position agrees well with some of the main points made in our discussion of standard setting in section 2.5 of this chapter. There, we distinguished between standard setting used as a means of defining functional specifications in public technology procurement and simple standard setting. The latter, we argued, can be viewed as an important demand-side policy instrument in its own right, since it will influence the buying decisions and technology procurement activities of private firms. We reviewed positions and arguments against and for 'interventionist' standard setting by public authorities. The position *against* holds that most important issues can and should be decided by negotiation among private economic actors (Rankine, 1995). In contrast, the Porterian perspective highlights those kinds of issues that are more readily decided by public authorities. The position *for* interventionist standard setting is, of course, very close to the Porterian one, and is supported by an extensive literature on the dynamics of technological competitions (Cowan, 1995).

Given the prevalence of 'bandwagons' (Arthur, 1988; Foray, 1989) it is important in terms of national competitive advantage that 'good' standards be set early, particularly in areas where a given country may hope to seize a strategic advantage for domestic firms by encouraging the appropriate kinds of technology development and procurement (OECD 1991: 95n). The standards most useful for this purpose will not be parochial or protectionist. And, to balance the concerns of developing 'national' industries and participating in a 'global' economy, standards favouring so-called gateway technologies' can be developed to resolve the trade dilemmas that public authorities may face (Cohendet and Llerena 1997: 236 - 238).

4. PUBLIC TECHNOLOGY PROCUREMENT REVISITED

Having dealt at some length in the preceding part with 'private technology procurement', we will now return to the discussion of our main topic – public technology procurement (or PTP). In the following sections of this chapter, we will relate the preceding overview of key aspects of private technology procurement to parallel aspects of public technology procurement. To this end we will organise the discussion under headings corresponding to those in part 3.

4.1 Organising Public Technology Procurement

In section 3.1 of this chapter we adopted a definition of private technology procurement that was highly similar to the initial definition of public technology procurement we gave in part 1. We then considered the nature of private technology procurement as a distinctive form of economic relations. In comparison with regular market relations, the special advantage of private technology procurement as a form of relations conducive to technological innovation was shown to be its potential for shifting a part of the burden of risk from sellers to buyers and achieving a general reduction of transaction costs. In comparison with vertical integration, it was shown that the main advantage of private technology procurement is that it allows for greater efficiency and flexibility than full vertical integration, thus avoiding problems such as bias towards internal sources of supply, organisational 'lock in', and loss of entrepreneurial initiative. These are often cited as 'faults' of public technology procurement (Dalpé, 1994).

One clear implication of the above analysis for public technology procurement is to warn against protectionist policies or favouritism leading to the establishment of an effective monopoly of supply. This is the functional equivalent in PTP of full vertical integration in private technology procurement. Such monopolies can be detrimental, over the longer term, to both technological innovation and economic efficiency. However, some analysts argue for 'protectionism' in public technology procurement when it shelters potentially important 'infant industries' – i.e., "provides an assured market for new products and processes, particularly early in product life-cycles when commercial possibilities are, at best, hazy" (Geroski 1990: 189).

Protectionism in PTP is not justified in maintaining national champions or protected national suppliers in what are essentially mature industries. This is often done on the grounds that a single, large established firm is supposedly the best organisational framework for the integration of any complex R&D project (Graham, 1988). But protected markets for established firms and industries erode competitive incentives for the 'champions' and provide public authorities with "no substantive basis upon which to evaluate their performance" (Geroski 1990: 194).

In subsection 3.1.2 of this chapter we considered a prominent example of private technology procurement that involved two large firms both situated in mature industries – steel and automobile manufacturing – in the development of a new technology (Hellman, 1993). In the case examined, the procurer worked extremely closely with the supplier until the technology had been successfully developed, but then insisted on entering into a contractual arrangement that would grant the supplier an effective monopoly of supply for only a limited period of time and assurances of contracts for only partial supply of the product thereafter. The contract also stipulated that the supplier should make the technology available to other potential suppliers through patent licensing. This ultimately benefited the supplier through market growth based on the overall expansion of its industry.

Policies such as these in public technology procurement could help to eliminate some well known problems. First, they would eliminate preferential treatment leading to entrenched monopolies of supply for large, established firms. This would encourage the entry of smaller, rival firms whose efficiency and capacity for successful commercialisation are demonstrably greater, creating competitive conditions that would spur (capable) champions on to higher levels of technological and commercial achievement (Baldwin & Scott, 1987; Cohen & Levin, 1989; Kamien & Schwartz, 1982; Scherer, 1980). Second, they would encourage substantial inter-firm 'spillovers' in supplier industries, resulting in industry and market growth through the exploitation of new commercial possibilities (McKinsey and Company 1988; Rothwell and Zegveld 1985: 119 - 120). And third, they would encourage the development of more flexible organisational arrangements than the giant, vertically integrated firm in supplier industries, through such means as decentralisation, specialist subcontracting and the formation of precompetitive consortia (Ergas 1987; Maddock 1983: 7).

There is, however, a caution against taking the prescription for the restoration of competition too far and resorting to 'perfect' market relations. Public technology procurement, as we have defined it, would be extremely difficult if not impossible to accomplish under these conditions, since it would preclude public agencies taking an active role in the innovation process as buyers. This would make interactive learning between organisations impossible – although it is crucial for most innovation processes.[43] It will be recalled that the special virtue of technology procurement as a form of economic relations is that it distributes risk more evenly between buyer and seller, and establishes special forms of communication between them, thus facilitating innovation that would otherwise not be carried out by the sellers.

There are alternative mechanisms by which public authorities could ease the burden of risk borne by technology sellers or suppliers. They are, for the most part, supply-side solutions, such as R&D subsidies. While they are certainly more consistent with market relations, however, supply side solutions such as these have their own serious drawbacks. Thus, survey evidence indicates that R&D subsidies are unduly expensive (Folster 1989; Stoneman 1987: 202) and are often used to finance projects of an inferior quality that are unable to attract private sector financing (Rothwell and Zegveld 1982: ch. 6). Moreover, the rate of return on publicly funded R&D tends to be extremely uncertain and is often exceedingly low (Stoneman, 1987). In contrast, the evidence for public R&D contracts indicates clearly that "government procurement as

[43] Innovations in products for consumer markets may be an exception.

a whole has a positive and substantial effect on private R&D investment" (Lichtenberg, 1988). Thus, there is a strong, though qualified, economic case to be made for the effectiveness of public technology procurement as an instrument of innovation policy. That case is related, as we have emphasised, to the active role played by public agencies as buyers. This provides, among other things, for quality control and "the expression of a clear, consistent set of needs towards which innovative efforts can be directed" (Geroski 1990: 189).

4.2 Demand Structures and Public Technology Procurement[44]

In section 3.2 of this chapter we discussed the structure of demand in private technology procurement. The discussion was organised on the basis of the well known classification of 'monopsony', 'oligopsony' and 'polypsony' as demand structures. This classification was originally applied to PTP by Rothwell and Zegveld (1982).

4.2.1 Monopsony

In our discussion of monopsony, in subsection 3.2.1 of this chapter, we made the point that this condition – domination of the buyer market by a single large buyer – seldom occurs, except in the case of certain "natural monopolies". Though these natural monopolies have usually fallen under public ownership, we were able to examine at least one important case of effective private-sector monopsony (AT&T). We demonstrated that monopsony provides one essential condition for effective technology procurement – i.e., an extremely high concentration of buying-power or 'demand-pull'. But it has the drawback of encouraging the development of rigid and non-innovative relations between buyers and sellers, due to the lack of competitive incentives and market opportunities for the development of new technological products or systems. This has been a key argument for the deregulation of mature industries, such as telecommunications.

The literature on the practice of public technology procurement makes similar points about the structure of so-called 'natural monopolies'. Some of the main concerns about monopsony in PTP seem to revolve around the tendency for monopolists (public or private) within these sectors not only to 'lock-in' to one or another technological trajectory but also to reproduce the same effects in the suppliers towards whom they act as monopsonists (Kanz, 1993). The fairly common practice in public technology procurement of sole-source buying can thus lead to the creation of monopoly on the supply side (Ponssard & Pouvourville, 1982; Williams & Smellie, 1985). And, "If there is only one supplier in a certain market there is considerable danger of an incestuous relationship between it and government, leading to a lack of competitive pressure on this

---

[44] We acknowledge here the contribution of our colleague Christopher Palmberg (Group for Technology Studies, Technical Research Centre of Finland) who put forward many of the ideas discussed in this section of the chapter.

firm and, quite likely, lack of technological performance" (Rothwell & Zegveld, 1982). This is the well-known problem of 'national champions' discussed above in section 4.1 of this chapter. Aversion to risk-taking, a penchant for 'quick fixes', and the desire to avoid electoral backlash from 'picking winners' can exacerbate this problem (Cohen & Noll, 1991).

Notwithstanding these considerations, the high concentration of public demand has been clearly recognised as a strong potential catalyst in technological development (Dalpé, 1987). The condition of monopsony in public demand for new technology is common in large scale projects (Faucher & Fitzgibbons, 1993), which are characterised by high levels of cost and risk together with long-term schedules (Cohendet & Lebeau, 1987) and also by their indivisibility (Shulman, 1980) and structuring effects on the productive system (Faucher, 1990). Examples include: hydro-electric and other energy projects, communications satellites, and new aircraft.

A Canadian survey of such projects found technological innovation to be "most dynamic in projects with high demand concentration", the most technologically innovative projects to be those occurring "fairly early in the development of the industry", and the highest levels of innovative activity to be associated with "projects for which there existed a critical mass of technical, financial and organisational resources" (Faucher and Fitzgibbons 1993: 182). Critical mass could be achieved through a variety of arrangements, including not only complete vertical integration within single organisations but also "strategic alliances" between the public and private sector. However, the pattern with respect to demand structure was an unambiguously positive relation between innovativeness and effective monopsony, complemented by "quasi-market ... arrangements ... organised to control the economic and technical risk associated with the projects" (Faucher and Fitzgibbons 1993: 183).

Monopsonistic conditions in PTP, then, may be justified from the standpoint of innovation under certain circumstances, including an early stage of industry development, a high potential for innovation, and a high level of technical risk. Properly managed, they can be used to develop important sources of competitive advantage within national economies, as a comparison of the UK and Norwegian experiences with North Sea oil has indicated (Cook & Surrey, 1982).[45] The best-known example of this remains the role played by the US Department of Defence (DoD) in the development of the US semiconducter industry. Key aspects of the DoD's role included a combination of R&D subsidies and informed technology procurement. In technology procurement, "policies of dual sourcing helped in the diffusion of semiconductor technology and the DoD's favourable attitude toward purchasing from new technology based firms accelerated NTBF [New Technology Based Firm] formation and growth rates" (Rothwell 1994: 645). Through the adoption of similar policies, public sector monopsonists can avoid the creation of monopolistic or protected supplier industries.[46]

---

[45] See the earlier discussion of this example in section 1.3 of this chapter.

[46] In such industries, "Well entrenched vested interests have an almost congenital propensity to support the status quo and their often well financed attempts to maintain it can ossify industry structure and cause performance to stagnate" (Geroski 1990: 190).

4.2.2 Oligopsony

Oligopsony, in which there are a number of large buyers in the market, was identified in subsection 3.2.3 of this chapter as possibly the ideal structure for technology procurement. In public technology procurement, oligopsony is a fairly common condition. This is partly due to the fact that different public authorities may make different procurement decisions. A notable example of 'oligopolistic' tendencies within the public sector is the US federal government's attempt to achieve complete interchangeability of all computer equipment it purchased through the creation of a standard input/output interface (Rankine 1995: 188 - 189). There was strong industry opposition to this policy, based on fear of losing the competitive advantage vested in each system's distinctive architecture. Against the proposed strategy of declaring only one of these competing interfaces to be the standard for all US federal government agencies, the manufacturers argued that this would prevent them (the agencies) from purchasing the most advanced technologies. This view was "shared by many of the agencies" (Rankine 1995: 189).

Oligopsony in public technology procurement is a feature of many 'infrastructural' sectors where a public authority or public agency may share the market with private firms. Rothwell and Zegveld (1982: 93) point to the example of the US electric utilities, which comprise an oligopsonistic market for capital goods related to electric power generation and transmission. They then go on to discuss the TVA (Tennessee Valley Authority). TVA's role in purchasing goods of this nature has exemplified how "analogous to the role of the 'price leader' in the theory of oligopoly, one of the oligopsonists can play the role of 'quality leader'" (Rothwell and Zegveld 1982: 93).

The role of 'quality leader' can, however, be lost by public agencies due to conditions of industry maturation and globalisation under which the needs of public buyers begin to diverge from those of private ones. This has apparently been the case for UK defence spending in the electronics market, of which it was observed some ten years ago that it could no longer serve as the "engine of electronics industry growth" in the private sector (McKinsey and Company 1988: 75). This was because "Only when Defence electronics procurement is so massive and so 'leading edge' as to create fundamental new technologies ... is there a chance that defence spending can find civic market success" (McKinsey and Company 1988: 75).

Similarly, in the US, the Department of Defence (DoD), which in the 1960s and 1970s played a leading role in the development of the US electronics industry, has in the context of industry maturity and internationalisation "evolved from being a microelectronics technology leader ... to adopting a largely technology follower position" (Kanz 1993: 71). Due to specialisation and obsolescence, DoD's current requirements are "seriously mismatched with commercial priorities", and even the establishment of a new component supplier base will involve "further isolation from mainstream micro-electronic technologies and commercial business methodologies" (ibid.: 71). Suggested remedies include: planning based on the recognition of "long-term, probably permanent, interdependencies among components makers" within an increasingly global industry; "increasing use of commercial standards, practices, and components, wherever feasible"; and revision of the field support service and logistics system "to permit flexibility ...and ... enhance the substitutability at higher assembly levels of components" (Kanz 1993: 72). Through such means, DoD might be able to maintain its previously successful strategies of 'technology leveraging' and 'force multipliers' (ibid.: 72). As this example shows, an influential public sector buyer within an oligopsonistic market needs to address a wide range of issues concerning product standards in order to maintain or re-gain the role of 'quality leader'.

4.2.3 Polypsony

In subsection 3.2.2 of this chapter we pointed out that technology procurement as such is often a practical impossibility under polypsony, the type of demand structure characterised by many buyers. In section 2.3 of this chapter, however, we drew a distinction between cases of public technology procurement where the procuring agency is also the end user and cases where the procuring agency acts as a catalyst, co-ordinator and resource to the benefit of other users. Some of the most successful examples of this type of PTP have been situated in highly polypsonistic consumer markets where it has been possible to identify important societal needs that were not being met by normal market mechanisms. There appears to be an emerging role for this type of PTP which 'organises' the demand of customer groups in fragmented markets (Nilsson, 1996).

Shorter product life-cycles mean that producers for "fragmented" markets will have to "interact with a broad and representative range of users during the product specification and development processes, in order to be able to identify functional segmentation at an early stage and plan accordingly" (Rothwell 1994: 637).[47] Public technology procurement of the 'catalysing' kind is one mechanism whereby user needs which are not prominent at the current time but which might soon constitute important sources of functional segmentation can be articulated and brought into the process of product design and development. It can also be used to improve the competitiveness of industries at an early stage in their development. One example of this is Japan's creation of a quasi-governmental rental agency, the JECC, that aided the growth of a domestic market for computers by rapidly extending the range of domestic users, thereby assisting the development of the Japanese computer industry (Anchordoghy, 1988).

In section 2.3 of this chapter, we referred to the experience of NUTEK, a Swedish public agency which has "empowered users" in a number of instances where products for mass markets were involved. The NUTEK experience also provides examples of the economic case for the use of this type of PTP as a policy instrument (Nilsson, 1996). As we pointed out in subsection 3.2.2 of this chapter, a need for this kind of intervention can arise in situations where a "pay-back gap" occurs within markets (Nilsson, 1994).

---

[47] The term, 'functional segmentation', used here, is explained as follows: "markets can ... be differentiated on a functional basis, and market segmentation can increase as technology develops, i.e., as the number of options the supplier is capable of offering increases and as new users and new uses emerge" (Rothwell 1994: 637).

A prominent example of the 'pay-back gap' occurs in the case of energy efficiency. It may be possible to reduce energy consumption more cheaply than to increase energy production. However, the different economic perspectives of suppliers and end-users create a situation in which "the suppliers calculate and act with a lower rate of return than the users" (Nilsson, 1994: 12). This 'pay-back gap' takes the form of a tendency within the market "to overspend on supply and to disregard the demand opportunities" (ibid.). The risk involved in taking the latter course becomes lower as the degree of supply monopoly becomes more pronounced, so suppliers' incentives to participate in the enhancement of energy efficiency might be decreased by measures such as deregulation that would emphasise their role as profit-maximisers (ibid.: 12). For this reason, organising the demand side is a more viable policy intervention than deregulating or privatising the supply side. Under such conditions, the 'catalysing' kind of PTP is justified in theory, though its degree of success remains to be proven in practice.

4.3 Technology Life Cycles and Public Technology Procurement[48]

In section 3.3 of this chapter we discussed technology procurement in relation to theories and models of the technology life cycle. Our discussion was organised according to a distinction, which we first made in section 2.2 of this chapter, between 'developmental' and 'adaptive' technology procurement. Although the discussion in section 3.3 of this chapter did not concern public authorities as procurers of new technology, it is possible to make this application.

4.3.1 'Developmental' public technology procurement

In subsection 3.3.1 of this chapter we discussed the well known conventional model of the technology life cycle (Utterback & Abernathy, 1975; Utterback, 1994). According to the conventional model, the decisive influence exercised by the demand side in the development of a given technology occurs at a fairly early stage in the life-cycle. A 'bandwagon' is formed as the consequence of the influential buying decision(s) made by one or a few major adopters to procure one of a number of competing 'product-concepts'. Subsequently, product innovation declines and the focus is on refining process techniques for large scale production of the product. The initial winners of the technology-competition and the early adopters who decide its outcome are presumed to reap the greatest economic rewards from subsequent development of the technology (Utterback & Suarez, 1993).

The conventional model of the technology life cycle thus presents a strong theoretical rationale for the use of public technology procurement at an early stage, rather than at a later stage, in the development of a given technology. There is also abundant

[48] We acknowledge here the contribution of our colleague Lena Tsipouri (Centre of Financial Studies, University of Athens) who contributed many of the ideas discussed in this section of the chapter.

empirical evidence to support this outlook. A comprehensive survey of large scale projects involving the development of new technology in one country has found that "public procurement is generally more effective as a positive factor in technological production in earlier stages of the product cycle when technical uncertainties are highest", while in later stages "the shift in emphasis focuses on economic risk, often resulting in less innovative activity" (Faucher and Fitzgibbons 1993: 182).

Other work has shown that public technology procurement has been especially influential in the development of pioneering innovations in technology: "This seems to suggest that government orders are oriented towards high performance products, quite often in an early stage of their product cycle" (Dalpé and DeBresson 1988: 11). For these reasons, Rothwell (1994: 646) ends an extensive review of the literature on public technology procurement with the unambiguous conclusion that PTP "has its greatest impact when the technology is newly emerging".

The diminishing impact of PTP in later stages of an industry's development would appear to be one of the main historical lessons of the US Department of Defence's evolution from 'technology leader' to 'technology follower' in the now mature US electronics industry (Kanz, 1993). However, the growing divergence between military and commercial technological trajectories may be as important a factor (in such cases) as the late developmental stage of the industry. This is what the evidence from other US industries owing their initial 'take-off' to military public technology procurement also suggests (Geroski 1990: 188). Indeed, as Rothwell (1994) observes, "there has been some concern that emphasis in the SDI programme on developing highly sophisticated radiation-proof semi-conductor devices might be leading to a divergence between US civilian and defence technological trajectories in this area" (Rothwell 1994: 646).[49] He goes on to conclude that PTP's "potential for stimulating the growth of infant industries seems significant in areas where evolving public and private sector technologies are not too divergent" (ibid.: 647).

4.3.2 'Adaptive' public technology procurement

These considerations bring us to the case to be made for 'adaptive' public technology procurement. In subsection 3.3.2 of this chapter we considered later theoretical revisions to the conventional model of the technology life cycle – in particular, Dosi's (1982) formulation of 'technological paradigms' and 'technological trajectories' and Silverberg's (1990a) 'self-organisation' approach to the analysis of diffusion processes. The Dosi formulation indicates that the demand side can have a continuing influence on the development of technologies, giving rise in later stages of diffusion to new evolutionary 'branchings'.

Silverberg's (1990a) approach provides techniques for modelling such processes and emphasises the need for especially high levels of 'collective learning' about pay-

[49] The acronym, SDI, used here, refers to the Strategic Defence Initiative launched in the US by the Reagan administration.

backs to resolve the entrepreneurial dilemma of choice between early and late adoption. Early adoption can be highly beneficial to the entrepreneur and essential to broader processes of economic growth and development, but where it involves genuine technological innovation it requires exceptionally high levels of 'absorptive capacity' and the ability to bargain with suppliers from a position of strength. Herein lies a strong justification for what we have called 'adaptive' public technology procurement.

Less-developed countries in Europe (and elsewhere) are generally interested in the early procurement of technologically complex products and systems that can improve their infrastructure and contribute to economic growth and the enhancement of social conditions. Particularly within the EU, transfer payments make it possible for them to make such purchases at a relatively early stage in the development and diffusion of these products or systems. This relatively early adoption of diffusing technologies, together with the inevitable requirements for modification and adaptation to local circumstances, generates strong potentials for technological product innovation in what would normally be considered the mature stage or 'specific phase' of technological development.

Situations such as this deserve close study and analysis, not only because of their high degree of interest from the standpoint of theory and research on innovation, but also because of their strong relevance for public policy. As the 'self-organisation' model described in subsection 3.4.2 of this chapter suggests, countries that engage successfully in 'adaptive' public technology procurement will need to assess their technological opportunities very closely, in order to achieve the correct timing. If they can do so, however, they will stand to gain favourable prices from suppliers and a leading position among competitor nations, due to improved infrastructure. There may be other benefits as well.

4.4 Stimulating Early or Sophisticated Demand through PTP

In section 3.4 of this chapter, we considered the main elements of Porter's (1990) policy perspective on the 'demand-side' and its importance with respect to innovation and the development of 'competitive advantage'. Throughout, we maintained a focus on the relation between various aspects of the demand side and (non-public) technology procurement. Here, we will not review that entire discussion. Rather, we will focus on the one recommended demand side policy instrument that we have hitherto left undiscussed. We refer, of course, to public technology procurement.

In Porter's (1990: 644 - 647) view, public technology procurement is the most direct means that public authorities have of influencing demand conditions, but it can, according to Porter, work either for or against national competitive advantage. Protected markets, usually associated with public or private monopolies that also act as monopsonies, generally work against innovation and competitiveness. Among the domestic firms that benefit from such protection, "innovation and upgrading ... slow down" and "products and services diverge in quality and cost from those demanded internationally (Porter 1990: 645). We have made similar points throughout this discussion – for example, in sections 3.2, 3.4, and 4.2 of this chapter.

Porter (1990: 646) also makes the argument that defence procurement, though it represents a major and early market for some types of sophisticated goods, is often a mixed blessing: "Not only is defence demand often distracting but the goals of a defence agency diverge from those best suited to upgrade industry" (ibid.: 647). We have shown with reference to the US experience that this was not the case in the earlier development of some major "sunrise industries" in the US (subsection 4.2.1, this chapter), but has become increasingly true of the later stages of their development (subsection 4.2.2, this chapter).

However, the perspective we outline here extends well beyond taking a critical view of public sector practices commonly associated with PTP, such as protectionism and defence contracting that disregards civil needs. Public technology procurement, according to Porter (1990: 645 - 646), can be a "positive force" for national competitive advantage under several circumstances. In what follows, we will briefly relate the conditions he specifies to relevant aspects of theory and research contributing to a 'systems' perspective on public technology procurement. In doing so, we will simultaneously elaborate the 'rationale' for PTP. Integrating Porter's 'management' principles for the organisation of public technology procurement with conceptual approaches that can provide deeper insight into the practices he recommends, we will develop a 'systems' view of linkages between 'competitiveness' and 'innovation'.

4.4.1 Providing Early Demand

The first of Porter's (1990) conditions for an effective contribution to national competitive advantage by public technology procurement is that it should "provide early demand for advanced new products or services, pushing its local suppliers into new areas" (ibid. 645). We outlined the basis of this argument in subsection 3.4.3 of this chapter. Earlier, in subsection 3.2.2 of this chapter, we considered how 'network analysis' can be applied to the condition of 'polypsony' that often occurs in markets for advanced new products and services. Such analysis can indicate a need for the involvement of public agencies as entrepreneurs in the formation of new markets.

Network analysis (Håkansson, 1990; Håkansson & Lundgren, 1993) has developed certain implications of the concept of 'organised markets' (Lundvall, 1985) by pointing to the dynamic tension between vertical integration and horizontal technological collaboration among 'learning' organisations (users, producers, and others) in the creation of new product markets. Some analysts have used this approach to argue that a focal organisation providing vertical linkages between users and producers is often necessary to overcome a 'low-level equilibrium trap' facing networks formed around emergent technologies (Teubal et al., 1991). Markets, this argument holds, are only likely to allow the emergence of a focal 'entrepreneur' at a relatively late point in the development of a network based on new technology, given the many *ex ante* co-ordination problems that would have to be solved.

For these reasons, and especially where the further development of a given technology can be shown to have beneficial consequences for economic welfare, a case can be made for public intervention to resolve the co-ordination problems involved in market creation. This would involve an organisation or articulation of demand resulting in, or equivalent to, the formation of a core of innovative users sufficiently large to overcome the problem of otherwise inadequate 'critical mass'. A public agency could perform this role through PTP either directly, as a 'lead user', or indirectly, as a 'catalyser'. In both cases, the public agency, as a focal entrepreneur, would have to represent both the broad range of diversity in user demand and a high level of sophistication in user competence.

4.4.2 Setting Stringent Product Specifications

Porter's (1990) second condition is that "government agencies should set stringent product specifications and seek sophisticated product varieties rather than merely accept what domestic suppliers offer" (ibid.: 645). We made this case in section 2.5 of this chapter and returned to it in subsection 3.4.4 of this chapter. Earlier, in subsection 3.1.1 of this chapter, we pointed out that technology procurement, both as a practice and as a special form of economic relation (Granstrand, 1984), can be usefully explained from the perspective of 'interactive learning' theory (Lundvall, 1988; Lundvall, 1985). This theory's appraisal of the role played by the demand side in product innovation suggests reasons for public 'articulation of demand'.

Interactive learning theory, as we have seen, is a demand-oriented perspective that places more emphasis on the *quality* of demand than upon its simple *quantity* (Lundvall, 1985). Policy-oriented work in this tradition suggests that the important role played by demand should be taken as a basis for strategic intervention in the innovation process (Lundvall, 1988). First, it considers that the tendency of producers to dominate the innovation process might be as great a problem as any lack of user competence. Second, it argues from such reasons that satisfactory innovation may depend on initiatives to establish or restructure user-producer relationships – for example by sustaining existing linkages or by supporting organisations formed to promote new linkages. Third, a closely related consideration is that in periods of sudden technological change, established user-producer relationships may have to be transformed to overcome inertia reinforced by vested interests in existing organised markets.

These aspects of interactive learning theory point very directly to public technology procurement as a policy instrument enabling public agencies to establish new patterns of demand by acting either directly as a user or indirectly on behalf of users in the development of innovations and the creation of 'organised markets' conducive to innovation. This is not only relevant for "traditional" PTP but also for the 'catalysing' kind of PTP, where the public agency does not represent final demand but serves instead as a co-ordinator and supporter. (See section 2.3 and subsection 4.2.3, this chapter.)

4.4.3 Anticipating International Demand

Porter's (1990) third condition is that "Government specifications should be set with an eye to what will be valued in other advanced nations" (ibid.: 645). This was argued with reference to issues concerning standards in section 2.5 and subsection 3.4.4 of this chapter. In addition, we considered in our discussion of 'oligopsony' in subsection 3.2.3 of this chapter an example drawn from theory and research contributing to a systems-oriented 'distributed process' model of innovation (von Hippel, 1988). It showed how public policies apart from standard-setting could be used to encourage domestic firms to act as 'lead users' in setting leading-edge standards for domestic producers, thus contributing to international competitiveness of domestic industry. But the 'distributed process' perspective also suggests a more direct role for public agencies as standard-setting lead users.

As we have noted, the distributed process model (von Hippel, 1988) has identified different potential sources of innovation and revealed that the locus of product innovation certainly does not always reside with product manufacturers. Instead, it is 'moveable' and can be strategically shifted by altering incentive structures and re-ordering relationships between users and producers. This can be done indirectly, by inducing a 'lead user' strategy to restore or advance the innovative competitiveness of producers. But 'lead user' strategies can have a direct application in public technology procurement. It should be recognised in this connection that both approaches depend on the modification of existing incentive structures, and both imply a wider restructuring of existing relationships between users and producers.

4.4.4 Facilitating Innovation

Porter's (1990) fourth condition is that "Government procurement that makes innovation easier works to the benefit of a nation's industry" (ibid.: 645). This is, of course, one of the main arguments of this discussion. It is also one of the main emphases of the systems-oriented 'network', 'interactive learning' and 'distributed process' perspectives considered immediately above. In addition to these, another systems oriented perspective that points to an important role for public authorities in facilitating innovation is the 'chain-linked' model of innovation (Kline & Rosenberg, 1986), a predecessor of 'systems' approaches that we referred to in section 3.1 of this chapter.

The chain-linked model, as previously observed, has placed a strong emphasis on the crucial role played by the 'demand-side' in the innovation process, developing a main focus on product markets and product innovation (Kline & Rosenberg, 1986). It advances the perspective that the 'management of innovation' (including public policy) should recognise the complementary strengths of different types of firms and seek to co-ordinate their efforts through the creation of viable 'chains of innovation' involving linkage structures among firms and other actors. These principles can be directly applied in order to 'ease' innovation in public technology procurement.

4.4.5 Including Competition

Porter's (1990) fifth condition is that "Government procurement must include a strong element of competition if it is to upgrade the local industry" (ibid.: 645). We have made the same argument throughout this discussion – for example, in sections 2.5, 3.2 and 3.4 of this chapter. In subsection 4.2.1 of this chapter, we made this point with explicit reference to public agencies acting as 'monopsonists' in public technology procurement. Earlier, in our discussion of private sector monopsony in subsection 3.2.1 of this chapter, we related this perspective to the systems-oriented theory of "development blocks" (Andersen, 1992; Dahmén, 1970; Edquist & Lundvall, 1993; Stenberg, 1987).

Development block theory refers to the essential importance of competitive dynamics in the form of innovation-induced 'structural tensions' giving rise, through disequilibria, to the formation of a broad set of interconnected producers and users of products, developing interactively, often with the aid of knowledge-producing organisations (Dahmén, 1988). A development block (DB) will include in its most complete form a whole chain of production, as well as independent sources of knowledge. It is a large-scale framework for 'interactive learning'.

Essentially, development blocks are vertically integrated (but still horizontally evolving) user-producer networks formed around radical technological innovations. They differ from 'networks' *per se* primarily in that they have successfully overcome the problem of 'critical mass'. Significantly, 'development blocks' can be discerned *ex ante* as well as *ex post*. Their identification and analysis can therefore be used to guide public policies for economic development based on technological innovation. Competition-oriented public technology procurement (or PTP) is, of course, one important means through which this might be accomplished where no 'natural' entrepreneur is present – especially where the technology in question is extremely complex and the demand for resources and influence is large.

5. SUMMARY

In the later parts (3 and 4) of this discussion, we have considered various aspects of the 'demand-side' and its importance for innovation and technological development. We have also related this discussion to public technology procurement. The policy-oriented discussion of PTP in part 4 paralleled a similar treatment of private sector technology procurement carried out in part 3. We also linked the discussion in part 4 with our earlier discussion, in part 2, of different kinds of public technology procurement. We have shown in the course of this discussion that public technology procurement is not necessarily linked with protectionism, monopoly and favouritism. We also hope to have demonstrated that, as Porter (1990) argues, public technology procurement can be a "positive force" for innovation and the development of competitive advantage. We have indicated the conditions under which PTP is most likely to play this role as a powerful instrument of demand-side innovation policy. We have also indicated where this use of PTP may be inappropriate, and other policy instruments – or none – may be used instead.

In these respects, the perspective that we have developed on public technology procurement is quite different from the perspective underlying the current regime of EU directives governing public procurement. We described this regulatory regime and its theoretical rationale in part 1. There, we argued that the EU procurement rules have restricted public technology procurement, both in terms of limiting the scope of action available to public agencies and in terms of diminishing the possibilities of interactive learning between public agencies and their suppliers. These arguments concerned the *regulatory* aspect of EU policy. We also drew attention to the fact that EU policy on public procurement has neglected the *strategic* aspect of public technology procurement (i.e., its use as an instrument of innovation policy). The historical reasons for this 'oversight' have been reinforced with a theoretical rationale provided by mainstream economics in the form of 'auction theory'. But 'auction theory' is a body of work that does not consider public technology procurement as a process of innovation.

The 'innovation theory' perspective that we have elaborated on public technology procurement not only points to the severe limitations of 'auction theory' as a basis for the regulation of PTP but also draws attention to the organisational requirements of effective public technology procurement. In these respects, the 'innovation theory' perspective provides important foundations for the development of a strategic approach to public technology procurement – something that is now absent from European Union policy regarding public procurement. We will return to this issue after having complemented the theoretical outlook elaborated here with the insights provided by empirical case studies of public technology procurement as an innovation policy instrument.

REFERENCES

Akerlof, G. (1970). The market for lemons: Quality uncertainty and the market mechanism. *Quarterly Journal of Economics, 84*(3), 488 - 500.

Allen, R. C. (1983). Collective invention. *Journal of Economic Behavior and Organization, 4*(1 - 24).

Anchordoghy, M. (1988). Mastering the market: Japanese government targeting of the computer industry. *International Organization, 42*, 509 - 543.

Andersen, E.-S. (1992). Approaching national systems of innovation. In B.-Å. Lundvall (Ed.), *National systems of innovation: Towards a theory of innovation and interactive learning* (pp. 95 - 115). London: Pinter Publishers.

Arrow, K. (1962). The economic implications of learning by doing. *Review of Economic Studies, XXIX*(80).

Arrow, K. (1974). *The limits of organization*. New York: W. W. Norton & Company.

Arrow, K. J. (1985). The economics of agency. In J. W. Pratt & R. J. Zeckhauser (Eds.), *Principal and agents: The structure of business* (pp. 37 - 51). Cambridge, Massachusetts: Harvard Business School Press.

Arthur, W. B. (1988). Competing technologies: An overview. In G. Dosi, C. Freeman, R. Nelson, G. Silverberg, & L. Soete (Eds.), *Technical change and economic theory*. London: Pinter Publishers.

Baldwin, R. (1987). Preferential government purchasing policies. In K. Jones & J. Kenen (Eds.), *Handbook of international economics*. Amsterdam: Elsevier.

Baldwin, W., & Scott, J. (1987). *Market structure and technological change*. London: Harwood Academic Publishers.

Bannock, G., Baxter, R. E., & Rees, R. (1986). *Penguin dictionary of economics*. (Third ed.). Harmondsworth: Penguin Books.

Barfield, C. (1994). Flat panel displays – A second look. *Issues in Science and Technology, 1994 - 1995*(Winter), 21 - 25.

Bergström, E. (1980). *Projektorientierad marknadsföring*. Malmö: Liber Läromedel.

Blume, S. (1981). *Science policy research – The state of the art and implications for policy*. Stockholm: Forskningsrådsnämden.

Burns, T., & Stalker, G. M. (1961). *The management of innovation*. London: Tavistock.

Burton, J. (1983). *Picking losers? The political economy of industrial policy*. London: Institute of Economic Affairs.

Business International Ltd. (1991). *Public procurement in Europe: How EC rules will open up new cross-border markets*. London: Business International Ltd.

Carlsson, B. (Ed.). (1995). *Technological systems and economic performance: The case of factory automation*. Dordrecht: Kluwer Academic Publishers.

Carlsson, B., & Stankiewicz, R. (1995). On the nature, function and composition of technological systems. In B. Carlsson (Ed.), *Technological systems and economic performance: The case of factory automation*. Dordrecht: Kluwer Academic Publishers.

Choffray, J.-M., & Lilien, G. (1978). Assessing response to industrial marketing strategy. *Journal of Marketing, 42*(2), 20 - 31.

Clark, K. B. (1985). The interaction of design hierarchies and market concepts in technological evolution. *Research Policy, 14*, 235 - 251.

Cohen, L. R., & Noll, R. G. (1991). *The technology pork barrel*. Washington, D.C.: The Brookings Institution.

Cohen, W., & Levin, R. (1989). Empirical studies of innovation and market structure. In R. Schmalensee & R. Willing (Eds.), *Handbook of industrial organization*. Amsterdam: North Holland.

Cohen, W. M., & Levinthal, D. A. (1990). Absorptive capacity: A new perspective on learning and innovation. *Administrative Science Quarterly, 35*(March), 128 - 152.

Cohendet, P., & Lebeau, A. (1987). *Choix stratégiques et grands programmes civils*. Paris: Économica.

Cohendet, P., & Llerena, P. (1997). Learning, technical change and public policy: How to exploit and create diversity. In C. Edquist (Ed.), *Systems of innovation: Technologies, organizations and institutions* ((In Press.) ed.,). London: Pinter Publishers/Casell Academic.

Cook, L., & Surrey, J. (1982). *Government policy for the offshore supplies industry* (Occasional Papers Series). Sussex: Science Policy Research Unit, University of Sussex.

Cowan, R. (1995). The informatisation of government: From choice of technology to economic opportunity. *STI Review*(16), 195 - 223.

Dahmén, E. (1970). *Entrepreneurial activity and the development of Swedish industry, 1919-1939*. Homewood: American Economic Association Translation Series.

Dahmén, E. (1988). 'Development blocks' in industrial economics. *Scandinavian Economic History Review, 1*.

Dalpé, R. (1987). *Politique d'achat et développement technologique*. Quebec: Conseil de la science et de la technologie, Quebec.

Dalpé, R. (1994). Effects of government procurement on industrial innovation. *Technology in Society, 16*(1), 65 - 83.

Dalpé, R., & DeBresson, C. (1988). *Le secteur public comme premier utilisateur d'innovations* (88 - 10). Montréal: Centre for Research on Development of Industry and Technology.

Dertoutzos, M. L., Lester, R. K., & Solow, R. M. (1989). *Made in America*. Cambridge, Massachusetts: MIT Press.

Dosi, G. (1982). Technological paradigms and technological trajectories. *Research Policy, 11*, 147 - 163.

Eads, G., & Nelson, R. R. (1971). Government support of advanced civilian technology: Power reactors and the supersonic transport. *Public Policy*.

Edquist, C. (1994). Technology policy: The interaction between governments and markets. In G. Aichholzer & G. Schienstock (Eds.), *Technology policy: Towards an integration of social and ecological concerns* (pp. 67 - 91). New York: Walter De Gruyter.

Edquist, C. (1996). Government technology procurement as an instrument of technology policy. In M. Teubal & Associates (Eds.), *Technological infrastructure policy* ((In Press.) ed.,). Netherlands: Kluwer Academic Publishers.

Edquist, C. (1997). Systems of Innovation approaches: Their emergence and characteristics. In C. Edquist (Ed.), *Systems of innovation: Technologies, organisations and institutions* ((In Press.) ed.,). London: Pinter Publishers/Casell Academic.

Edquist, C., & Hommen, L. (1998). Public technology procurement and innovation theory. In C. Edquist & F. Texier (Eds.), *ISE: Innovation systems and European integration [CD-ROM]*. Linköping, Sweden: Systems of Innovation Research Programme, Department of Technology and Social Change, Linköping University.

Edquist, C., & Hommen, L. (1999). Systems of innovation: Theory and policy for the demand side. *Technology In Society, 21*, 63 - 79.

Edquist, C., Hommen, L., Johnson, B., Lemola, T., Malerba, F., Reiss, T., & Smith, K. (1998). *The ISE policy statement: The innovation policy implications of the 'Innovation Systems and European Integration' research project*. Linköping, Sweden: Unitryck (University of Linköping Press).

Edquist, C., Hommen, L., & McKelvey, M. (1998). Innovations and employment in a systems of innovation perspective. In C. Edquist & F. Texier (Eds.), *ISE: Innovation systems and European integration [CD-ROM]*. Linköping, Sweden: Systems of Innovation Research Programme, Department of Technology and Social Change, Linköping University.

Edquist, C., & Johnson, B. (1997). Institutions and organizations in systems of innovation. In C. Edquist (Ed.), *Systems of innovation: Technologies, organizations and institutions* (pp. 41 - 63). London: Pinter Publishers/Casell Academic.

Edquist, C., & Lundvall, B.-Å. (1993). Comparing the Danish and Swedish national systems of innovation. In R. Nelson (Ed.), *National systems of innovation: A comparative study*. Oxford: Oxford University Press.

Ergas, H. (1987). The importance of technology policy. In S. P. & P. Dasgupta (Eds.), *Economic policy and technological performance*. Cambridge: Cambridge University press.

European Commission. (1992). Communication to the Council and European Parliament on the principle of subsidiarity. *European Documents, 1804*(05), 14.

European Commission. (1997). *Public procurement in the European Union: Exploring the way forward*. Brussels-Luxembourg: DG XV.

Faucher, P. (1990, 27 May). *Large scale projects and technological development: An institutional approach.* Paper presented at the Annual Conference of the Canadian Political Science Association, Victoria.

Faucher, P., & Fitzgibbons, K. (1993). Public demand and the management of technological risk in large-scale projects. *Science and public Policy, 20*(3), 173 - 185.

Finsinger, G. (1988). Non-competitive and protectionist government purchasing behaviour. *European Economic Review, 32*.

Flamm, K. (1987). *Targeting the computer: Government support and international competition*. Washington, D.C.: The Brookings Institution.

Flamm, K. (1988). *Creating the computer*. Washington, D.C.: The Brookings Institution.

Folster, S. (1989). *The efficiency of R&D subsidies*. Stockholm: Industrial Institute for Social and Economic Research.

Foray, D. (1989). Les modeles de compétition technologique: Un revue de la littérature. *Revue d'Économie Industrielle, 48*.

Foray, D. (1997). Generation and distribution of technological knowledge: Incentive structures, norms and institutions. In C. Edquist (Ed.), *Systems of innovation: Technologies, organizations and institutions* ((In Press.) ed.,). London: Pinter Publishers/Casell Academic.

Freeman, C. (1991). Networks of innovators: A synthesis of research issues. *Research Policy, 20*(5), 499 - 514.

Fridlund, M. (1993). *The 'development pair' as a link between systems growth and industrial innovation: Cooperation between the Swedish State Power Board and the ASEA company* (Trita HST Working Paper 93/9). Stockholm: Royal Institute of Technology, Department of History of Science and Technology.

Gelsing, L. (1992). Innovation and the development of industrial networks. In B.-Å. Lundvall (Ed.), *National systems of innovation: Towards a theory of innovation and interactive learning* (pp. 116 - 128). London: Pinter Publishers.

Geroski, P. A. (1990). Procurement policy as a tool of industrial policy. *International Review of Applied Economics, 4*(2), 182 - 198.

Glassman, J. (1993, October 8). NAFTA and trade are tiny parts of a job revolution. *The Washington Post*, pp. G1.

Goodman, A., & Saunders, W. (1985). US federal regulation of foreign involvement in aviation: Government procurement and national security. *Journal of World Trade, 54*.

Graham, M. (1988). R&D and competition in England and the United States. *Business History Review, 62*, 261 - 285.

Granstrand, O. (1984). *Technology procurement as a special form of buyer-seller interaction in industrial markets: CIM report no. 84:06* (CIM Report 84:06). Göteborg: Chalmers University of Technology, Department of Industrial Management.

Granstrand, O., & Jacobsson, S. (1983). *Innovation economy and business development: A study of innovation take-overs: CIM report no. 83:08* (CIM Report 83:08). Göteborg: Chalmers University of Technology, Department of Industrial Management.

Granstrand, O., & Sigurdson, J. (Eds.). (1985). *Technological innovation and industrial development in telecommunications: The role of public buying in the telecommunication sector in the Nordic countries*. Lund: Research Policy Institute.

Grant, B., & Gadde, L.-E. (1984). *Automotive component supply strategies: CIM report no. 84:03* (CIM Report 84:03). Göteborg: Chalmers University of Technology, Department of Industrial Management.

Grübler, A. (1989, 14 - 16 June). *Diffusion: Long term patterns and discontinuities*. Paper presented at the International Conference on Diffusion of Technologies and Social Behavior, Laxenburg.

Ham, R. M., & Mowery, D. C. (1995). Enduring dilemmas in US technology policy and the Clinton administration. *California Management Review*.

Hansen, R. (1988). Auctions with endogenous quality. *Rand Journal of Economics, 29*(1), 44 - 58.

Harris, M., & Townsend, R. M. (1985). Allocation mechanisms, asymmetric information and the 'revelation principle'. In G. Fiewel (Ed.), *Issues in contemporary micro-economics and welfare* (pp. 379 - 394). Albany: State University of New York Press.

Hartley, K. (1987). Public procurement and competitiveness: A Community market for military hardware and technology? *Journal of Common Market Studies, XXV*(3), 237 - 247.

Hayek, F. A. (1948). Economics and knowledge and the use of knowledge in society. In F. A. Hayek (Ed.), *Individualism and the economic order*. Chicago: University of Chicago Press.

Hellman, J.-L. (1993). *Supplier-customer interaction in product development: A case study of Nippon Steel's cooperation with Toyota*. Kunitachi, Tokyo: Hitotsubashi University, Institute of Business Research.

Hidefjäll, P. (1997). *The pace of innovation: Patterns of innovation in the cardiac pacemaker industry*. Unpublished Ph.D. Dissertation, Linköping University, Linköping, Sweden.

Holmes, J. (1986). The organization and locational structure of production subcontracting. In A. J. Scott & M. Storper (Eds.), *Production, work, territory: The geographical anatomy of industrial capitalism* (pp. 80 - 106). Boston: Allen & Unwin.

Håkansson, H. (Ed.). (1987). *Industrial technological development: A network approach*. London: Croom Helm.

Håkansson, H. (1989). *Corporate technological behaviour: Cooperation and networks*. London: Routledge.

Håkansson, H. (1990). Technological collaboration in industrial networks. *European Management Journal, 8*(3), 371 - 379.

Håkansson, H., & Lundgren, H. (1993). *Industrial networks and technological innovation*. Uppsala/ Stockholm: Network for Industry and Market Studies.

Jeanrenaud, F. (1984). Marchés publics et politique économique. *Annales de l'Économie Publique, Sociale et Coopérative, 72*(2).

Jeanrenaud, F., & Meyer, L. (1984). L'incidence a moyen et longe terme des commandes de matériel de télecommunication: Deux études de cas. *Annales de l'Économie Publique, Sociale et Coopérative, 72*(2).

Kamien, M., & Schwartz, N. (1982). *Market structure and innovation.* Cambridge: Cambridge University Press.

Kanz, J. W. (1993). Technology, globalization and defence: Military electronics strategies in a changing world. *International Journal of Technology Management, 8*(1/2), 59 - 76.

Keyzer, M. (1968). L'importance et la politique des achats publics. *Hommes et Techniques*(278).

Kline, S. J., & Rosenberg, N. (1986). An overview of innovation. In R. Landau & N. Rosenberg (Eds.), *The positive sum strategy: Harnessing technology for economic growth* (pp. 275 - 304). Washington, D. C.: National Academy Press.

Levin, R. C. (1982). The semiconductor industry. In R. R. Nelson (Ed.), *Government and technical progress – A cross-industry analysis* (pp. 9 - 100). New York: Pergamon Press.

Lichtenberg, F. (1988). The private R&D investment response to federal design and technical competitions. *American Economic Review, 78*, 550 - 559.

Lundvall, B.-A. (1988). Innovation as an interactive process: From user-producer interaction to the national system of innovation. In G. Dosi, C. Freeman, R. Nelson, G. Silverberg, & L. Soete (Eds.), *Technical change and economic theory* (pp. 349 - 369). London: Pinter Publishers.

Lundvall, B.-A. (1991). Networking and interaction for innovation: The organized market and the productivity slowdown. In OECD (Ed.), *TEP: The challenge for economic policy*. Paris: OECD.

Lundvall, B.-Å. (1985). *Product innovation and user-producer interaction.* Aalborg: Aalborg University Press.

Lundvall, B.-Å. (Ed.). (1992). *National systems of innovation: Towards a theory of innovation and interactive learning*. London: Pinter Publishers.

Maddock, I. (1983). *Civilian exploitation of defence technology*. London: National Economic Development Office.

Mansfield, E., & Rapoport, J. (1977). Social and private rates of return from industrial innovations. *Quarterly Journal of Economics*(May), 221 - 240.

Martin, J. F. (1996). *The EC public procurement rules: A critical analysis*. Oxford: Clarendon Press.

Mattson, L.-G. (1978). *Systemförsäljning*. Stockholm: Marknadstekniskt Centrum.

McAfee, R. P., & McMillan, J. (1986). Bidding for contracts: A principal-agent analysis. *Rand Journal of Economics, 17*(3), 326 - 338.

McAfee, R. P., & McMillan, J. (1987). Auctions and bidding. *Journal of Economic Literature, 15*(June), 699 - 738.

McCrudden, C. (1994). *Public procurement and equal opportunities in the European Community: A study of 'contract compliance' in the member states in the European Community and under Community law.* Brussels: EC.

McKinsey and Company. (1988). *Strengthening competitiveness in UK electronics.* London: National Economic Development Office.

McLachlan, L. (1985). Discriminatory public procurement, economic integration, and the role of bureaucracy. *Journal of Common Market Studies, 23.*

Milgrom, P. (1989). Auctions and bidding: A primer. *Journal of Economic Perspectives, 3*(3), 3 - 22.

Milgrom, P., & Weber, R. J. (1982). A theory of auctions and competitive bidding. *Econometrica, 50*(5), 1089 - 1122.

Mowery, D., & Rosenberg, N. (1979). The influence of market demand upon innovation: A critical review of some recent empirical studies. *Research Policy, 8*, 102 - 153.

Mowery, D. C. (1995, January 11 - 12). *US postwar technology policy and the creation of new industries.* Paper presented at the Conference on Creativity, Innovation and Job-Creation, Oslo, Norway.

Mowery, D. C., & Rosenberg, N. (1982). The commercial aircraft industry. In R. R. Nelson (Ed.), *Government and technical progress – A cross-industry analysis* (pp. 101 - 161). New York: Pergamon Press.

Mowery, D. C., & Rosenberg, N. (1993). The US innovation system. In R. Nelson (Ed.), *National systems of innovation: A comparative study*. Oxford: Oxford University Press.

Nelson, R., & Winter, S. (1977). In search of a useful theory of innovation. *Research Policy, 6*(1), 36 - 76.

Nelson, R., & Winter, S. (1982). *An evolutionary theory of economic change*. Boston, Massachusetts: The Bellknap Press of Harvard University Press.

Nelson, R. R. (1984). *High-technology policies: A five-nation comparison*. Washington, D.C.: American Enterprise Institute.

Nilsson, H. (1994). *Technology procurement for market transformation – a customer tool for sustainable design and manufacturing* (Discussion Paper.). Stockholm: NUTEK.

Nilsson, H. (1996,). *Looking inside the black box of market transformation*. Paper presented at the ACEEE.

OECD. (1991). *Information technology standards: The economic dimension*. Paris: OECD.

OECD. (1996a). *Technology, productivity and job creation – Vol.1: Highlights*. Paris: OECD.

OECD. (1996b). *Technology, productivity and job creation – Vol.2: Analytical report*. Paris: OECD.

OECD/ICCP. (1996). *Mobile cellular communications: Pricing strategies and competition*. Paris: OECD.

Parker, P. (1990). The 1988 local government act and compulsory competitive tendering. *Urban Studies, 5*.

Parkinson, S. T. (1982). The role of user in successful new product development. *R&D Management, 12*(3), 123 - 131.

Parsons, H. (1988). Economic principles in the private and public sectors. *Policy and Politics, 16*.

Pavitt, K. (1984). Sectoral patterns of technical change: Towards a taxonomy and a theory. *Research Policy, 13*, 343 - 373.

Perroux, C. (1969a). Note sur la notion de la pole de croissance. In C. Perroux (Ed.), *L'economie du XXé siecle* (Vol. 3d,). Paris: Presses Universitaires de France.

Perroux, C. (1969b). Les espaces economiques. In C. Perroux (Ed.), *L'economie du XXé siecle* (Vol. 3d,). Paris: Presses Universitaires de France.

Phillips, A. (1980). Organizational factors in R&D and technological change: Market failure considerations. In D. Sahal (Ed.), *Research, development and technological innovation* (pp. 105 - 205). Lexington, Massachusetts: Lexington Books.

Ponssard, J. P., & Pouvourville, G. D. (1982). *Marchés publiques et politique industrielle*. Paris: Économica.

Porter, M. E. (1990). *The competitive advantage of nations*. London: MacMillan.

Rankine, L. J. (1995). The role of users in information technology standardisation. *STI Review*(16), 177 - 194.

Riley, J. G., & Samuelson, W. F. (1981). Optimal auctions. *American Economic Review, 71*(3), 381 - 392.

Robinson, P., Faris, C., & Wind, Y. (1967). *Industrial buying and creative marketing*. London: Allyn & Bacon Ltd. and The Marketing Science Institute.

Roll, L. (1982). *The mixed economy*. London: MacMillan.

Rosenberg, N. (1972). Factors affecting the diffusion of technology. *Explorations in Economic History, 10*.

Rosenberg, N. (1976). *Perspectives on technology*. Cambridge: Cambridge University Press.

Rothwell, R. (1994). Issues in user-producer relations in the innovation process: The role of government. *International Journal of Technology Management, 9*(5/6/7), 629 - 649.

Rothwell, R., & Zegveld, W. (1982). *Industrial innovation and public policy*. London: Frances Pinter.

Rothwell, R., & Zegveld, W. (1985). *Reindustrialization and technology*. London: Longmans.

Sahal, D. (1981). *Patterns of technological innovation*. New York: Addison Wesley.

Sahal, D. (1985). Technology guide-posts and innovation avenues. *Research Policy, 14*, 61 - 82.

Scherer, F. M. (1980). *Industrial market structure and economic performance*. (Second ed.). Chicago: Rand McNally.

Schmidt, S. (1992). *Negotiating technical change through standards: Technical coordination in market and committees*. Cologne: Max Planck Institute.

Sheth, J. (1973). A model of industrial buyer behavior. *Journal of Marketing, 37*(4), 50 - 56.

Shulman, P. R. (1980). *Large scale policy-making*. Elsevier: New York.

Silverberg, G. (1987). Technical progress, capital accumulation, and effective demand: A self-organization model. In D. Batten, J. Casti, & B. Johansson (Eds.), *Economic evolution and structural adjustment*. New York: Springer-Verlag.

Silverberg, G. (1990a). Adoption and diffusion as a collective evolutionary process. In C. Freeman & L. Soete (Eds.), *New explorations in the economics of technological change* (pp. 177 - 192). London: Pinter Publishers.

Silverberg, G. (1990b). Dynamic vintage models with neo-Keynesian features. In OECD (Ed.), *Technology and productivity: The challenge for economic productivity*. Paris: OECD.

Silverberg, G., Dosi, G., & Orsenigo, L. (1988). Innovation, diversity and diffusion: A self-organisation model. *Economic Journal, 98*, 1032 - 1054.

Snyder, F. (1990). Ideologies of competition: Two perspectives on the completion of the internal market. In F. Snyder (Ed.), *New directions in European Community law*. London: Croom Helm.

Stenberg, L. (1987). *Utvetlingsblock i förnyelse av svensk industri* (Ds I 1987:3). Stockholm: Industridepartementet.

Stigler, G. J. (1964). A theory of oligopoly. *Journal of Political Economy, 72*(1), 44 - 61.

Stoneman, P. (1987). *The economic analysis of technology policy*. Oxford: Oxford University Press.

Teubal, M., Yinnon, T., & Zuscovitch, E. (1991). Networks and market creation. *Research Policy, 20*, 381 - 392.

Tovias, S. (1990). The impact of liberalizing government procurement policies of individual EC countries on trade with non-members. *Weltwirtschaftliches Archiv*(722).

Townsend, J. (1976). *Innovation in coal mining: The Anderton shearer-loader*. Brighton: Science Policy Research Unit, University of Sussex.

Tushman, M. (1977). Special boundary roles in the innovation process. *Administrative Science Quarterly, 22*, 587 - 605.

Tushman, M., & Anderson, P. (1986). Technological discontinuities and organizational environments. *Administrative Science Quarterly, 31*, 439 - 465.

Utterback, J., & Abernathy, W. (1975). A dynamic model of process and product innovation. *OMEGA, 3*(6), 639 - 656.

Utterback, J. M. (1994). *Mastering the dynamics of innovation: How companies can seize opportunities in the face of technological change*. Boston, Massachusetts: Harvard Business School Press.

Utterback, J. M., & Suarez, F. F. (1993). Innovation, competition and industry structure. *Research Policy, 22*, 1 - 21.

Utterly, J., & Harper, B. (1993). The political economy of competitive tendering. In D. Clarke & H. Pitelis (Eds.), *The political economy of privatization*. London: MacMillan.

Vickrey, W. (1961). Counterspeculation, auctions and competitive sealed tenders. *Journal of Finance, 16*(March), 8 - 37.

von Hippel, E. (1988). *The sources of innovation*. Oxford: Oxford University Press.

Webster, F., & Wind, Y. (1972). A general model of organizational buying behaviour. *Journal of Marketing, 36*(2), 12 - 19.

Westling, H. (1996). *Cooperative procurement: Market acceptance for innovative energy-efficient technologies* (1996 - 3). Stockholm: NUTEK B.

Williams, R., & Smellie, R. (1985). Public purchasing: An administrative Cinderella. *Public Administration, 63*(1), 23 - 39.

Williamson, O. E. (1975). *Markets and hierarchies: Analysis and anti-trust implications*. New York: Free Press.

WS Atkins Management Consultants, & Associates. (1988). *The "cost of non-Europe" in public sector procurement*. (Vol. 5). Luxembourg: Office for Official Publications of the European Communities.

Zysman, J. (1977). *Political strategies for industrial order*. Berkeley: University of California Press.

OVERVIEW OF CASE STUDIES

C. Edquist, L. Hommen and L. Tsipouri

INTRODUCTION

Part II of this volume, because of both its size (by far the greatest part of this book) and the central importance to our work of empirical evidence, merits its own introduction. We provide this in the form of the present Overview of Case Studies.

The Overview first discusses (in part 1) the rationale for our selection of cases. Subsequently (in part 2), it addresses questions concerning generalisation from cases. Finally (in part 3), it presents a synopsis of the case studies. After stressing the importance of differentiating between 'developmental' and 'adaptive' public technology procurement, this part of the discussion presents summaries of two groups of case studies.

The final part of this Overview is also its most extensive. Our two groups of case studies – three from Sweden and six on telecommunications – represent 'national' and 'sectoral' contexts respectively. Each case is briefly summarised here. The summaries all highlight one or another aspect of the distinction between 'developmental' and 'adaptive' public technology procurement. This key distinction also figures importantly in our later discussions, in Part III, of Conclusions (Chapter 11, part 4) and Policy Implications (Chapter 12).

1. SELECTION OF CASES

The case studies that we present in this part (Part II) of the volume are cases of innovative public technology procurement carried out in response to perceived societal needs, combined with perceived requirements for 'demand side' intervention by public agencies responsible for meeting these needs. These cases occurred in a number of countries that are now EU member states. Namely, these countries are: Austria, Finland, France, Greece, Italy and Sweden.

Three cases are drawn exclusively from Sweden and concern the development of different technologies at different times: the procurement by Swedish public agencies of high speed trains and school computers in the 1980s, and of high voltage, direct current (HVDC) transmission systems in the middle decades of this century. An additional six of our case studies represent all of the countries mentioned above and concern the development of digital switching systems (as well as Finnish and Swedish mobile telephony) in the telecommunications sector during the later decades of the post-war period.

As this list shows, all of our cases deal with large-scale (i.e., national – level) public sector infrastructural projects.[1] A possible exception to this rule is the case of "The Computer in the School". From a systems of innovation perspective, however, school computers can be regarded as a technical component of the 'knowledge infrastructure'. This term, according to Smith's (1997) definition, refers to "a complex of private and public organisations whose role is the production, maintenance, distribution, management, and protection of knowledge", and which possesses "technical and economic characteristics that are not dissimilar to those of physical infrastructure" (Smith, 1997: 94 - 95).

The theoretical discussion in Chapter 1 provides a basic rationale for this type of case selection. It points out that effective technology procurement depends on a high concentration of buying power and a comprehensive 'articulation of demand'. These conditions often tend to occur in conjunction with certain 'natural monopolies' – which have all, or nearly all, been publicly controlled in most countries (see Chapter 1, subsection 3.2.1). Sectors of this kind are, of course, precisely those in which specifically *public* technology procurement has played an important role in the past and can be expected to do so in the future. According to existing theory and research, this is especially likely to be the case with the development of new technology in certain contexts. Specifically, these are in large scale projects characterised by high levels of cost and risk, long term schedules, indivisibility, and structural effects on the productive system (ibid., subsection 4.2.1). Our cases of public technology procurement conform to these criteria.

Another criterion met by the selection of cases is the representation of both 'developmental public technology procurement' and 'adaptive public technology procurement' – terms defined briefly in our Introduction and explained more fully in Chapter 1 (section 2.2). As shown in the following synopsis of case studies (part 3 of this Overview), many of our cases concern developmental public technology procurement. However, some deal with cases of adaptive public technology procurement. In most there is a mix of 'developmental public technology procurement' and 'adaptive public technology procurement' elements. Thus, our selection of cases indicates a broad scope for public technology procurement policies.

As mentioned in our Introduction, the selection of cases also includes instances of both "success" and "failure". However, it is important to note in this connection that "success" and "failure" were not determined according to a uniform standard applied to all cases. Although we contemplated this strategy, we rejected it because a common

---

[1] A widely accepted definition of 'infrastructure' refers to "structural elements of an economy which facilitate the flow of goods and services", such as "communications, and transport ... housing, sewerage, power systems, etc.". It recognises that such facilities "are usually, though not necessarily, provided by public authorities and may be regarded as a prerequisite for economic growth in an economy" (Pearce, 1986: 204).

definition of 'success' could not be arrived at. In the absence of a common definition (of success), we judged each case to be a success or failure 'on its own terms'. That is, every case of a public technology procurement process was considered in relation to the particular objectives that were established for it. The evaluative question asked was, 'did the procurement process fail to realise its self-declared objectives or not?'. On this basis, it became possible to recognise comparable cases of failure and success without imposing a normative standard on any of the cases.[2]

Finally, as already mentioned, the selection of cases also covers several different countries. This group includes both larger and smaller national economies (e.g., France *vs.* Finland) and more-developed and less-developed ones (e.g., Sweden *vs.* Greece). It also encompasses a wide variety of institutional settings. Thus, common themes, conclusions and implications emerging from the analysis of the case studies can be expected to be applicable across a similar diversity among member states within the European Union as a whole.

2. GENERALISATION FROM CASE STUDIES

Case study research and analysis is commonly confronted with objections to its limited capacity for "generalisation". Such objections are, more often than not, based on a misunderstanding of case study methods, combined with a view of "generalisations ... as an end in themselves" that is tied to "a conception of social science as the search for order and regularity" (Sayer, 1992: 100). In other words, what is often (wrongly) demanded are *statistical* generalisations. These are statements of frequency with respect to relationships among variables occurring within a 'representative sample' that can be readily generalised to a larger 'population' or 'universe'.

Case studies are not, however, properly intended to make statistical generalisations. Rather, case studies produce *analytical* generalisation, in which "the investigator is trying to generalise a particular set of results to some broader theory" (Yin, 1994: 36). Thus, the object of case study research is not to select 'representative' cases from which it will be possible to generalise directly to a wide universe of other cases. Instead, cases are selected for the purpose of generalising findings to 'theory', after testing to determine whether the results of initial cases can also be found to occur in comparable cases. This use of *replication logic*, as it applies to our analysis, will be further discussed in Chapter 11 (in Part III of this book).

Case study procedures involving further analysis based on structured comparisons among similar cases are particularly well suited for the development of 'typological' theory. Such theory is not generated solely by means of induction. Rather, it proceeds from the empirical 'grounding' or particularisation of general explanatory theories in

[2] In comparative policy studies, such use of "context" to compensate for the absence of "rules" for making comparisons is particularly appropriate where "situations may be so different that existing behavioural methods do not apply, but at the same time there may be striking similarities and differences of behaviour that we want to examine systematically" (Ashford, 1992: 13).

analytically relevant cases. This leads to a clearer specification of "typical patterns and ranges of variation" (Diesing, 1971: 195 - 196). What results from this analytical-inductive approach to theory development is thus "rich, differentiated theory", which is cast in the form of contingent generalisations.[3] Thereby, it possesses a capability for more discerning explanations than those offered by formal deductivist theories. Moreover, this kind of theory "also has far greater practical value for policy-makers, because it enables them to make more discriminating diagnoses of emerging situations" (George, 1979: 59).

The generalisations from the case study analysis that we present in Part III of this book are thus intended to further theory and policy developed from a Systems of Innovation perspective. The general theoretical perspective 'grounded' in the cases is outlined in considerable detail in Chapter 1 (Part I of this book). The particular themes or dimensions drawn from that perspective for the systematic comparison of public technology procurement across cases are subsequently identified and explained in Chapter 11 (in Part III of this book).

The general 'type' of case that we consider here concerns a recent process of public technology procurement taking the form of a large-scale project in public infrastructure development. It also deals with a procurement process occurring within a country that is now an EU member state. The 'contingencies' addressed are identified in the case studies themselves and further discussed in Chapter 11 (Part III of this book), which analyses and presents our main findings and conclusions from the case studies.

3. SYNOPSIS OF THE CASES

As mentioned above, our cases fall into two main groups. First, there are three cases drawn exclusively from the country of Sweden. Taken together, they provide varied perspectives on the process of public technology procurement within one 'national' system of innovation. These studies provide a basis for sectoral comparisons within a national context. The second group of case studies deals exclusively with the telecommunications sector but documents public technology procurement in a number of different countries. The telecommunications case studies thus make possible national comparisons within a common sectoral context. We summarise the two groups of cases below, in sections 3.2 and 3.3 of this chapter, respectively.

Prior to presenting the summaries, we draw attention to a further important distinction that can be made among the cases. This is the differentiation that we have made, conceptually, between developmental public technology procurement and adaptive public technology procurement. (These terms are defined both in our Introduction [part

---

[3] 'Contingent generalisations' indicate how general rules, or propositions, are modified by different kinds or 'types' of circumstance, as in the following illustration: "Contrast, for example, a general explanatory theory such as 'war is the result of miscalculation' with a richer, more differentiated theory comprised of contingent generalisations that identify the different conditions under which different types of miscalculations lead to different types of war outbreaks" (George, 1979: 59).

2] and in Chapter 1 [section 2.2].) As indicated in our later discussion of Policy Implications (Chapter 12 in Part III of this book), the distinction between these two types of public technology procurement proves to be a significant basis for comparison. It assumes great practical importance in relation to key differences in 'systems of innovation' such as, for example, their relative degree of 'maturity'.

3.1 'Adaptive' versus 'Developmental' Public Technology Procurement

The evidence and analysis presented in the several case studies of public technology procurement in telecommunications and other sectors show that including both 'development-oriented' (developmental public technology procurement) and 'adaptation-oriented' (adaptive public technology procurement) cases is rewarding. They also point to analytic and empirical problems involved in constructing and maintaining a rigid distinction between developmental public technology procurement and adaptive public technology procurement.

Many of the cases address adaptation- rather than development-oriented procurement. In particular, the Austrian and Greek cases deal with clear instances of 'adaptive' public technology procurement. The French and Italian cases deal with processes involving elements of both development-oriented and adaptation-oriented public technology procurement – the former predominating in the French case, the latter in the Italian. The Nordic cases are also mixed, though they deal primarily with 'developmental' technology procurements in various sectors. The following summaries consider the balance between 'development' and 'adaptation' in the various studies.

3.2 Swedish Cases

The Swedish X2000, a tilting high speed train, is discussed as an adaptive public technology procurement case. The design was, in some respects, unique and it incorporated some major innovative components, with a potential to generate new dominant designs, but it lost out in an international technology competition. The project would have been a case much closer to developmental public technology procurement than adaptive public technology procurement, had it been implemented as early as originally envisaged by the Swedish railway industry in 1969. But the very long time elapsed between this vision and the first deliveries turned it into a more conventional project from the point of view of technological change.

Overall the X2000 project was a moderate success: It succeeded in technical terms and industrial delivery, and fulfilled its role in increasing the competence of the supplier, ASEA (later ABB-ADtranz, and now ADtranz). Nevertheless, it failed to create a strong Swedish presence in the international high speed train market. The case study demonstrates the importance of time in turning a potential developmental public technology procurement into adaptive public technology procurement. It also shows that within adaptive public technology procurement there are different grades of advance of knowledge, since the X2000 had a more innovative character than the typical adaptive public technology procurement telecommunication cases in Austria and Greece.

The procurement of High Voltage Direct Current (HVDC) transmission technology in Sweden belongs to the most successful cases studied and constitutes an ideal developmental public technology procurement with follow-ups leading into a broad sectoral and geographical technology transfer. This makes it a case with additional adaptive public technology procurement elements over and above the initial developmental public technology procurement success. Over a very long period of time, 'adaptive' technology procurement both preceded and followed the 'developmental' procurement of HVDC.

After a long proto-procurement phase between 1919 and 1939, the actual technology procurement phase began in 1940 and the first procurement contract was signed in 1949. The decision for the final construction was taken in 1953 and the first part of the transmission started in 1954. Ultimately, it wasn't until 1961 that the HVDC technology matured and took its first steps in the European and international market. Developmental public technology procurement was justified by the need for two major technological solutions. After long experimentation during the 1940s, the producer, ASEA, could face both critical problems and combine them into a reasonable HVDC transmission for the first time.

The adaptive public technology procurement that followed was related to demands of the user (a public utility) for reduction of cost, improved efficiency and better regulation. The HVDC case being both developmental public technology procurement and adaptive public technology procurement suggests in its description that a successful developmental public technology procurement is often followed by possibilities for adaptive public technology procurement. Derived benefits were high for both the supplier, who gained an international reputation and major follow-up and export contracts from the 1950s onwards, and the procurer, who initiated at the same time a period of lower transmission cost that outweighed by far the initial budget increase. The supplier, in particular, had a practical monopoly on HVDC technology world-wide and could apply profit margins up to 20% until 1970.

The Compis project, aimed at developing a Swedish school computer, is another case of a development-oriented project and therefore a developmental public technology procurement. At the time the project was conceived, i.e., 1981, the personal computer for use in schools did not exist in the Swedish market and, with the specifications suggested, it did not exist in the international market either. Additional and new technology development was necessary so it really constituted an ideal type of developmental public technology procurement.

Time was also a crucial factor in the Compis project. The relative maturity of the basic technology and the existence of parallel design efforts by US producers, notably Apple, meant that the project was associated with a high level of uncertainty and a strong need to anticipate correct technical solutions, entering the market not only early enough but also with the right standards. In the end, Compis failed to realise its potential, due to problems with the selection of technical standards and problems of co-ordination and decision making that led to late delivery and premature discontinuation of the product.

3.3 Telecommunications Cases

There is a clear focus on developmental public technology procurement in the Swedish telecommunications study, but this is set in the context of a much longer history of 'switching relations' dominated by adaptive public technology procurement. Over a period spanning the greater part of this century, recurrent collaboration in the context of public technology procurement formed a 'Development Pair' between the public utility, STA (the Swedish Telecommunications Administration), and its main supplier, LME (the L.M. Ericsson company, later renamed Ericsson). In the formation of this Development Pair, therefore, 'adaptive' technology procurements of an incremental character and informal collaboration laid foundations for more formal 'developmental' procurement. Similarly, the Finnish study demonstrates that the analyst's selection of the period of time to be used as a frame of reference for the study of a particular instance of public technology procurement influences whether the procurement should be viewed as adaptive public technology procurement or developmental public technology procurement.

The French and Italian cases deal with processes of public technology procurement occurring over very long periods of time. They stress not the objective, but the subjective, aspect of time: how the possession of a short-term or long-term strategic perspective, or 'vision', on the procurer's part can decisively affect whether a series of related technology procurements will turn out to have a primarily 'adaptive' or 'creative' character. Both France and Italy set out to develop unique 'national' digital switching systems. Though France had a much earlier start than Italy, Italy's CSELT was in fact able to keep pace with the French CNET's development of the time-division switching technology. Yet Italy lagged far behind France in the modernisation of its telecommunications network during the 70s and early 80s. Moreover, the later development of the Italian system was based to a remarkable extent on the adaptation and diffusion of knowledge and expertise acquired from external sources. In France, however, the result was clearly a 'developmental' technology procurement, involving a much stronger development of original 'national' competence. This was due to the fact that France, unlike Italy, began with (and adhered to) a long term perspective or 'vision' of ending the domination of its domestic telecommunications sector by foreign subsidiaries and replacing them with a 'national champion' capable of competing on international markets with a superior technology.

The Greek and Austrian cases follow the pattern of the Italian one, at least insofar as they are narratives about 'catching up', rather than 'forging ahead'. However, they also point out in different ways that a lack of 'vision' is not the sole factor of explanation in accounting for failures to realise the full developmental potential of public technology procurement. Both of these public technology procurement processes were, from the outset, conceived of as 'adaptive' projects. However, they were also motivated by the desire to provide opportunities for 'development' rather than restricting their aspirations to 'adaptation' only.

The Austrian case study describes, for the most part, a 'success story' but a failure "to foster exports". This failure is attributed, primarily, to the over-riding conception of the Austrian public technology procurement process as 'export policy', rather than 'technology policy', witnessed by the failure to co-ordinate the procurement with other policy instruments. The Greek study deals with a case in which there was a better fit and co-ordination of policies, particularly in the design of the procurement mechanism. Yet, at the same time, the Greek case was also one in which long delays in the implementation turned the switches into a standard product procurement and the buyer lost every advantage in terms of timing. The delays are attributed to discontinuity and friction between formal and informal institutions, which stalled the agencies responsible for implementing the procurement.

The conflict between formal and informal institutions referred to in the Greek case was based on the desire of the administration responsible for implementing the procurement to avoid public controversy and notoriety. Thus, official public technology procurement initiatives and the formal rules and procedures for conducting them were stymied by Greece's informal rules for public procurement. These consisted of procedures for delaying or avoiding non-unanimous decisions in order to avoid publicity and thereby safeguard the integrity and public image of civil servants. In contrast, no such pronounced conflict between formal and informal institutions, but rather a complementarity between them, was evident in the Austrian case.

REFERENCES

Ashford, D. E. (1992). Introduction: Of cases and contexts. In D. E. Ashford (Ed.), *History and context in comparative public policy* (pp. 3 - 23). London: University of Pittsburgh Press.

Diesing, P. (1971). *Patterns of discovery in the social sciences*. Chicago and New York: Aldine-Atherton.

George, A. L. (1979). Case studies and theory development: The method of structured, focused comparison. In P. G. Lauren (Ed.), *Diplomacy: New approaches in history, theory and policy* (pp. 43 - 68). New York: The Free Press.

Pearce, D. W. (Ed.). (1986). *The MIT dictionary of modern economics*. Cambridge, Massachusetts: MIT Press.

Sayer, A. (1992). *Method in social science: A realist approach*. (2nd ed.). London: Routledge.

Smith, K. (1997). Economic infrastructures and innovation systems. In C. Edquist (Ed.), *Systems of innovation: Technologies, organizations and institutions* (pp. 86 - 106). London: Pinter Publishers/ Casell Academic.

Yin, R. K. (1994). *Case study research: Design and methods*. (2nd ed.). London: Sage.

2. PUBLIC TECHNOLOGY PROCUREMENT IN SWEDEN: THE X2000 HIGH SPEED TRAIN

C. Edquist, P. Hammarqvist and L. Hommen

INTRODUCTION

This chapter describes and analyses the development of Sweden's high-speed train (HST), the X2000, as a case of public technology procurement. The case study is based mainly on primary data collected in recent (1997) interviews with some of the main actors involved in the procurement process, but it also draws upon relevant secondary sources. The analysis employs evaluative criteria defined from the standpoint of public policy (as elaborated below) and develops a critical perspective on the case that leads to more general policy implications.

To situate and characterise the case briefly, the procurement of the X2000 occurred during the mid-1980s, culminating the development of proposals, plans and technical preparations for a Swedish HST over a period of some 20 years (Flink & Hultén, 1993; Pålsson, 1987; SJ, 1969). Throughout this period, there was close collaboration between the Swedish state railway company, SJ, and the Swedish railway equipment manufacturer, ASEA, in joint research concerning a future HST (Andersson, 1983; Flink, 1992; Flink & Hultén, 1993). The eventual procurement was a case where the only existing user – SJ – also represented final demand for the trains. The market was also very concentrated on the supplier side and it can be questioned whether the chosen supplier, ASEA (now AD Tranz), had any real competitor, because of the way the procurement process was organised. The previous very close interdependence between SJ and ASEA is the main basis for this scepticism about competition.

Since the procurement of the X2000, there have been several changes in the corporate identity of the supplier. Originally, the supplier of the X2000 was the Swedish corporation ASEA and, more specifically, its division for the manufacture of railway equipment, ASEA Traction. However, shortly after the procurement of the X2000 was finalised in 1986, ASEA merged with the Swiss corporation, Brown Boveri, in 1988, to form ABB, a multinational corporation of which ASEA owned 50%. ASEA Traction then became ABB Traction. Several years later, in 1995, ABB together with Daimler-Benz founded the joint venture, AD Tranz, which was responsible for all business of ABB within the rail transportation sector. Thus, the supplier of the X2000 became AD Tranz, of which ABB owned 50%. In the beginning of 1999, ABB sold its part in AD Tranz to Daimler, which had (almost simultaneously) merged with Chrysler to form Daimler-Chrysler. Hence, AD Tranz is now wholly owned by Daimler-Chrysler.

One of the most notable aspects of the X2000 procurement was the length of time required for its completion. As early as 1964, the R&D department at SJ (Swedish State Railways) had considered the benefits of a high speed train (Flink & Hultén, 1993: 89). SJ's interest in developing a Swedish HST was due to the increasing competition from road and air transport during the 1960s. Furthermore, there was a general pessimism about the future of railways. This concern, together with a forecast indicating an increasing demand for fast reliable passenger transport, legitimised the vision of a high speed train within SJ. However, SJ took a very long time to realise this vision. The analysis presented here suggests that this delay was based in large part on SJ's initial lack of the competence required to carry out the train's procurement. The critical evaluation supported by this analysis points to certain negative competitive and innovative consequences for the producer of the X2000, AD Tranz (formerly ASEA). However, it also recognises some positive effects of the procurement overall.

One of the main problems experienced by the X2000 was the failure of the train as originally designed to win a significant share of the export market for high speed trains. The analysis presented here indicates that a major reason for this failure was poor timing. The X2000 did not fare well in this market because of its late development and introduction. In Europe, at least, the X2000 lost possible foreign sales most consistently to a competitor, the Italian 'Pendolino', that had the advantages of much earlier development and introduction to the market. It is reasonable to hypothesise that the competitor's greater success has mainly been due to these advantages. This argument can be linked directly to the explanation of delay on the part of SJ as being the result of an initially low 'buyer competence'. There are, of course, counter-arguments to this explanation of the X2000's lack of export success. These are addressed in the analysis that follows in later sections.

The societal and industrial aspects of the X2000 procurement case will here be addressed in terms of three main kinds of criteria. These are criteria related to: 1) Innovation Policy, 2) Industrial Policy and 3) Costs, Benefits and Profits.

Evaluation in terms of 'Innovation Policy' entails establishing what kind of technology procurement this case represented and determining whether or not it was successful in altering the direction of technical change, increasing technical diversity, or influencing the pace of technical change. The X2000 procurement process was a case of 'adaptive public technology procurement'. Although adaptive public technology procurement can have significant effects on technological change, in this case it did not result in any radical change in the speed or direction of technical change. Most of the elements in the train system existed previously, and a high-speed train with similar functions – the Italian 'Pendolino' – existed well before the X2000 was developed (Agerberg, 1996a; Giuntini, 1993). Thus, the procurement was not altogether successful from an innovation policy point of view.

Evaluation in terms of 'Industrial Policy' involves establishing whether or not the competitiveness of the supplying firms was enhanced. It is true that the procurement of the X2000 did mean the development of a somewhat different train than the Pendolino – one drawn by a locomotive rather than consisting of motor coaches. However, this concept has not proven successful in other countries and the locomotive version of X2000 has not been exported to any other country.[1] This means that the X2000 procurement has not functioned well as an industrial policy instrument either.

The appraisal of 'Costs, Benefits and Profits' concerns whether or not the buyer and seller, respectively, 'gained' or 'lost' in strict accounting terms from the procurement process, and (if this can be established) whether or not there were any corresponding gains or losses in general economic welfare. The assessment presented here indicates that in these terms the X2000 procurement was for the most part successful.

The negative judgements given above can be qualified by an historical perspective. The X2000 case was far from a complete failure. Although the train itself did not prove to be a success on export markets, its development did result in significant improvements to the 'system competence' (i.e., capacity to design, build and maintain entire technical systems) of the supplier firm, AD Tranz (formerly ASEA). In this respect, it improved the firm's competitiveness. Moreover, the X2000 design incorporated some important technical innovations which have since been transferred to other, more successful, commercial applications. In what follows, we develop this perspective more fully and end our discussion by drawing some general conclusions and policy implications.

1. INDUSTRIAL POLICY

Although the X2000 procurement can be considered, overall, to have been a 'partial failure' in terms of innovation policy, it can be deemed to have been at least a 'limited success' in terms of industrial policy. There is general consensus among informed sources that the X2000 procurement improved the competitiveness of ASEA (now AD Tranz). This is certainly true so far as the development of the supplier firm's competences is concerned.

Historically, state railway companies (i.e., railway operators) in Europe have held the overall 'system competence' with respect to railway equipment. 'System competence', in this context, means the capacity not only to design but also to assemble and maintain or guarantee the operation of entire rail transportation systems. In contrast to the state railway companies, railway equipment manufacturers have tended to have only more limited – i.e., specialised – technical competences. That is, they have dealt primarily with the development of specific components for rail systems, and not with the systems as such. Yet at the end of the procurement process for the X2000 – the terms

[1] The train later sold by AD Tranz to Norway was a motor coach version of X2000 (Agerberg, 1996b: 44). So was the train sold more recently — in August, 1997 — by AD Tranz for operation on the future bridge now under construction between Malmö, Sweden, and Copenhagen, Denmark.

of which required the supplier to demonstrate such a capacity – ASEA (now AD Tranz) had acquired this kind of system competence (L. Pålsson, 1997, interview). In this respect, the competitiveness of the supplier was greatly enhanced (B. Andersson, 1997, interview; J. Lundgren, 1997, interview).

The acquisition of 'system competence' was particularly vital to the continuing survival and commercial success of the supplier, given that there is now a definite trend – in Sweden as elsewhere – toward deregulation and privatisation of the rail transport sector (B. Andersson, 1997, interview). This trend implies that equipment suppliers will have the primary responsibility for designing and maintaining entire systems in the future (J. Lundgren, 1997, interview; R. Persson, 1997, interview). Another implication is that there will be more operators and fewer suppliers of railway equipment, so that emerging 'system problems' within the sector will have to be addressed first by suppliers (Nilsson & Wallner, 1993). These aspects of restructuring in the rail transport sector have made the acquisition of 'system competences' crucial to firms like AD Tranz (formerly ASEA).

For these reasons, an important part of the story of the X2000 procurement process concerns the transfer of 'system competences' from SJ to ASEA. While the transfer was more or less successful, it was not without its problems. On balance, it can be regarded as illustrating both the strengths and the weaknesses of the conventional 'Swedish model' of public technology procurement, in which 'development pairs' such as SJ-ASEA have played a prominent part (Fridlund, 1993). As this case study shows, one of the strengths is the ability of development pairs to develop a high level of competence through interactive learning between buyer and supplier, even when both begin from a position of relatively low competence. However, one of the weaknesses is the inordinately long periods of time that may be needed to accomplish this. In this case, the competence of SJ remained problematic, even after a formal procurement process had been initiated. During this process, almost all functional requirements were altered (Flink & Hultén, 1990: 21). This would not have occurred if SJ had possessed the required competence in HST technology at the outset. The fact that SJ was slow to acquire this specialised technical competence had serious consequences, not only for ASEA, but also for Sweden. Arguably, the X2000 could have had greater success on export markets if procurement of a Swedish HST could have been accomplished more quickly.

There are reasons to believe that in the development of the X2000 the Swedish state railway company, SJ, might have had a preference for working with ASEA (now AD Tranz) because it was a Swedish firm. Part of SJ's traditional mandate was to 'care about' Swedish industry (E. Andersson, 1997, interview; A. Hörberg, 1997, interview). This meant caring about ASEA, since the Swedish market for railway equipment was characterised not only by buyer monopsony (SJ) but also by producer monopoly (ASEA) (Flink and Hultén, 1990; 1993). This situation – and the long history of 'collegial' relations and shared 'tacit knowledge' between SJ and ASEA that it entailed – gave ASEA certain competitive advantages in bidding on the X2000 procurement (E. Andersson, 1997, interview; A. Hörberg, 1997, interview). It may also have deterred serious competition from outside of Sweden (L. Pålsson, 1997, interview).

Moreover, the decision criteria of the first investigation into the possibilities of a Swedish HST gave priority to the development of domestic industry (SJ, 1969: 163).

Thus, it is not hard to discern some kind of protectionism in the initial phase of the procurement. Although at a later point the entire process of bidding and negotiation was open to foreign firms (J. Lundgren, 1997, interview; R. Persson, 1997, interview) and SJ was careful to evaluate all offers in 'a particularly honest way' (L. Pålsson, 1997, interview), there appears to have been lingering suspicion among potential foreign suppliers that 'they could not receive fair judgement' (A. Hörberg, 1997, interview). In other words, they were concerned that SJ's evaluation of tenders would be influenced by covert protectionism favouring the 'national' firm.

'Protectionism', however, does not fully account for SJ's highly 'professional' handling of the X2000 procurement through several rounds of bidding and procurement occurring between the first tender request in 1982 and the final contract negotiations in summer, 1986 (Flink and Hultén, 1990; Gunnarson, 1994; J. Lundgren, 1997, interview; Pålsson, 1987:8 - 10). In that process, a tender from at least one foreign supplier was apparently accepted and considered as an alternative to ASEA's offer even during the second round of bidding (B. Andersson, 1997, interview; E. Andersson, 1997, interview; R. Persson, 1997, interview; L. Pålsson, 1997, interview). This suggests that foreign competitors to ASEA *were* considered seriously.

Certainly, the legal and formal institutional context in which the procurement was carried out worked against protectionism. The law under which the procurement was conducted stressed the exclusivity of business criteria as the sole basis for evaluating tenders (SFS 1973:600, 1973). The law, which was scrupulously observed, emphasised openness, transparency and accountability in tendering and contracting procedures, and it insisted upon maintaining a 'competitive environment' (Linder, 1996: 13 - 14). Moreover, and not least importantly, SJ itself was never fully committed to a 'protectionist' solution – i.e., choosing a Swedish supplier in all cases and at any cost. SJ's first investigation of the options to acquire a high speed train ruled out *both* procurement of a complete train from abroad *and* procurement of a train originating from strictly domestic design, development and manufacturing (SJ, 1969: 163). The latter option was initially rejected because of the length of time expected. That it was later accepted and acted upon may have had less to do with 'protectionist' impulses on the part of SJ than with a reappraisal of ASEA's improving competence (R. Persson, 1997, interview). It may have had even more to do with SJ's attempts to remedy its own lack of competence.

SJ recognised, early on, that it would have to improve its own knowledge regarding high speed train systems and their various components in order to act as a competent procurer of high speed trains (SJ 1969: 171; A. Hörberg, 1997, interview; J. Lundgren, 1997, interview; R. Persson, 1997, interview). This thinking was in line with the responsibilities conventionally assumed by state railway companies for developing and maintaining 'system competence' (Flink and Hultén, 1990: 21; Pålsson, 1987: 3 - 7). The rationale was also consistent with the logic of 'best practice' in private technology procurement, where no 'political considerations' normally apply. In that context, too, effective procurement of new technology requires effective 'demand pull', and buyers must be not only strong enough to 'articulate demand' forcefully but also competent enough to make the correct demands. In this connection it has been noted that most potential suppliers were sceptical of SJ's competence (A. Hörberg, 1997, interview).

For these reasons, there was an imperative – not only in terms of 'industrial policy' but also in terms of 'innovation policy' – for SJ to develop its own lagging technological competence. There were limited options for doing this within Sweden, given the highly concentrated structure (monopsony/monopoly) of the Swedish market for railway equipment. Moreover, it is likely that SJ would have suffered serious disadvantages in trying to enhance its own knowledge by co-operating with foreign suppliers possessing technical expertise superior to its own. Under 'low trust' conditions, SJ would have had to pay a high price for the acquisition of competence from sources that it could have demanded little accountability from. Additionally, SJ might never have been able to develop truly superior expertise vis-á-vis potential suppliers. In the event of a formal bidding process, then, SJ would have remained unable to evaluate the offers thoroughly enough to arrive at the most rational decision regarding the award of a contract. It was therefore 'natural' for SJ to turn to ASEA as a development partner for the mutual upgrading of 'system competences' essential for the development of a high speed train. This was one main object of the lengthy collaboration between SJ and ASEA during the early and mid-1970s (Andersson, 1983; Flink, 1992; Flink and Hultén, 1993; Gunnarsson, 1994).

The actual procurement of the X2000 experienced a long delay while SJ (together with ASEA) endeavoured to raise its level of competence to a level where SJ (and, presumably, ASEA) could participate credibly in a high speed train procurement process open to foreign bidders and aimed at the development of a technologically advanced high speed train. SJ first expressed interest in acquiring a high speed train in 1968 - 1969, but it spent the subsequent ten years improving its knowledge of high speed train technology in collaboration with ASEA, through development projects such as the X15 prototype (Flink, 1992; Gunnarsson, 1994).

In 1975, when the X15 project had been completed, the Swedish parliament passed a Bill for the speedy introduction of high speed trains (Flink, 1992). But SJ did not respond by trying to secure financing for such a project (E. Andersson, 1997, interview). Instead, its response was to undertake further preliminary investigations (SJ, 1980). Acquiring sufficient 'buyer competence' to procure a high speed train from a position of strength appears to have been a continuing concern for SJ. Conversely, SJ's reason for further delay is not likely to have been that it was concerned with the basic competence of potential suppliers.

The basic capability to build a high speed train existed in Sweden by 1975, due to the completion of the X15 project. It would also have been possible at this time to obtain from a foreign supplier a high speed train that fulfilled most of SJ's initial (1969) functional requirements – such a train, the Pendolino, had already been developed and came into full operation in 1976 (Agerberg, 1996a; Giuntini, 1993). However, even after 1981, when SJ received the necessary 'political support' for procuring the X2000 through a government decision to acquire and finance a fleet of high speed trains (Flink and Hultén, 1993: 98), SJ was still in the process of acquiring the necessary 'critical competence' to act effectively in procuring a high speed train (R. Persson, 1997, interview). Thus, for example, the X2 study was conducted by SJ in 1982, in order to test and develop newly acquired expertise in digital technology, control engineering and the techniques of 'Life-Cycle-Costing' (LCC) analysis to be used in the evaluation of alternative HST systems

(J. Lundgren, 1997, interview; Pålsson, 1987: 3 - 7). For its part, SJ is reported to have remained concerned at that time about the competence of ASEA (Flink and Hultén, 1990; A. Hörberg, interview) – with whom it was still collaborating as late as 1982.

Even after the X2 study had been completed and the formal bidding process for procurement of the X2000 had been initiated in 1982, the procurement remained a 'learning process' for SJ (B. Andersson, 1997, interview; A. Hörberg, 1997, interview). Nearly all functional requirements were changed during the bidding process, indicating that SJ did not initially have the competence required (Flink and Hultén, 1990: 21). Thus, one way of interpreting the result of the first round of bidding, where no suppliers were found capable of fulfilling SJ's functional requirements (Pålsson, 1987: 9), is to consider that SJ may not have been able at first to propose 'realistic' requirements. This problem took some years to remedy, and apparently was only resolved in 1985, when SJ sent out its supplementary request to the second invitation for tenders (Flink and Hultén, 1990: 21; Pålsson, 1987: 10). Subsequently ASEA (now AD Tranz) was awarded the contract in summer, 1986, and final contract negotiations were initiated (J. Lundgren, 1997, interview).

ASEA was clearly one of the main beneficiaries of SJ's learning, in much of which it had been a joint participant. Collaborative learning was most pronounced in the 1970s, when SJ and ASEA jointly developed and tested a tilting mechanism and 'soft bogie' and – later – the X15 train (Flink, 1992; Flink and Hultén, 1993; Gunnarsson, 1994). At this time, ASEA began to acquire extensive 'tacit knowledge' of SJ's requirements for a high speed train (E. Andersson, 1997, interview). Even at the outset of the procurement process that began in 1981 - 1982, ASEA and SJ continued to develop knowledge on a joint basis, and to share technical resources. For example, during the period when the X2 study was being conducted, SJ recognised, as a consequence of the study, that it lacked the necessary competence in digital control engineering (J. Lundgren, 1997, interview). SJ then hired a consultant expert in digital technology – who was subsequently hired by ASEA "in the last minute" before procurement (ibid.).

Once formal bidding on the X2000 procurement began in earnest in 1982, the formerly 'collegial' relationship between ASEA and SJ gave way to a more distanced kind of relationship. This shift in relations was a matter of policy for SJ, which was under political pressure to exhibit an arm's-length stance towards ASEA in order to develop credibility with potential foreign suppliers bidding on the X2000 procurement (J. Lundgren, 1997, interview; R. Persson, 1997, interview). The change in relations was accompanied by the requirement for LCC costing in all bids submitted on the X2000 procurement (Flink and Hultén, 1993: 98). Although SJ had developed knowledge and expertise in LCC analysis, it had not shared this competence with ASEA. Even in this context, however, SJ could be seen to have encouraged the development of ASEA by 'formalising professionalism', prompting the firm to acquire an important basis of 'competitive advantage', and instilling 'systems thinking' in ASEA (R. Persson, 1997, interview; Pålsson, 1987: 12).

Finally, even once it came to an end with the initiation of bidding, the earlier 'collegiality' between SJ and ASEA still had a continuing value for ASEA. While SJ was very strict in treating the procurement process as an open competition and was scrupulous about answering all requests for information, its reluctance to visit potential tenderers in order to explain its request probably favoured ASEA over other bidders (Pålsson, 1987: 8). Given that it had previously accumulated a large stock of 'tacit knowledge' about SJ's requirements, ASEA would have been able to understand SJ's request more fully than most, if not all, other tenderers (B. Andersson, 1997, interview; Andersson, 1983; R. Persson, 1997, interview).

A significant result of the X2000 procurement process was that ASEA developed full fledged 'system competence' as a designer and manufacturer of railway equipment (L. Pålsson, 1997, interview). SJ's fundamental requirement in this procurement was that the supplier could design, make, deliver and guarantee in all details the reliability of an entire system (J. Lundgren, 1997, interview; Pålsson, 1987: 3). ASEA (now AD Tranz) succeeded in meeting this requirement. Its accomplishment is encapsulated in the remark that "what SJ required in 1982 is what AD Tranz is just about able to accomplish today" (R. Persson, 1997, interview).

As mentioned earlier, it turned out that the 'system competences' developed by ASEA (now AD Tranz) were fairly specialised in terms of market suitability. However, this might not necessarily have represented a great competitive drawback for the firm. The fragmentary structure of the international 'market for trains' leads to regionally specialised demands for 'tailor-made' solutions (Agerberg, 1996a: 13; Agerberg, 1996b: 45; J. Lundgren, 1997, interview). Subsequent sales to Norway – and, more recently, to the Malmö-Copenhagen bridge project – of a motor-coach variant of the X2000 suggest that AD Tranz (formerly ASEA) was at least able to compete in a 'regionalised' international market (Agerberg, 1996b: 44).

The fact remains, however, that – as further discussed in the next section – the X2000 design as a whole (a single-locomotive, tiltable, high speed train) has proven to have little if any potential for export. In this sense, it can be said to have 'lost out' in the international technology competition to develop high speed trains. The X2000's lack of commercial success beyond Sweden has meant that the world market impact of this innovation has been effectively limited to alternative applications of the innovative subsystems or components incorporated into the original product design (Flink and Hultén, 1993: 96; Gunnarsson, 1994; R. Persson, 1997, interview). These might have helped to establish a competitive position for AD Tranz (formerly ASEA), (J. Lundgren, 1997, interview). However, ABB's recent sale of its part of AD Tranz to Daimler-Chrysler implies that few such benefits are now likely to accrue to Sweden.

In sum, the joint development and 'transfer' of competence between SJ and ASEA can be considered to have improved the competitiveness of AD Tranz (formerly ASEA). There is consensus among informed sources that this has been the case. However, to the extent that exports are an indicator of competitiveness, the positive impact was not all that large. In this respect, and others (discussed below), the procurement of the X2000 may be regarded as having had only very partial success as 'industrial policy'. The question remaining is whether it would have enjoyed greater success – and perhaps also succeeded as 'innovation policy' – if the procurement of a Swedish HST could have been accomplished more quickly. The crucial problem in this regard appears to have been one of competence – not just that of ASEA, but also and more fundamentally, that of SJ. We will return to this question and a discussion of the trade-off between 'timing' and 'competence' in drawing conclusions and policy implications.

2. INNOVATION POLICY

The procurement of the X2000 did not fundamentally alter the direction of technical change in rail transport, and it resulted in only minor increases in technical diversity. The X2000 was not the first high speed train to be developed, nor has it come to be widely used outside of Sweden. On the whole, then, SJ's procurement of the X2000 from ASEA (now AD Tranz) must be regarded as a case of 'adaptive public technology procurement' rather than 'developmental public technology procurement'. There is a fundamental distinction between the former, 'diffusion-oriented', kind of public technology procurement, and the latter, 'creation-oriented', kind. In the latter, new products, processes or systems are created. In the former, the product or system procured is not new to the world but only to the country of procurement.

In the analysis of cases of adaptive public technology procurement, emphasis is placed on the fact that the system needs adaptation to local circumstances – not simply that it is new to the country. This adaptation may or may not bring about true 'branchings' in the development of existing technologies. In cases where this does occur, however, the buyer – and more broadly, the country that the buyer as a public or government agency represents – has the possibility of realising certain economic benefits and advantages.

Early adoption of an emerging technology can contribute to broader processes of economic growth and technical development. At a later stage of development, the further adaptation of the technology to local conditions can still involve real technological innovation and generate substantial economic benefits. Successful adaptation of this kind, however, requires exceptional capabilities for 'collective learning' and a strong bargaining position with suppliers. If buyers can meet these conditions, they can obtain favourable prices from suppliers and a leading position among competitor nations, due to improved infrastructure. And, if their timing is appropriate, they may also share in innovation 'rents' from the sale of the (adapted) technology to other national buyers occupying the same market segment.

By the standards outlined above, the X2000 was neither a complete success nor a complete failure. This evaluation is elaborated below.

As an adaptation, the X2000 did represent a diversification of high speed train technology – the design was in some respects unique and it incorporated some innovative components. The X2000 also met national requirements more or less successfully. It thus resulted in significant improvements to infrastructure – and, hence, the conditions for economic growth.[2]

[2] The actual contribution to economic welfare remains difficult to establish, given insufficient data, but see the discussion that follows this appraisal of the X2000 as a technical innovation.

Certain components of the X2000's design were new, or at least original, technological solutions to enduring problems in the development of high speed train technology. This was particularly true of the 'soft bogie' developed by ASEA on the basis of earlier collaboration with SJ (Gunnarson, 1994) – which, in and of itself, could be regarded as having constituted a successful case of developmental, or 'creation-oriented' public technology procurement. This innovation was transferred to other commercial applications, as well as being combined in the X2000 with supporting innovations from other sources – such as the specialised computer technology developed to control the tilting mechanism and the GTO-thyristor which made the asynchronous engine feasible (Flink and Hultén, 1993: 96). These developments represented 'a series of large, innovative steps' with considerable commercial potential (J. Lundgren, 1997, interview; L. Pålsson, 1997, interview).

For the above reasons, the X2000 should not be dismissed in terms of its innovative value. Rather than simply characterising it as an unspectacular case of 'incremental innovation' (refinement of an existing technology), it may be more appropriate to regard the X2000 as exemplifying 'modular innovation' (new technology in a familiar context) (Abernathy & Clark, 1985). At a subsystem level – e.g., the 'soft bogie' and its use in combination with supporting components – the X2000 could also be said to have entailed 'configurative innovation', which has at least the potential to generate new dominant designs (Hidefjäll, 1997: 43).

From its lack of success on export markets, it is clear that the X2000 failed to establish a new 'dominant design' – the single-locomotive, tiltable, high speed train – in the international market for high speed trains. Instead, as indicated in the introduction, the main demand in this market has been for 'motor-coach' designs, and as reiterated in the preceding section, all export sales of the X2000 have been for a motor-coach variant, not the original design. In this sense, the original X2000 'lost' a technology competition. In so-called 'technology-competitions' markets tend to 'lock in' on one of several options in technology, due to positive feedback effects of initially influential choices (Arthur, 1988; Foray, 1989). However, as the case of the X2000 demonstrates, it remains important to consider the nature and extent of the competition.

We have noted that, as with railway equipment in general, there is no true 'world market' for high speed trains. Rather, there is a very fragmented and diverse market consisting of a series of national or regional markets that are essentially 'custom' markets demanding 'tailor-made' systems (Agerberg, 1996a: 13; Agerberg, 1996b: 45; J. Lundgren, 1997, interview). Thus, even though the Italian Pendolino out-competed the X2000 in parts of Europe and may thus be considered to have 'won' a limited technology competition (R. Persson, 1997, interview), this competitor still appears to have fallen well short of establishing a new 'dominant design'. For example, an altogether different competitor – the Canadian firm, Bombardier – managed to secure a significant part of the largest North American market (Agerberg, 1996a: 12 - 13: Agerberg, 1996b: 45). Accordingly, the fact that the X2000 'failed' in past competitions with the Pendolino did not necessarily imply that it would lose in future competitions (B. Andersson, B., 1997, interview; E. Andersson, 1997, interview; A. Hörberg, SJ, 1997, interview). Given the nature of the market, all such competitions are 'limited' and some might well favour the X2000 or its descendants.

In addition to the above mentioned characteristics of the 'market for trains', there is also an apparent trend towards increasingly fragmented national markets (B. Andersson, 1997, interview). In these markets the manufacturers of trains and railway equipment will be required to assume a much greater share of the responsibility for designing, delivering and maintaining – perhaps even operating – entire systems (J. Lundgren, 1997, interview; Nilsson and Wallner, 1993; R. Persson, 1997, interview).

In this context, it may be particularly significant that the X2000 was fairly successful in meeting national requirements. The X2000 may also be considered to have been at least a partial success in terms of winning national markets within its region – despite significant differences in national requirements within the region. The recent sale of a motor-coach version of the X2000 to Norway is reported to have been for a 'completely different' train (R. Persson, 1997, interview). This might nevertheless indicate that the excessive customisation that may have prevented the X-2000 from capturing markets elsewhere in Europe (ibid.) could also have been a source of regional advantage. That is, the high degree of customisation involved might have enabled the supplier to develop a design well suited to its own region, as well as the capability to modify this design (or to develop others) for specialised national markets.

In addition to the commercial viability of the X2000 on export markets, the question of whether genuine societal needs were met should also be considered as a criterion for evaluating the X2000's successfulness as an innovation at the national level. These criteria will be considered in more detail below. Here, we will only refer briefly to the estimations of possible societal benefits that formed part of the second (1980) investigation into the costs and benefits of building a Swedish high speed train (SJ, 1980: 9). The investigation projected socio-economic benefits due to decreased travel time of around 300 million SEK/year, decreased energy consumption on the roads resulting in savings of about 30 million SEK/year, plus additional improvements to social conditions and savings in infrastructure costs also related to decreased use of automobiles (ibid.). It has not been possible to ascertain whether all of these projected benefits have been realised, but they would not have had to have been met completely in order for one year's societal 'savings' to compensate for the actual development cost of the X2000, which turned out to be about twice the originally quoted figure of 100 million SEK (E. Andersson, 1997, interview). This does not take into account the benefits (versus costs) for either SJ or ASEA, which will be addressed subsequently.

A final issue for consideration is that of whether or not the procurement of the X2000 increased the speed of high speed train development. Formulating an answer to this question depends on the frame of reference. From an international standpoint, the answer is probably 'no'. Indeed, from an 'outsider's' perspective, it even appears that the pace of technological development may actually have been slowed down within Sweden, given the length of time it took to produce the X2000 after the X15 had already been developed and the first generation Pendolino had come into operation.

From a national viewpoint, however, the answer to this question may be different. The question might be restated as follows: Would SJ have saved time by buying the Pendolino? It is not clear that this would have been the case. Buying the Pendolino for use in Sweden would have involved an extensive adaptation of the existing design to local conditions, including differences in the electrical power supply that would have required, among other things, a major redesign of the train's propulsion system (E. Andersson, 1997, interview; R. Persson, 1997, interview).

Related to these technical considerations, there is a fundamental issue of cost – or, to be more precise, uncertainty about cost. Assuming that the Pendolino could be successfully adapted to Swedish conditions, could this have been done for a substantially lower cost than developing a Swedish train? We can not answer this question on the basis of the available data. We have, however, reviewed extensive evidence concerning the procurer's lack of adequate competence until a very late point in time. Given these circumstances, it is arguably the case that SJ was unable to make a proper evaluation of the costs until the time that the procurement process was formally initiated, and possibly for some time afterwards. Under these conditions, it would have been rational for SJ to minimise uncertainty by preferring a 'known' supplier with whom a relationship of 'trust' had been established. This point was also made in the preceding section.

3. COSTS, BENEFITS & PROFITS

We will now consider the strictly economic costs and benefits for SJ and ASEA – and Sweden. If it did not succeed entirely at the level of innovation policy, the X2000 procurement was at least minimally successful at the level of industrial policy. This procurement was effective in extending the life of a nationally based industry and enhancing the competitiveness of the leading national firm in that industry – even though it did not lead directly to export sales. But at what cost – or profit – to the procurer was this accomplished? What was the extent of the economic benefit to the supplier? And what losses or gains were there for Sweden as a whole?

3.1 Societal and Industrial Results

In the investigation of 1969, several calculations were made. A Swedish express train set of 1969 was running about 275,000 km a year or about 750 km a day. This is the same as today since not much has happened regarding the speed for express trains (SJ, 1969: 158).

If a high speed train was going to be used 300 days a year, it would probably have reached a usage of 600,000 km a year or about 1,600 km a day. This increased production capability of each train set would have allowed SJ to diminish the size of the train sets and increase the frequency of departures with a constant demand (SJ, 1969: 158).

An express train has a capacity of 450 passengers, while the predicted high speed train would have carried 200-250 passengers. The investigators also thought that this would require less personnel, since a high speed train only would require a crew of two persons, while an express train needs a crew of 4 persons. Therefore they thought, in the investigation of 1969, that if the current train sets were replaced by the same number of high speed trains astonishing results would emerge:

- the same number of passengers could be transported.
- the travel time could be cut to 50%.
- the frequency of the departures could be doubled.
- the need for personnel could be reduced to 50%.

(SJ, 1969: 158)

The investigation of 1980 did not make calculations similar to those above, but the travel time in 1980 was about the same as in 1969 (SJ, 1980: 23). SJ made corresponding calculations, to which we do not have access. However, we think that the above discussion was of relevance in 1980 as well, since actually not much had happened since 1969 with regard to travel time and maximum speed.

3.2 The Cost

Each train, consisting of five coaches and one locomotive, of X2000 that ASEA started to deliver to SJ in 1990, had a cost of about 100 millions SEK (Agerberg 1996: 16). The final sum of the contract in 1986 was about 1,5 billion SEK (Pålsson, 1987: 3). SJ bought 20 trains in 1986 and have by now had a total of 43 trains delivered. The first 20 trains were used on major passenger routes. The remaining 23 trains were used on routes supported by rural cities and regions (B. Andersson, 1997, interview).

3.3 Benefits & Profits

As the high speed train reduces travel time between cities, the real time distance decreases. Round trips to neighbouring cities that it was formerly not possible to make within one day are now possible. In Japan this has allocated business activities to other cities than Tokyo. It has furthermore enhanced business activities between cities and led to overall economic growth as well. For the hotel industry in places that have become closer to Tokyo in real time distance, the impact of the 'Shinkansen' in this respect has meant fewer guests, since it is not necessary to stay over night any more (Hultgren, 1981: 6).

We are unable to determine whether the X2000 has had similar effects in Sweden. However, such effects would be plausible.

There are some indications that the X2000 has enabled the procurer, SJ, to compete successfully with air and highway transportation. Between Stockholm and Gothenburg – the first and second largest cities in Sweden, respectively – the X2000 has reduced journey times by 25 per cent. This gain in efficiency has "increased rail's market share on the 285-mile route from 41 to 55 per cent" (Batchelor, 1996).

It appears, then, that at least on major routes the X2000 has met the expectations of SJ. By doing so, moreover, it will also have realised to some extent the societal benefits projected by the second (1980) investigation into high speed trains. These included not only increased commercial profitability for SJ but also reduced infrastructure costs for the building of roads and traffic routes, lower energy consumption costs from decreased use of automobiles, benefits due to shorter travel times, and reduced costs in terms of the accidents and pollution caused by highway traffic (SJ, 1980: 9). However, we can not quantify the extent of these benefits on the basis of data currently available.

It is hard to determine whether the X2000 project as a whole has been profitable for AD Tranz (formerly ASEA). However, according to Bernt Andersson, ASEA did not make any profit on the first 20 trains and, at best, the total was probably break even (B. Andersson, 1997, interview). Despite this, it has been important for AD Tranz from a strategic point of view. As argued previously, the X2000 project enhanced the general level of competence at AD Tranz (E. Andersson, 1997, interview).

4. MAIN CONCLUSIONS

In our introduction, we suggested that the procurement of the X2000 was not altogether successful as an instrument of industrial policy, due to the train's failure to win export markets. We also suggested that the X2000 was not a particularly successful example of innovation policy, since it resulted in little or no radical change to the speed or direction of technical change. Subsequently, we have reviewed evidence and arguments to support these statements. Here, we will offer further conclusions that follow from salient features of the case.

An important inference that can be drawn from the history of the train's development and its consequences concerns the management of the procurement process. It is a plausible interpretation of the evidence that the requirement of SJ that X2000 should be a locomotive-drawn train pushed the supplier, ASEA (now AD Tranz), into a dead-end street. In other words, the demand from SJ was perhaps not appropriate or visionary enough to function well as industrial policy. Although the new motor- coach version of the rapid train is partly designed on the basis of experience with the X2000, the locomotive requirement might, at the same time, have delayed the development of the motor coach. It might have been better if SJ had been looking more closely at the Pendolino design at the outset of the procurement.

One of the main problems faced by the X2000 in capturing a significant share of export markets for high speed trains was simply that of timing. The X2000 was a very late arrival to the international 'market for high speed trains', and fared poorly. In contrast, the competitor to which the X2000 lost possible foreign sales most consistently – the Pendolino – was one of the early arrivals. It is reasonable to consider that the greater success of the Pendolino may have been due, in large part, to better timing – i.e., earlier development of a viable, marketable train.

There is, of course, a counter-argument to this position. It has been explored at some length here. According to this counter-argument, the failure of the X2000 to gain export sales in Europe and elsewhere can be attributed largely to the highly fragmented and nationally or regionally divided nature of existing markets for railway equipment.

There is no true 'world market' for high speed (or other) trains – only a series of 'custom' markets, the differences among which are often exacerbated by political conditions. For these reasons, the counter-argument holds, the X2000 can not be considered to have lost a truly international technology competition. Moreover, the X2000 proved the supplier's competence as a 'custom' manufacturer of high speed trains. The manufacturer, if not the train, subsequently began to make foreign sales, and will likely continue to do so.

This counter-argument, however, ignores the fact that the X2000 usually lost export markets to the Pendolino. This is particularly true of competition within Europe. Thus, it appears that, at least in Europe, a 'bandwagon' was formed around the Pendolino and not around the X2000. This would not be possible in a completely fragmented European market. For this reason, the counter-argument is not fully convincing. Additional explanations must be sought, and of the other possible explanations available, the one concerning 'timing' is the most powerful.

In the preceding discussion we identified late development as a main flaw of the X2000, in terms of both innovation and industrial policy. We also referred to the theoretical argument, made elsewhere, that adaptive public technology procurement can alter the diversity, direction or pace of technological change. To do so, it must be conducted by a 'buyer' that is capable of developing exceptional competence through learning – thereby becoming able to deal with 'suppliers' from a position of strength. If the buying agency, moreover, can exercise good timing, both the public agency and the country it represents may also be able to benefit from subsequent sales of the (innovatively adapted) technology in other countries. Under these conditions, adaptive public technology procurement can achieve successes both in terms of innovation policy and in terms of industrial policy.

These conditions were not met in Sweden, with the result that the X2000 procurement can not be regarded as a 'success story'. As we pointed out above, it seems from an international perspective that the X2000 procurement actually slowed down the pace of technological development in Sweden, instead of speeding it up. Earlier, in our discussion of industrial policy, we reviewed extensive evidence indicating that one of the main causes of this effect was the initially low competence level of SJ, the state railway company, and the length of time it took for SJ to raise its competence to a level where it could function effectively as the 'buyer' in procuring a high speed train for Sweden. We returned to this theme in the discussion of innovation policy, when considering the poor export performance of the X2000. We have suggested, throughout, that in the case of the X2000 procurement there had been a trade-off between 'timing' (earlier development, increasing the probability of greater success on export markets) and 'competence' (in particular, the competence of SJ).

Closer scrutiny of the available evidence indicates that SJ executives involved in the procurement of the X2000 were quite conscious of this trade-off. From the mid-sixties onwards, SJ was aware that it was under considerable economic pressure to acquire a high speed train (Flink and Hultén, 1993), and the leadership of SJ contributed on several occasions to increasing the political pressure to make such an acquisition (Flink, 1992). Despite this, the development process headed by SJ before the procurement was initiated, as well as SJ's actual procurement of the X2000, proceeded very cautiously and slowly. SJ was concerned from the outset to acquire and demonstrate to potential suppliers 'a certain critical competence', and it remained unable to do this satisfactorily until the very late stages of the procurement process. After SJ became fully capable of 'matching' its functional requirements with the technical capabilities of potential suppliers, the procurement proceeded smoothly. However its conclusion had suffered a lengthy delay, even in terms of the schedule that SJ had originally set for itself (SJ, 1980: 25).

5. POLICY IMPLICATIONS

The evidence suggests that an insufficiently high level of 'user competence' was one of the primary reasons for the poor 'timing' of the X2000. What, then, are the main policy implications to be drawn? We can begin to answer this question by first considering the orthodox view.

According to this prevailing wisdom, the main problem with public technology procurement as it has been conventionally practised in most countries is that government involvement has tended, for political reasons, to interfere unduly with the operation of market forces. The main remedy, as conceived and enacted by the EU, has been to institutionalise in the legal and administrative regime governing public procurement a 'free market' orientation, based on classical economic efficiency principles and emphasising the exclusive priority of commercial criteria in the awarding of contracts (Martin, 1996: 41).

The stated goals of the EU policy regarding public procurement, including public technology procurement, have been not only to broaden competition and thereby achieve public savings, but also to bring about restructuring and adjustment leading to economic growth. The policy has sought to eliminate protectionist procurement practices on the part of public agencies with monopsonistic powers within nationally 'closed' sectors such as rail transport. Such practices were considered, in analyses that informed the current EU policy, to have resulted in supply-side overcapacity, an excessive number of firms, the duplication of research and marketing efforts, and inadequate scales of production. The solution recommended (and since attempted) in order to meet the challenges of increasing international competition, was (and remains) to create "viable producers" in Europe by widening competition. It is expected that more open competition will both enable and encourage existing firms "to merge, rationalise and invest in new technology" (WS Atkins Management Consultants & Associates, 1988: 14 - 15).

The orthodox view, it can be seen, focuses primarily on the supply side. Insofar as it is concerned with issues of competence it addresses only questions about producer competence. With respect to the demand side, there is only a preoccupation with opening markets and increasing the extent of demand. This is a perspective that takes little real account of user competence. Its main prescription, in the case of Sweden's acquisition of a high speed train, would have been for SJ to purchase an existing high speed train at the best possible price, as soon as possible.

We considered this proposition, when we posed the question, 'Would SJ have saved time by buying the Pendolino?'. Reviewing the evidence indicating that a major overhaul of the Pendolino would be required in order to adapt it to use in Sweden, we could not state that the answer to this question was a definite 'yes'.

We also drew attention to the issue of cost. Assuming that an adaptation of the Pendolino to Swedish conditions could have been made within a shorter period of time than it required SJ to take delivery of the X2000, would it also have cost less to buy the Pendolino? Our answer to this supplementary question was that SJ, because of its lagging technological competence, was in no position to know whether this would have been the case or not. There is ample documentation to show that, at least until a very late stage in the procurement process, SJ found itself unable to evaluate properly the costs, benefits, capabilities, and inadequacies of the Pendolino and, for this reason, felt itself to be at a considerable disadvantage in negotiating a contract with its producers.

During its procurement of the X2000, SJ was subject to a procurement regime similar in its essential features to that now in force in Sweden as a member state of the EU. SJ also acceded to political pressures to 'widen competition' to an international level. Yet SJ still confronted a fundamental problem of uncertainty due to insufficient knowledge. SJ remedied this problem by building upon a relationship of trust with a known domestic supplier, which it enlisted as a 'development partner' in acquiring the needed competence. This was a lengthy and difficult process and its eventual outcome was that the domestic supplier gained considerable competitive advantage over foreign rivals and, largely for this reason, won the procurement contract from SJ.

In this case, then, the simple 'widening of competition' provided no immediate solution to the fundamental problem of developing a sufficiently high level of 'user competence'. It is also questionable whether the current trends towards deregulation and privatisation in the rail transport sector will, over the longer term, result in a restructuring of the supply and demand side of 'the market for trains' that would eventually mitigate problems of 'user competence' such as those which held back development of the X2000. The apparent trend is towards consolidation and concentration of equipment manufacturers on the supply side of the market, and a proliferation of smaller-scale railway operators on the demand side. 'System competence' will no longer be held by the users, but rather by the producers. This is a potentially very negative scenario for the future of innovation in rail transport. As observed elsewhere, in markets dominated by producers there is a strong tendency towards "unsatisfactory innovations" arising out of deviation away from user needs in innovative activities that are producer-led (Freeman, 1982; Lundvall, 1988).

This possible trajectory of development would contribute to an emergent overall 'system problem' in the railway industry. Mainstream economic analysis does not fully recognise or appreciate all of the dimensions of such problems, including the one which is most fundamental from a Systems of Innovation perspective – namely, the *quality* of demand (Lundvall, 1988). Addressing this problem effectively in cases where users are dominated by producers and possess little capability to articulate their needs forcefully or well enough requires some form of public intervention, direct or indirect, aimed at establishing or restructuring user-producer relationships. Somewhat ironically, therefore, the current trend to institutionalise a 'free market' orientation in public procurement and elsewhere, which has been justified in part by arguments about 'non-market' barriers to innovation, may, over the longer term, generate strong demands for public intervention, justified by arguments about 'market-based' barriers to innovation. In the case of rail transport (and similar sectors) in Europe, however, the level of public intervention required may well be supra-national, rather than national.

If the EU finds itself called upon to restructure the demand side of the rail transport sector, the type of solution required might well be that of overcoming a 'low level equilibrium trap' by constituting a focal organisation capable of (re)creating an organised market through the development of an appropriate form of vertical linkage between users and producers (Teubal, Yinnon, & Zuscovitch, 1991). In that (admittedly hypothetical) case, the main focus of public activity would not be to consolidate demand by acting as an end-user, but rather to resolve the co-ordination problems involved in market creation by organising and, to some extent, leading existing end-users. This would involve an organisation or articulation of demand resulting in, or equivalent to, the formation of a core of innovative users sufficiently large to overcome the problem of otherwise inadequate 'critical mass'. The key requirement for the responsible public agency, in the role of a 'catalyser' or 'focal' entrepreneur, would be to represent both the broad range of diversity in user demand and a high level of sophistication in 'user competence'.

These are very general policy implications, based on the most general aspects of the Swedish experience in procuring a high speed train. With regard to more specific issues, connected with more detailed considerations about the case, it seems to us that the experiences accumulated in Italy by the Pendolino case could have been used more intensively in some way. Drawing upon the Italian experience could have decreased the long time period it took Sweden to develop the X2000.

We would add to this that the Swedish state railway company, SJ, and the Swedish government might also have studied the Italian case more carefully from the 'buyer's' perspective. What, for example, were the key features of user-producer relations in the Italian 'market for trains' that contributed to the early development and subsequent international market success of the Pendolino? What organisational aspects of the Italian state railway company, FS, contributed to its apparently superior capabilities for 'collective learning' in preparing to procure the Pendolino? And how did public policy operate in relation to these and other features of the Italian case? We can not, of course, answer those questions here. They are, however, of more than 'historical' interest. As we note above, there are reasons to believe that national – or supra-national – public technology procurement in rail transport has 'a future' in Europe.

INTERVIEWS

Andersson, Bernt – General Manager, Swedrail (interview, April 28th, 1997).

(B. Andersson was the Director of Technique within the Engineering Department at SJ between 1984 and 1988. From 1988 until October he was the Director of the Engineering Department at SJ.)

Andersson, Evert – Professor KTH & AD Tranz (interview, April 28th, 1997).

(E. Andersson entered the high speed train project in 1973 and worked half-time until 1981 with high speed train project. He was a specialist in mechanical systems. In 1981 he became the director of the office for mechanical system technique. He continued to work with the high speed train until 1990.)

Hörberg, Agne – Currently Director of Technique within the Engineering Department at SJ (interview, April 28th, 1997).

(A. Hörberg entered the high speed train project after the first tender request. He was project leader for the project at SJ.)

Lundgren, Jan – Director of The Service Division, ABB Atom (interview, May 12th, 1997).

(J. Lundgren took part in the final negotiations towards SJ in 1986. As the contract was written by midsummer 1986, he took over as the overall project leader. He was responsible for the development and the total delivery of 20 train sets. He left the high speed train project in 1992 when the trains worked satisfactorily and ASEA had delivered 10 train sets.)

Persson, Richard – AD Tranz (interview, May 12th, 1997).

(R. Persson entered the high speed train project in 1982. He made, e.g., calculations regarding the life-time of the train. From 1986 until 1993, he was responsible for the tilting mechanism and the coaches themselves.)

Pålsson, Lars – System Purchaser, SJ (interview, April 28th, 1997).

(L. Pålsson entered the high speed train project in 1982. He started at the finance and accounting department. Pålsson introduced LCC [the Life Cycle Cost concept] in SJ. He was a member of the group that defined the functional requirements in the technology procurement of the high speed train, X2000.)

REFERENCES

Abernathy, W. J., & Clark, K. B. (1985). Innovation: Mapping the winds of creative destruction. *Research Policy, 14*, 3 - 22.

Agerberg, M. (1996a). Pendelinos segertåg. *Ny Teknik*(45), 12 - 17.

Agerberg, M. (1996b). Nytt svenskt snabbtåg får motor i varje vagn. *Ny Teknik*(44).

Andersson, P.-G. (1983). *Snabbtåget X2 – teknikupphandling: Teknikupphandling dokumentation från ett symposium* (Rapport 23). Stockholm: Ingenjörsvetenskapsakademin.

Arthur, W. B. (1988). Competing technologies: An overview. In G. Dosi, C. Freeman, R. Nelson, G. Silverberg, & L. Soete (Eds.), *Technical change and economic theory*. London: Pinter Publishers.

Batchelor, C. (1996, October 17). Inclined to try again – On the revival of high speed trains in the U.K. *Financial Times*, pp. 23.

Flink, T. (1992). *The development of high speed trains: An industrial dynamics approach*. Stockholm: The Economic Research Institute at Stockholm School of Economics.

Flink, T., & Hultén, S. (1990). *Det svenska snabbtågsprojektet: De första 20 åren*. Stockholm: The Economic Research Institute at Stockholm School of Economics.

Flink, T., & Hultén, S. (1993). The Swedish high speed train project. In J. Whitelegg, S. Hultén, & T. Flink (Eds.), *High speed trains: Fast tracks to the future* (pp. 89 - 103). Hawes, North Yorkshire: Leading Edge Publishing Ltd.

Foray, D. (1989). Les modeles de compétition technologique: Un revue de la littérature. *Revue d'Économie Industrielle, 48.*

Freeman, C. (1982). *The economics of industrial innovation.* (Third ed.). London: Frances Pinter.

Fridlund, M. (1993). *The 'development pair' as a link between systems growth and industrial innovation: Co-operation between the Swedish State Power Board and the ASEA company* (Trita HST Working Paper 93/9). Stockholm: Royal Institute of Technology, Department of History of Science and Technology.

Giuntini, A. (1993). High speed trains in Italy. In J. Whitelegg, S. Hultén, & T. Flink (Eds.), *High speed trains: Fast tracks to the future* (pp. 55 - 65). Hawes, North Yorkshire: Leading Edge Publishing Ltd.

Gunnarsson, U. (1994). *Banor för snabba tåg i Sverige.* Malmö: Nordisk Banteknisk Ingenjörsutbildning.

Hidefjäll, P. (1997). *The pace of innovation: Patterns of innovation in the cardiac pacemaker industry.* Unpublished Ph.D. Dissertation, Linköping University, Linköping, Sweden.

Hultgren, K. (1981). *Höghastighetståg: Järnvägens tredje generation olika lösningar i olika länder.* Malmö: Frank Stenvalls Förlag.

Linder, G. (1996). *Handbok i offentlig upphandling.* Stockholm: Norstedts Juridik AB.

Lundvall, B.-A. (1988). Innovation as an interactive process: From user-producer interaction to the national system of innovation. In G. Dosi, C. Freeman, R. Nelson, G. Silverberg, & L. Soete (Eds.), *Technical change and economic theory* (pp. 349 - 369). London: Pinter Publishers.

Martin, J. F. (1996). *The EC public procurement rules: A critical analysis.* Oxford: Clarendon Press.

Nilsson, C., & Wallner, J. (1993, March 3). *Dagens Nyheter.*

Pålsson, L. (1987). *Upphandling av snabbtåg.* Stockholm: SJ.

SFS 1973:600. (1973). *Kungliga Majestäts upphandlingskungörelse.* Stockholm: Swedish Law.

SJ. (1969). *Höga hastigheter i SJ persontrafik tekniska förutsättningar.* Stockholm: SJ Centralförvaltning Utvecklingsavdelningen.

SJ. (1980). *SJ snabbtågsprojekt: Utredning 1980.* Stockholm: SJ Centralförvaltning.

Teubal, M., Yinnon, T., & Zuscovitch, E. (1991). Networks and market creation. *Research Policy, 20,* 381 - 392.

WS Atkins Management Consultants, & Associates. (1988). *The "cost of non-Europe" in public sector procurement.* (Vol. 5). Luxembourg: Office for Official Publications of the European Communities.

3. PROCURING PRODUCTS AND POWER: DEVELOPING INTERNATIONAL COMPETITIVENESS IN SWEDISH ELECTROTECHNOLOGY AND ELECTRIC POWER

M. Fridlund

1. INTRODUCTION

Competitiveness comes in mysterious ways. Today, the Swedish state, through its power utility, Vattenfall, is a competitor on the European power market and the private Swedish-Swiss firm, ASEA Brown Boveri (ABB), is one of the major international electrotechnical manufacturers. The aim of this study is to investigate the development of one of the tools behind this competitiveness of the Swedish state and industry. The tool in question was the High Voltage Direct Current (HVDC) transmission technology.[1]

The first commercial HVDC plant was inaugurated in 1956 and was the outcome of a public technology procurement project between the Swedish electrotechnical company, ASEA, and the Swedish public power agency, the Royal Board of Waterfalls (Vattenfall). However, the HVDC development neither started nor ended with this project. Therefore the central global and local processes that influenced the prospective HVDC user and producer before and after the formal procurement project are described. The study closes when HVDC reached commercial maturity locally and globally as well as having gone through its first transfer abroad. The procurement process is divided into the phases of Proto-Procurement, Procurement, and Post-Procurement. Note however that this linear division is made in retrospect for analytic purposes and does not resemble the actual perception of the actors involved, who on the contrary were uncertain all through the procurement project – and long after – whether HVDC actually had a commercial future.

[1] ASEA was in 1988 merged with the Swiss Brown Boveri & Co (BBC) into ASEA Brown Boveri (ABB). Vattenfall, formerly the Royal Board of Waterfalls, was in 1992 divided into the state-owned power producing stock company, Vattenfall, and the grid administration agency, Svenska Kraftnät.

2. PROTO-PROCUREMENT: IN SEARCH OF POWER AND PRODUCTS

In the interwar period high voltage alternating current (AC) transmission was the dominant technology for electric power transmissions. That had not always been the case however. The end of the 19th century had seen the "battle of the currents" between proponents of direct current (DC) and AC technologies about which was best for long-distance power transmissions. The battle was decided by the first large-scale transmission project, the American Niagara Falls project. This included building the world's largest power plant and experts had first recommended DC transmission but as the experience of AC increased swayed in favour of AC. In 1896 Niagara started producing hydro power and became a powerful example of the value of AC transmission in utilising "remote waterfalls hitherto running to waste" (Hunter and Bryant, 1991: 254). This decided the battle of the currents and gave AC an hegemony in long-distance power transmission that was to keep its hold until the interwar period and the "second battle of the currents" (Fridlund and Maier, 1996).

2.1 New Beginnings

The interwar period can with the benefit of hindsight be seen as being formative for several crucial procurement resources globally as well as locally by identifying obstacles of a new dimension for AC transmissions. The interwar period saw a rising demand for power reflected in discussions concerning several international "super power transmissions" projects – long-distance transmissions of large amounts of electric power – that were suitable for HVDC. This generated a *proto-demand* for an HVDC innovation-to-be, and created a familiarity of the future user Vattenfall with the critical problems with such a project. Concerning supply, globally novelty and diversity was created through the emergence of several HVDC *proto-technologies* and locally an "absorptive capacity" for transmission technologies was created in the future producer, ASEA. Last but not least, the period saw the structuring of relations in the Swedish electrotechnical sector in that the *development pair* between ASEA and Vattenfall had matured. After a 20-year long period of social learning in different user-producer projects a mutual trust had been developed that became important in the following Procurement Phase.

2.2 Shaping the Needs: Developing Proto-Demand

The technical characteristics of the AC technology made it unsuitable for several super-power projects. Mainly, there were critical problems with the bad electrical stability of long AC transmissions; these could cause power failures and black-outs, and made impossible the building of long underwater cable transmissions using AC. As a solution, it was proposed in the interwar period to use DC transmission, which did not suffer from these problems and also promised to be less expensive. Internationally AC and DC were once again pitched against each other.

2.2.1 Global visions: plans for international interconnections

The largest and most seriously contemplated super power transmission project concerned electric power transmission from Norway to the European continent. WWI had demonstrated to Denmark its large dependency on coal-imports and since Norway had one of the world's largest hydro power resources with only a limited domestic market potential Denmark proposed a project concerning a super power transmission from Norway through Sweden to Denmark. A Nordic commission investigated critical technological and economic problems of such an Inter-Scandinavian power transmission and concluded that it was feasible with either AC or HVDC. However, the world economic crisis of the 1930s made the project unrealistic because of a lack of demand in industry (Svenska Vattenkraftsföreningen, 1921: 68; Fridlund and Maier, 1996).

Large transmission projects were also discussed in Germany, which after the Nazi take-over began to prepare its national power grid for a new war. In connection with this the German military became interested in super power and cable transmission projects (Maier, 1993: Ch. 4.1.2). In Russia, also, there were similar discussions of national super power transmissions.

2.2.2 Local visions: plans for a national grid

In Sweden, the State through the public agency, Vattenfall, was the largest actor in the electric power system. There was no public monopoly on ownership or generation of hydro power. Nevertheless, Vattenfall and the large private or municipal power utilities, because of exclusive regional transmission networks, enjoyed de facto regional monopolies. Vattenfall was in charge of exploiting the waterfalls of the state and it was the largest Swedish power producer with four large power plants. Aside from Vattenfall, three other large power utilities owned interests in inter-regional AC transmission trunk lines and through this had interests in future Swedish super power transmissions. Among the smaller utilities was the ASEA subsidiary, SEV, which was going to be important in the development of the HVDC technology.

During the interwar period Vattenfall interconnected the networks of its two hydro power plants in central Sweden. This was the first step in a plan to interconnect all of the State's power plants from the upper North to the South into one national grid and through this use the plants more efficiently. The last of the State's three regional systems was centred around a power plant in the very far north at Sweden's largest waterfall. The system was isolated from the southern systems and in the 1910s it was seen as a utopia to have super power transmissions across the thousand kilometres to central Sweden. In 1930 Vattenfall expanded on its earlier vision of a national grid and put forward a plan with *all* Sweden's power plants – private, municipal and state – interconnected into one national power system. But this got thwarted in the mid-30s when two large private power companies decided to interconnect their systems in the north and south of Sweden with an AC-transmission link. Vattenfall protested to the government, which decided that further development of the long-distance transmission grid was to be carried out in co-operation between Vattenfall and private utilities. The majority of the state's large hydro power resources in the very far North had to wait for demand and technology to progress enough for it to be possible to build new power plants and transfer the power south. And this was seen as coming sometime in the 1960s.

2.3 Shaping the Means: Developing Proto-Technologies

The critical technological obstacle to HVDC transmission problem was the lack of a *converter* between high voltage DC and low voltage AC, which was needed since the majority of customers used low voltage AC electricity to power motors and machines. A competitive development atmosphere was created globally as well as locally in Sweden, as electrotechnical firms and independent inventors competed to present prototype converter technologies to solve the HVDC problem.

HVDC is, as a technology, basically an application of previous technological knowledge and therefore more an example of 'technology as applied technology' than 'technology as applied science' (Rosenberg, 1982; Rosenberg, 1992).[2] In the proto-procurement phase this application of technology came from the know-how accumulated in the industrial processes of incremental improvements on an already commercial technology.[3]

2.3.1 Global search: creating technological diversity

There already existed a DC transmission technology – the Thury system – that in 1906 with a 180 km power link had shown the possibility of long distance DC transmission. However, the Thury system used a very large number of electromechanical converters which were regarded as too costly and clumsy. Starting in the 1920s more promising solutions to the converter problem began to appear. In the 1920s two different experimental prototype converters intended for HVDC transmissions were presented by the large international electric manufacturers English Electric and the American giant General Electric Co. (GE). In 1931 GE stated that there existed no theoretical reasons against HVDC transmissions with their new converter and the press reported that the battle between AC and DC "seems to be entering upon a new and most interesting phase". The same year the German Siemens company reported that their commercial Mercury-Arc Rectifiers now could also be used to convert from DC to AC. Siemens saw this as the solution to the HVDC problem and meant to be ready to start building HVDC transmission systems. The other German giant, AEG, also said they were working on HVDC converters. In May 1932 yet another German converter was presented by the famous scientist and independent inventor Erwin Marx. A devoted German national-

---

[2] As shown by Lindqvist (1994), HVDC also provides a good example of "science as applied technology".

[3] This is a good example of how strategic absorptive capacity is developed as a by-product of a firm's normal manufacturing. See Cohen and Levinthal (1990).

ist, Marx wanted to use his converter to further the war effort. HVDC transmission with underground cables was of great military interest because AC power lines disturbed airfields and made industrial plants easy to find by aircraft. Marx's converter therefore became a secret defence project and the focus of the German firms. But after lack of progress it came to be seen in 1937 as a dead-end and the German firms restarted their other converter research (Direct Current, 1963; Maier, 1993: 19, 106 - 8).

In addition to these proto-converters, two Swedish prototypes were presented in the 1930s. The first – the Glesum system – was developed by a famous Swedish independent inventor with help from a professor at the Royal Institute of Technology in Stockholm. When the converter was presented at a meeting in 1934, ASEA "gave the sensational announcement" that it was also working on its own converter technology (Svenska Dagbladet, 1934: 3).

In the mid-30s GE conducted the first field tests of HVDC transmission with a small-scale proto-converter. The experiments were seen as a success and GE planned larger experiments. In 1939, a short HVDC transmission was publicly demonstrated in Europe when the Swiss company, Brown Boveri (BBC), used a converter of their own to transmit DC. Even though the transmitted power was small, BBC claimed that "doubt can no longer exist" about super power transmissions by HVDC and intended to go ahead with larger transmissions. The Swiss demonstration was, however, stopped with the outbreak of WWII (Direct Current, 1963: 4 - 5).

2.3.2 Local search: the factory as learning laboratory

ASEA was the most important Swedish manufacturer of electric power technology and, after the 1930s, the other only independent firm was a cable manufacturer controlled by the Ericsson company. Also the German and Swiss giants Siemens, AEG, and BBC were active in Sweden through Swedish subsidiaries and representatives.

In 1925, ASEA decided to start manufacture of a Mercury-Arc Rectifier for converting low-voltage AC to DC. The use of low-voltage DC had expanded with the growth of city lighting and tramways. Traditionally electromechanical converters had been used for converting AC to DC but since 1904 the electronic mercury-arc rectifier had become a serious alternative.[4] The other large European electrotechnical manufacturers sold rectifiers with BBC as the leader. ASEA being a laggard, first considered acquiring a license on a foreign construction but eventually decided to develop a more independent construction (ABBCA, 1925: 1). In 1927 ASEA wrote a contract with a Hungarian consulting engineer for the construction of a rectifier. The engineer had worked for BBC and supplied ASEA with know-how and drawings in what "most likely was a plagiarism from Brown Boveri" (Håkansson, 1983, 1:5, 1:17; TMA, 1977: 17,

[4] The mercury-arc rectifier is a vacuum tight vessel filled with mercury gas and a cathode in form of a mercury pool and anode of iron or graphite rods and with a continuous current in the form of an electric arc burning in the mercury gas between cathode and anode. *Alternating* current is transformed to *continuous* direct current because an anode gets "blocked" when the AC-phase becomes negative and the positive current flows in the anode that is experiencing positive AC-phases.

26). In 1928, an experimental rectifier was finished. However, it did not function satisfactorily because of critical material problems and random short circuits, so called "back-fires". ASEA set up a group to solve the critical problems and construct a commercial rectifier. They soon found the engineer's construction not to be "capable of development in many respects" and instead went on and developed an independent rectifier construction (Wollard, 1988: 44; TMA, 1975: 2).

The next step towards a commercial product was to construct a prototype for internal use. This was not a totally independent construction since ASEA had acquired a lot of know-how from the failed experimental rectifier. This was later recognised through royalties to the consulting engineer. As the prototype and confidence progressed, ASEA in 1930 turned to the municipal power utility in Stockholm and asked to be able to install a prototype rectifier to try it out in real use. The utility accepted and a rectifier was delivered in 1932 and worked so well that it was accepted for operation one year ahead of contractual stipulations. Meanwhile, ASEA had also given a tender and delivered a commercial rectifier to the municipal utility in Gothenburg. However it suffered from severe problems and had to be taken back for adjustments. In retrospect these problems served an important purpose in providing important know-how for the further development of low-voltage rectifiers and – later on – for the first HVDC mercury-arc rectifier. Parallel to this ASEA's management worked hard to get an international reference order. The first came in 1932 to the Danish State Railways (DSB) and it was important because of its large size – six rectifiers – but also because it was the first export order. More so, DSB did not dare to give ASEA the whole order because its lack of reference-projects and also ordered three rectifiers from BBC, the leading manufacturer. This meant that ASEA's new rectifier construction was to be measured against its foremost competitor, which especially pleased ASEA. ASEA considered it important to deliver a construction that could match BBC's (ABBCA, 1932: 2 - 3). This was successful and in 1933 ASEA's low-voltage rectifiers had reached technical and commercial maturity.

As ASEA's group had worked on the solution to the problems of the low-voltage rectifiers they had also thought about how to construct a converter for the 50 - 100 times higher voltages of super power transmissions. And in 1933 the first experiments on a experimental HVDC converter were "sneaked in" between the work on the low-voltage rectifiers. The practical HVDC development had been initiated because the invention of the Glesum system (Lamm, 1983a: 37; Lamm, 1976: 4). A new company had been formed to exploit the invention, which since 1932 had been further developed by a professor at the Royal Institute of Technology. ASEA had been approached about investing to acquire patents and know-how. But although ASEA's management considered the Glesum system to have solved the converter problem and to be the best solution to the HVDC transmission issue, the international prospects of a commercial HVDC project in the next 10 - 20 years looked extremely small and non-existent in Sweden. This was because an eventual Swedish transmission from upper Norrland would "wait a very long time" and probably use AC. The only possible project that could be conceived in the foreseeable future was transmission of Norwegian hydro power to the European continent although it was also considered too uncertain (Alm, 1935: 1; ABBCA, 1933b; ABBCA, 1933a: 3 - 4). But the Glesum company did not give up their efforts and in 1933 it gave a successful demonstration of its new converter. ASEA's management discussed acquiring a license to manufacture the new technology, but eventually decided to wait (ABBCA, 1933c: 1).

ASEA's rectifier group wanted to convince ASEA's management that an HVDC mercury-arc converter – an Ion Valve – was a better development path and sped up their development in 1934, in connection with an announcement that the Glesum system would be presented at a public meeting. ASEA's project manager attended the presentation, and when he left ASEA's laboratory on the day of the meeting a new experimental ion valve was finished but hadn't yet been tested. There were great expectations of the Glesum system, and several hundred engineers attended the presentation. During the presentation the eager project manager phoned the laboratory and was told that the ion valve worked fine with a voltage 10 times that of commercial rectifiers. Back at the presentation he announced that ASEA had also worked out an HVDC solution. But the next day he found out that his phone call had come at the "right" time since 20 minutes later "the whole thing collapsed, and after that we never got it going again" (Berneryd, 1992: 10; Svenska Dagbladet, 1934; Lamm, 1976: 6). Although the first ion valve had broken down after a very short time, it seemed to prove that the construction was correct in principle. The construction was based on a patent from 1928 that ASEA's project manager had come up with when trying to improve the back-fire problem of the Hungarian rectifier. ASEA's group continued with more tests but they all suffered from the critical problem of an excessive amount of back-fire that, together with more urgent issues concerning commercial rectifiers, stopped the HVDC development in the mid-30s.(Lamm, 1947: 309; Lamm, 1946: 28).

In 1937, ASEA continued the development on the ion valve solution. The group was now less busy with problem-solving in connection with the commercial rectifier orders and had gained extensive experience and training in solving DC converter problems (ABBCA, 1938a: 2). Crucial to the further development of low-voltage rectifiers and high-voltage ion valves were a number of such critical problems concerning choice of materials. A series of experiments was conducted on a new ion valve construction that succeeded, and in 1940 ASEA came up with a construction that worked longer than a couple of hours and that could be tried in more realistic large-scale HVDC test transmissions (ABBCA, 1938b: 11; Lamm, 1949: 189).

2.4 Shaping the Relations: Developing Informal Institutions

In this period Vattenfall and especially ASEA developed international and national relations that became important in the coming procurement phase. Internationally, ASEA gained a respect among its international competitors and, nationally, ASEA and Vattenfall formed a long-term and intimate user-producer relation around the development of new technologies. Both these relations led to ASEA asserting and strengthening its independence and developmental capacity.

2.4.1 Global institutionalisation: gaining an international respect

From its foundation ASEA was set on competing with the major electrotechnical firms and being a respected international company. In the 1920s and 1930s ASEA established itself as a major electrotechnical enterprise and was accepted on a more or less equal basis into a market cartel with the German giants. It managed to keep its independence from the foreign giants which included staving off an attempted take-over by the American giant General Electric. In the 1930s, when industrial R&D became institutionalised among the electrotechnical giants, ASEA followed this example and began considering R&D as a tool for international competitiveness. That led to the establishment of a high voltage laboratory and a new will to invest in more strategic and long-term R&D such as HVDC.

2.4.2 Local institutionalisation: establishing a national development pair

The joint HVDC development between Vattenfall and ASEA was not unique but rather one link in a series of previous intimate and informal user-producer development collaborations around advanced technologies. This joint history of long-term collaborations in uncertain projects on high-risk technologies had matured in the 1930s and established the trust relation between the two partners that I call a Development Pair. This relation goes back to 1909, when Vattenfall and ASEA conducted joint experiments to improve ASEA's circuit breakers. These collaborations were taken up again in the 1920s, when Vattenfall erected a power line of 130.000 volts to interconnect their two regional systems in central Sweden. This was the first European project to use such a high voltage and through this it provided incentives to Vattenfall and ASEA to initiate their a large joint strategic development project. ASEA's high voltage transformers were not up to international standards. When ASEA got a procurement order on transformers for the power line, it got the chance to catch up. Through joint tests ASEA and Vattenfall could develop transformers comparable to foreign ones. After this Vattenfall initiated a new strategic project to improve ASEA's circuit breakers. With the use of Vattenfall's power plants for advanced experiments ASEA in the 1930s could improve their constructions and also develop a completely new type of breaker (Fridlund, 1995; Glete, 1984a).

This collaboration was characterised by a "spirit of national engineers" among Swedish electrical engineers. The historian Jan Glete argues that when discussing the collaboration between Vattenfall and ASEA one must "take into account a positive interest in encouraging Swedish industry and Swedish technology". There were nationalistic sentiments connected to Swedish technology concerning national prestige and the "guarding of the national technology and ambitions to foster it were strong ambitions in Swedish industry". Sweden was a country that stood outside the wars and the struggle for national eminence came to be fought on the industrial battlefield instead of the military.[5] This "spirit of national engineers" characterised the electric power sector up to the late 1950s (Glete, 1984b: 43, 69).

3. PROCUREMENT: HVDC IN THE MAKING

3.1 Developing a Problem: Putting HVDC on the Agenda

In 1940 DC transmissions once again became an issue globally as well as locally in Sweden. The global interest was shown by Soviet plans for exploiting large waterfalls in Siberia and Caucasus and to use HVDC to transmit 600 megawatts (MW) over 900 km (VA, 1940b: 1, 15). The local interest was shown when a Swedish HVDC project proposal of about less than one-hundredth the size of the Soviet plans was put forward to Vattenfall. This was in a conversation about Swedish transmission projects between the Director General of Vattenfall and the general manager of ASEA's subsidiary power company, SEV. ASEA through SEV owned several smaller Swedish utilities and waterfalls. This involvement in the power sector made ASEA into a customer and sometimes competitor to Vattenfall. Usually this was not a problem, but on this occasion in 1940 it was. ASEA owned a utility on the island of Gotland that it wanted to supply with power from one of its waterfalls 450 km away on the Swedish mainland. The power could be transferred via Vattenfall's existing power lines but since Gotland's distribution net was isolated from the mainland it would also require an undersea power cable to Gotland. SEV's manager proposed this power transfer should be with HVDC. The amount of power transmitted should be rather small and the Director General found the project interesting but nevertheless objected because it might strain Vattenfall's capacity in times of crisis (VA, 1940a: 3).

This market-directed interest in HVDC projects from ASEA is also shown through two 'market surveys' conducted in 1940. The first concerned what previously had been written about possible HVDC transmissions and gave ASEA a view of the potential global demand for HVDC. Based on this ASEA did an internal survey of the technical and economic reasons for and against HVDC. It concluded that there could be no doubt that HVDC was going to "obtain a footing" and also be used in Sweden. Also two critical problems for a successful future HVDC transmission were identified. The first was technical and was to develop converters for high voltages; the second was social and concerned establishing a development project "in collaboration with those power companies concerned to try to get a plan for the gradual testing of equipment on a practical scale for larger and larger power." To develop the social collaboration was considered "as important" as developing the technology since

> "those on the first hand economically justifiable projects are so large that their realisation must be preceded by the execution of smaller DC-transmissions, which in themselves might not generate any eventual economic profit, but would give experience necessary to determine if it will be possible to risk the realisation of the larger projects. [...] When one goes about executing projects involving such large investments, one usually avoids large technical risks, in that one seeks to apply constructions and principles that have proven their trustworthiness and operational reliability in smaller projects" (ABBCA, 1940: 9).

---

[5] The view that nations battled each other through their technological development was prevalent and that nationalism was connected to technological development from the late 19th century onwards is supported by a lot of historical research. See Eriksson (1978), Fritzsche (1992), Ekström (1993), Björck (1993), Samuels (1994), Fridlund and Maier (1996), Elam (1997), Fridlund (1997).

In 1941, Vattenfall's Director General and ASEA's project manager met to discuss Swedish HVDC projects. The Director was an enthusiastic expert on the issue, as he had worked on the previous Inter-Scandinavian project and on a plan to use HVDC to export power from one of Vattenfall's power plants to Denmark. He had several practical ideas and suggestions and offered ASEA the use of Vattenfall's transmission network for HVDC experiments as, in a similar way, it had been allowed to use it to test their circuit breakers (VA, 1941: 1 - 2). The project manager stressed the risks with an HVDC transmission to Gotland but despite this the Director was very optimistic about possibly finishing a transmission within two years. If Swedish cable manufacturers – to which an ASEA subsidiary belonged – would be willing to sell a cable without any profit, an immediate purchase could be possible (VA, 1941: 2 - 4). But despite Vattenfall's optimism, it did not become an 'immediate purchase' and HVDC came instead to be considered for a much larger transmission project, a Swedish super power project.

3.2 Enter Procurement: Formalising the Process of Problem-Solving

In 1942 Vattenfall started preparations for a future super power transmission from the northern Norrland region of Sweden. In central Sweden, only 20 % of the exploitable hydro power remained untapped and was rather expensive to develop, while in lower and upper Norrland around 70% and 90% was still unused (Rusck, 1945: 155). To keep the number of large transmission lines from Norrland from becoming too many, though, it would be necessary to develop a technology with higher transmission capacity than before.

The previously mentioned state inquiry had concluded that the future large power transmissions should be built in collaboration between the major power utilities. In 1942 the Swedish Collaborative Committee for Superpower Transmissions was formed to investigate the problems concerning a future transmission from Norrland. It consisted of representatives from Vattenfall, the three other large utility companies owning power lines of 220.000 volts, and the potential Swedish manufacturers, ASEA and a cable company. Its main task was to investigate the alternative possibilities of using HVDC or AC transmission with much higher voltages than before. This project was much larger than those previously discussed and the Director General was less optimistic about HVDC and more doubtful concerning the assurances made about its usefulness. Nevertheless, he found it "very laudable" that ASEA had started developing HVDC converters, even if he doubted whether ASEA with its limited resources would manage to solve a problem "where the major international firms have not succeeded" (VA, 1942: 1 - 2; VA, 1943: 1).

ASEA had constructed an improved Ion Valve prototype and after the first committee meeting Vattenfall investigated the possibility of using one of its power lines for large-scale and long-term construction experiments on the new prototype. The problem with constructing the Ion Valve was that it was too complex a construction to make possible its development on the basis of known laws of electromagnetism and theoretical models. Instead, it had to be based on results from empirical trial-and-error testing of prototypes. This had to be done in realistic full-scale tests with almost operational voltages. Furthermore, it was necessary to test each modification of the prototype during a very long period of operation because of the possible critical problems of "ageing" of the valve material and random and unpredictable back-fires. All this made the development very time consuming and expensive, as well as the fact that it needed access to large amounts of electric power (VVA, 1942; Direct Current 1960/61: 224; Lamm, 1976: 6 - 7). The result of the committee's main investigation about AC versus DC confirmed that HVDC should give lower transmission cost than AC, estimated as 2/3 the cost of AC. Following this, ASEA and Vattenfall decided to go ahead with more formalised co-operation (Borgquist, 1968: 19; Lamm, 1976: 2).

3.2.1 Contracting to procure joint know-how

In 1943, ASEA and Vattenfall signed an agreement to "jointly establish and operate" an experimental power transmission for HVDC to give the "most complete experience basis for the consideration of a possible use of HVDC for future super power transmissions from Norrland to central Sweden." This experimental transmission set-up would consist of a power line connecting two converter stations, one to be built in Vattenfall's hydro power plant Trollhättan and the other 50 km away in a small substation belonging to Vattenfall. Except for supplying the power for the experiments and building the two converter stations Vattenfall supplied personnel to operate the power line while ASEA supplied the Ion Valves and necessary personnel and equipment for the experiments. All other expenses should be divided equally between the two parties. According to the contract the results from the experiments should be freely available to both parties although they could not be communicated to outside parties (VVA, 1943: 1, 3). The transmission was estimated to cost 1.3 million Crowns to build and the experiments were planned to commence during the end of 1944 and be finished in 1948. While waiting for the experimental transmission link to be set up, ASEA continued its experiments and in late 1943 they performed the first successful long-term experiments on their Ion Valve prototype (Sylwan, 1944: 57; [Helén], 1957: 105; Lamm, 1949: 189). But crucial tests still missing were to actually transmit power over longer distances and to try out how and if the converters also worked outside the laboratory, out in the field.

3.2.2 Confronting critical technical problems

A great advantage for HVDC transmissions would be the possibility of using 'earth-return'. This meant that, rather than using a closed DC circuit of two power lines for each transmission, the returning DC current could instead be transmitted through the earth between the two converter stations (or through the sea, in the case of undersea transmission). This would mean that only one power line was needed instead of two without earth-return or three with AC. Since a large cost of the transmission link usually consisted of the cost for lines or cables, this would mean very large savings for DC compared to AC.

However, an important critical problem with using earth-return was that it might damage or disturb other large technical systems such as telephone cables or signal systems of the railway networks. To investigate this Vattenfall and ASEA in 1944 started a series of large-scale experiments in collaboration with the Swedish Telecommunication Administration (STA), the State Railways and Chalmers Institute of Technology. First earth-return was investigated at the joint experimental transmission link at Trollhättan in a 50 km low-voltage DC transmission. After this the scale of the experiments gradually increased until it was almost spanning the whole length of the country. The first of the more large-scale field experiments was conducted during a Saturday night in November, when DC was transmitted more than 300 km via one of Vattenfall's long power lines. During the five minutes the experiment lasted all railway traffic in central Sweden was stopped because of the danger of faulty signalling. The experiment also showed that a real super power transmission with earth-return could cause wrong railway signals within 150 km of each electrode. The critical problem still remained unsolved. During 1945 these 'macro-experiments' were continued with "one of the most exceptional electrotechnical experiments performed in recent time". This was a DC transmission of 300 km along the Norrland coast but this time using the Baltic Sea as a return conductor (sea-return). This was a continuation of a previous failed test to transmitting power with sea-return more than 450 km. By using sea-return, it was hoped to avoid disturbing the mainland telephone and railway systems (Blomqvist 1946: 21; Lundholm, 1947: 321). This experiment succeeded in the sense that this seemed to be the case for the major part of the current.

Meanwhile, the completion of the experimental stations at Trollhättan had been delayed by a year because of a nation-wide strike at the end of the war. The transmission set-up was finished in the end of 1945 but then the preconditions for the HVDC experiments were altered because of the earth-return issue that had changed from a possible advantage to an actual obstacle. This was because the large risk of disturbing the railways and telephone cables around the Trollhättan power plant made it impossible to conduct any long-distance transmission tests. What could be done, however, was to simulate such a long-distance power transfer in the Trollhättan power station (with an AC/DC-converter connected over a large resistance – 'the power line' – to a DC/AC-converter).

The most important part of the experiments was however the testing of the Ion Valves which now could be moved from ASEA's laboratory to the Trollhättan power plant. This empirical testing was done through 'life-time' experiments wherein teams of ASEA's specialists together with Vattenfall's operators worked around the clock in Trollhättan for several months of time. A vivid description of the reasons for this long and arduous engineering work has been provided by ASEA's project manager:

> "[We] have tested [...] different modifications of the interior design of the valves, many of them in two or more samples. The more fruitful modifications have been run for a period of about half a year, or longer, while quite a number have been run unchanged for several years. One would like, of course, to finish the test on one modification before one goes on with the design of the next, but as the time that elapses between building a new modification from the first sketches and obtaining the first test results can generally not be cut

> down below 9 months, it is quite obvious that one must work with a great number of modifications 'in parallel'. The setting out of a design, therefore, often has to include a lot of guesswork regarding the outcome of previous designs which are still under construction in the works" (Lamm, 1956: 10).

3.2.3 Eliminating an institutional obstacle

Another obstacle for future Swedish super power transmissions was of an institutional nature and concerned the ownership and operation of the future super power lines. Except for the state, private companies owned hydro power in Norrland and it would not be considered practically and economically justifiable that every owner of power in Norrland that wanted to transfer it south should build its own super power transmission. Therefore they had to transfer their power on transmission lines owned by other companies. Since it was decided that the future power system was going to be developed in co-operation between private and public utilities it was predicted that several critical problems would arise concerning ownership of the power transmitted, etc. As a solution to this institutional obstacle, Vattenfall in early 1945 proposed to the three other main power companies owning power in Norrland that the four of them should form a joint company that should take over all old transmission lines and build all new super power transmission lines. None of the other companies was really interested in the proposal, but after hard negotiations Vattenfall managed to get them to agree on the proposed company. This company should own, build and operate all large Swedish transmission lines in the future. During the spring of 1945 Vattenfall asked the government to approve the proposed company. Opinions differed among the parties of the coalition government and the decision had to wait for the change of government following the end of WWII. In November the new Social Democrat government decided that the State – i.e. Vattenfall – should own and be responsible for any new Swedish super power transmission lines. This also meant that it became the sole responsibility of Vattenfall to decide what technology to use (Bjurling, 1982: 140).

3.2.4 Increasing demand and de-selecting supply

In 1946, the HVDC experiments continued with smaller power transmissions between the two experimental stations. Furthermore, Vattenfall performed a series of joint experiments with STA on radio disturbances from HVDC power lines. In September, 1946, yet another large-scale experiment was performed with earth-return. This was a full-scale experiment in its true sense since it was a transmission of 1,000 km. The north electrode of the transmission line was placed down in a mine in upper Norrland and the south in a fjord outside Gothenburg. But the experiment gave rather few positive results and strengthened the negative ones from previous experiments that had shown HVDC super power transmission to cause serious disturbances of the railway and telephone systems (Lundholm, 1952/53: 82; Rathsman and Glimsted, 1949a: 43).

During the autumn of 1946 the plans for developing the large waterfall at Harsprånget were suddenly brought forward by the Parliament. The anticipated post-war depression did not come and instead there was an increase in the demand for electric power. Therefore it was seen as necessary to start exploiting the hydro power in upper Norrland earlier than foreseen. Vattenfall already had to decide what technology to use for the super power transmission already during 1946. Both AC transmissions with 300 - 400,000 volts and HVDC were considered risky options. HVDC did not exist and the highest AC voltage previously used was 287,000 volt. ASEA did not dare to take the risk with the untried HVDC system and stated "that there are *no* chances to get the DC system fully ready and sufficiently tested in time for the first stage of the exploitation of Harsprånget, while the prospects look good to have it ready in time for it to be applied to the major part of the [further] developments in upper Norrland" (Lamm, 1947: 311). In December 1946 Vattenfall decided to chose AC technology for the development of Harsprånget and entered into a public technology procurement project with ASEA around developing a 380,000-volt AC technology. The project was successful and the first power line from upper Norrland was inaugurated in 1952. But this had not meant the end for HVDC.

3.3 Reconstructing Procurement: Return of an Old Project

Vattenfall and ASEA in 1947 decided to continue the HVDC collaboration with the aim of developing Ion Valves for future Swedish super power transmissions from upper Norrland. The old experimental station in Trollhättan would be too small for the new experiments and in 1948 it was decided to build another larger joint laboratory next to the old one. The new laboratory would make possible experiments on Ion Valves of up to eight times the power (25 MW) as that of the old laboratory (Lamm, 1947: 314; Rathsman, 1954b: 12).

At the same time an older much smaller HVDC project had been reopened when, barely a month after the decision to chose AC for Harsprånget, the question was raised anew in the Parliament about the possibility of supplying Gotland with electric power from the mainland. Gotland was the only Swedish region without its own hydro power or connection to the national power grid. Because of this 'power isolation' and high fuel costs, the electricity prices on Gotland had been much higher than in the rest of Sweden during WWII. The government approved a state inquiry and ordered Vattenfall to investigate it. Vattenfall conducted the inquiry jointly with the manager of the ASEA subsidiary that owned the only power plant on Gotland. Such an HVDC transmission to Gotland would be a very suitable first reference project for ASEA, in that it would be of a 'lagom' – Swedish for 'not too big, not too small' – scale and difficulty. The amount of power transmitted was only a tenth of that of a super power transmission. Furthermore, in case the project failed, there was no risk of law suits or claims of damage from dissatisfied customers since ASEA owned the island's one and only power company.

As for the super power transmission, it was crucial that the HVDC transmission did not disturb or damage surrounding infrastructure systems and as a part of the inquiry several experiments were performed. With the help of Chalmers Institute of Technology, potential disturbances to the telephone cables to Gotland were investigated. Furthermore, the transmission was not allowed to damage sea life because "possible demands from the fishing industry" might bring to nothing the economic gain of using sea-return. In the summer of 1949 Vattenfall together with the Royal Board of Fisheries conducted model experiments on the effect of DC sea-return on the flora and fauna of the sea. In these experiments a small (1000 m) scale-model of a transmission link with sea-return was used. The investigations showed that any disturbances could easily be avoided. The joint Vattenfall–ASEA inquiry was finished in January 1949 and recommended an HVDC cable transmission because it did not suffer the same large loss of power as an AC cable transmission. It stated the primary advantage of the transmission to be lower electricity prices on Gotland, and furthermore, an operational HVDC transmission "for reasonable power" would also be positive for the development of larger super power transmissions (Rathsman and Glimstedt, 1949a: 45, 60 - 62; Deines, 1949: 108).

In 1949 Vattenfall followed the inquiry's recommendations and presented the government with a proposal to build an HVDC power transmission to Gotland. The government and Parliament followed the proposal and in 1950 the procurement contract between ASEA and Vattenfall for a 20 MW Gotland transmission could be signed. According to Lamm, the technology was not considered as completely safe since it was still under development and all involved knew that the project meant "considerable technical risks" (Lamm, 1950: 146). The new Ion Valve laboratory was finished in 1951 and all experiments now became focused on developing the Ion Valves for the Gotland transmission. The two converter stations should each have two converter groups of 10 MW each with every group consisting of six normal Ion Valves and an extra 'by-pass' Ion Valve. This by-pass valve was ASEA's second solution to the problem of back-fires. It had been possible to decrease but not to eliminate completely the back-fire problem. Since it had been found impossible to completely eliminate them, even in commercial low-voltage rectifiers, the group had found it to be "sound realpolitik" to accept them and, instead, 'invent around' the problem. This was done by using a method developed for the commercial rectifiers. When a back-fire occurred, all the Ion Valves were automatically blocked and the faulty current was passed through a by-pass valve. When the back-fire had died out, the current came back as normal (Lamm 1947: 311 - 2). After the decision one and a half years were spent on further development of the Ion Valves. This time was used to test different modifications. When the decision was taken in 1953 to freeze the Ion Valve construction, around 140 different Ion Valve constructions had been tried, with several in two or more versions.

3.4 Supplying the Product: Laying the Link

The set up of the Gotland transmission began in 1952. The whole transmission consisted of two converter stations and one underwater cable. In the mainland converter station power was taken from the national grid and converted to DC and then transmitted through the cable to the converter station on Gotland. Here the power was converted back to AC before it was sent out to consumers on Gotland. The transmission used only one cable and the circuit was closed by sea-return. The station buildings were built so that it would be possible to increase the power to 40 MW at a later stage by doubling the number of converters and by laying an extra cable. The 96 km long cable was put out in 1953 and the transmission started in 1954. During the try-out transmission that followed, it was possible to tune in the equipment and to do "some measurements under realistic conditions" and, based on this, construct and manufacture some supplementary details (Rathsman and Svidén, 1956: 12). According to the contract the try-out transmission should continue until 1957. However, since Vattenfall was very satisfied with the trial results it was decided to bring forward the take-over one year and on January 1st, 1956, the transmission was taken into commercial operation by Vattenfall.

The whole transmission had originally been estimated to cost 9.5 million Crowns but had increased to 19 million because of an additional build-out of the converter stations. This was, however, cheap, considering that Vattenfall got the equipment for the same price as ASEA's manufacturing cost (Rathsman, 1954a: 7; Glete, 1983: 175). The price of 19 millions can also be compared to a Vattenfall assessment from 1959, which estimated Vattenfall's total costs for the HVDC development work in Trollhättan 1943 - 54 to be 2.6 million Crowns.[6] The hydro power from the mainland transferred through the new transmission link lowered the price of power on Gotland to half of what it had been before. Furthermore, ASEA also gave up the right to control the operations of its Gotland steam plant to Vattenfall (Rathsman and Glimstedt, 1949b: 99). A lagom cost for a lagom first HVDC order. The next step was to go from "lagom" to "big".

4. POST-PROCUREMENT: RE-MAKING PRODUCTS AND POWER

In 1957, the English and French public power utilities, British Electricity Authority (BEA) and Electricité de France (EdF), placed an order with ASEA for Ion Valves for an HVDC transmission under the English Channel. The transmitted power was to be eight times the amount in the Gotland project and the total cost was 70 million Crowns, including ASEA's share of 16 million Crowns (Vi Aseater, 1960: 20). Although the formal procurement process began in 1957, this project – which was of an adaptive procurement kind – also had its proto-procurement phase.

It started in 1950, when a joint French-English committee began to investigate the possibility of establishing a power exchange between the two countries. This would save building power plants of 300 MW in the two countries (Teknisk Tidskrift, 1954: 556). The distance over the Channel was rather short (40 km) which made both AC and HVDC possible. The committee's official report came only three weeks after the Gotland link started trial transmission. The report proposed using AC. Concerning HVDC, for which "so far no tangible success has been commercially demonstrated", the report declared that it was not immediately practicable because of its immature state

[6] Calculations are based on information in handwritten memos VVRA (1959a) and VVRA (1959b).

of development (ABBCA, 1954: 4). It nevertheless stated that an experimental HVDC transmission under the Channel merited serious consideration. When the report was due to be publicly presented in London, ASEA's project manager travelled there together with Vattenfall's Director General and vice Director General. During the meeting they provided information about the Gotland transmission and showed pictures from its opening. Although the report had settled the technical preconditions in favour of AC, ASEA with the help of Vattenfall lobbied to convince EdF and BEA of the advantages and trustworthiness of HVDC. The committee had envisioned an AC transmission using four cables, wherein the fourth cable was a spare cable that would use sea-return to experiment on HVDC transmission. If that worked well, a change-over would be made from AC to HVDC. However, this was unacceptable to the powerful British Admiralty, because the sea-return would create a small error in compass inclination in the heavily trafficked English channel. Although the error was insignificantly small, EdF and BEA thought it "would take 10 years to convince the Admiralty" about that (Berneryd, 1992: 86). Therefore, it was impossible to make a gradual change-over from AC to DC and the choice between HVDC or AC had to be taken from the outset. The great advantage of DC was that the expected economic savings would be around 4 million Crowns compared to 1.3 millions with AC; moreover, it promised to be more reliable (Vi Aseater, 1960: 20; Aseas Tidning, 1962: 17).

The discussions continued between ASEA and EdF and BEA in 1954 and 1955. After a while, ASEA's engineers were allowed to participate informally in the committee meetings. Later still, these meetings turned into serious negotiations. Finally, in 1957, the power companies placed their orders with ASEA. The transmission had been changed to an HVDC cable transmission with Ion Valves of twice the current and voltage of the Gotland transmission, and with eight times the power. This was also a procurement project of the adaptive variety, wherein the major adaptive innovation activities concerned the further development of the ion valves and the regulation system of the link. The transmission was brought into operation in December, 1961. The HVDC technology had matured and taken its first steps out onto the international market.

5. CONCLUSIONS: PRIVATE AND PUBLIC COMPETITIVENESS

The Channel project had been the first project in which ASEA managed to export its new product – the HVDC transmission technology. This project was followed in the 1960s by other export projects for long underwater transmissions in Italy, Sweden-Denmark (Konti-Skan, see below) and New Zealand. HVDC also found application in its originally intended Swedish use – as an overland transmission link for super power transmission. This was in 1965, when ASEA and GE got a joint order for the so-called Pacific Intertie. This was a more than 1.300 km long transmission of 1.440 MW from the Colorado River in Oregon to Los Angeles in California. Thanks to this, HVDC proved its superiority over AC for long super power transmissions. This was also a public procurement project, as the major buyers were two federal and municipal power companies. And, like the Channel project, this too was technology procurement of an adaptive kind, with Ion Valves developed to more than 15 times the power of those at Gotland (Lamm, 1983b).

Up until the 1970s ASEA had a practical monopoly on HVDC technology world-wide, with a profit margin of 20 %. In the 1970s the HVDC Ion Valve converter technology met harsh competition from new Semiconductor HVDC Thyristor converters. ASEA switched to the new technology trajectory and thanks to its experience in the Ion Valve technology managed to keep its leading position. The economic result of ASEA's HVDC Ion Valve technology was orders of 545 million Crowns in 1954 - 73 with net profits of around 74 million Crowns. The latter amount accounted for 13.5 % of the HVDC turnover (Glete, 1983: 287). Up until 1951, ASEA had spent 5 million Crowns on the HVDC development work, to be compared with the around 2.2 million Crowns spent by Vattenfall. In 1960, ASEA had spent around 22 million Crowns, to be compared to around 6.4 million Crowns by Vattenfall. But Vattenfall profited from its investments in its next large super power transmission project, which it started in 1960.

HVDC was never used by Vattenfall for its intended purpose, to transmit power from Norrland directly south. Further transmissions in the 1950s and 1960s used the existing AC technology and, instead, HVDC came to be used in another large power project, whereby Vattenfall redefined its demand from local Swedish to European in the 1960s. This was a project of building a power link between the Scandinavian peninsula and the European continent. The large development of hydro power in Norrland had created a power surplus that could be used for a "profitable" export to West Germany (Lalander and Gradin, 1962: 1189; Vi Aseater, 1963). In the early 1960s, discussions started concerning a power exchange between Sweden and Denmark and West Germany which would mean great savings because of the complementary diversity of different national consumer patterns and energy resources (hydro power vs. steam power). A joint committee of the involved power companies concluded that HVDC was the most suitable technology for this "Konti-Skan" link.

The super power transmission was for 250 MW and the whole project cost around 100 million Crowns. Vattenfall in 1963 ordered equipment for 50 millions (equalling around 400 million Crowns in 1997) from ASEA (Vi Aseater, 1963: 4). This project was seen as meaning cost reductions and profits of the same order of magnitude through reduction of the needs of thermal power in Sweden and export of surplus power. This cost can be compared with the sum of 6.4 million Crowns, which was the estimate of Vattenfall's cost until 1960 for the joint HVDC development work in Trollhättan (Konti-Skan, 1962: 35, 51; VVA, 1959a; VVA, 1959b). The Konti-Scan project was successfully inaugurated in 1965. It was described as "the European highway of electric power" and as opening the gate for "a complete European exchange of power" (Vi i Vattenfall, 1963). Through this, both ASEA and Vattenfall had succeeded on the European market with the help of their new HVDC technology.

REFERENCES

ABBCA (1925). "Protokoll hållet vid konferens den 29 och 30 december 1925", *ABB Central Archives, Västerås (ABBCA),* D:N7 02-005 [361/D-1002], Direktionskonferenser: Protokoll 1921 - 1932 (Ing. Hambergs).

ABBCA (1932). Letter Lindén to Edström, 21 July 1932, *ABBCA*, D, 526/1448, Diverse extern korrespondens: 1930, 1932.

ABBCA (1933a). R. Liljeblad, "Utlåtande angående den av von Platenska uppfinningen av anordningar för genererande och utnyttjande av högspänd likström", 4 October 1933, *ABBCA*, D, 544/1472, Diverse extern korrespondens: 1933 - 1934.

ABBCA (1933b). Letter Lindén to C.G. Langenskiöld, 7 October 1933, *ABBCA*, D, 544/1472, Diverse extern korrespondens: 1933 - 1934.

ABBCA (1933c). "Protokoll fört vid konferens 18/12 1933", *ABBCA*, D:N7 02-001 [401/D-1110], Diverse sammanträdesprotokoll 1906 - 1938.

ABBCA (1938a). U. Lamm, "Redogörelse för verksamheten under år 1937", 18 February 1938, *TM 1578*, 2, *ABBCA*, Tekniska Meddelanden, H:C1 03-018, TM 1550 - 1589.

ABBCA (1938b). Å. Karsberg, "Undersökningar av anodkonstruktioner för högspända likriktare: *Konfidentiellt*", 23 March 1938, *ABBCA*, Tekniska Meddelanden, H:C1 03-018, TM 1550 - 1589.

ABBCA (1940). U. Lamm, "Tekniskt-ekonomiska skäl för och emot kraftöverföring med högspänd likström", 11 December 1940, *TM 1794*, 9, *ABBCA*, D, 500/1387, Diverse Extern och Intern korrespondens: 1940.

ABBCA (1954). D.P. Sayers, M.E. Laborde & F.J. Lane, "The Possibilities of A Cross-Channel Power Link between the British and French Supply Systems", (Proof to be published in *Proceedings of the Institution of Electrical Engineers*, London, received 21 January 1954), 4, *ABBCA*, D, 1072/2059, Högspänd likström 1943 - 1957.

Alm, E. (1935). Report 347: le système Glesum: un système électromagnétique noveau pour la production et l'utilisation du courant continu a haute tension. In *Conférence Internationale des Grands Réseaux Électriques a Haute Tension: Compte-rendu des travaux de la 8ieme Session,* vol. 3. Paris.

Aseas Tidning (1962). Likströmsöverföringen för Engelska Kanalen invigd. *Aseas Tidning*, 54, 17.

Bemeryd, J. (1992). Uno Lamm: Framgångar och baktändningar. *Polhem: Tidskrift för teknikhistoria*, 10, 82 - 91.

Bjurling, O. (1982). *Sydkraft–samhälle: En berättelse i text och bild.* Malmö: Sydkraft.

Björck, H. (1993). *Teknikens art och teknikernas grad: Föreställningar om teknik, vetenskap och kultur speglade i debatterna kring en teknisk doktorsgrad, 1900 - 1927.* Stockholm Papers in the History and Philosophy of Technology TRITA-HOT 2025. Stockholm: Kungliga Tekniska Högskolan.

Borgquist, W. (1968). Svensk elkraftförsörjning. In *En del om el: Minnesskrift utgiven till Sveriges Elektroindustriförenings 50–årsjubileum 1968*. Stockholm.

Cohen, W. M. and Levinthal, D. A. (1990). Absorptive capacity: a new perspective on learning and innovation. *Administrative Science Quarterly*, 35, 128 - 52.

Deines, W. (1949). Elströmmens inverkan på havsfaunan. *ERA*, 22, 108 - 11.

Direct Current (1960/61). The development of a valve. *Direct Current*, 6, 224 - 34.

Direct Current (1963). The history of d.c. transmission: part IV. *Direct Current*, 8, 2 - 5 & 27.

Ekström, A. (1993). *Den utställda världen: Stockholmsutställningen 1897 och 1800-talets världsutställningar*. Uppsala: Nordiska Museets förlag.

Elam, M. (1997). National imaginations and systems of innovation. In C. Edquist (ed.), *Systems of Innovation: Technologies, Institutions and Organizations*. London: Pinter.

Eriksson, G. (1978). *Kartläggarna: Naturvetenskapens tillväxt och tillämpningar i det industriella genombrottets Sverige 1870 - 1914*. Uppsala Studies in the Humanities 15. Umeå: Acta Universitatis Umensis.

Evans, P. (1995). *Embedded Autonomy: States and Industrial Transformation*. Princeton: Princeton University Press.

Fridlund, M. (1993). The 'development pair' as a link between systems growth and industrial innovation: cooperation between the Swedish State Power Board and the ASEA company. Working papers from the Department of History of Science and Technology 93/9. Stockholm: Department of History of Science and Technology, Royal Institute of Technology.

Fridlund, M. (1995). Ett svenskt utvecklingspar i elkraft: Aseas och Vattenfalls FoU-samarbete, 1910 - 1980. *BI Forskningssenteret forskningsrapport* 1995/2 *& Senter for elektrisitetsstudier* 1995/312/7. Sandvika: Handelshøyskolen BI.

Fridlund, M. (1997). Teknikens konstruktion av nationen: svensk nationalism och industrialisering kring sekelskiftet 1900. In B. Brenna and K. M. Fjeldstad (eds.), *Kollektive identiteter, ting og betydninger*. TMV skriftserie 24. Oslo: Pensumtjeneste.

Fridlund, M. and Maier, H. (1996). The second battle of the currents: a comparative study of engineering nationalism in German and Swedish electric power, 1921 - 1961. Working Papers from the Department of History of Science and Technology 96/2. Stockholm: Department of History of Science and Technology, Royal Institute of Technology.

Fritzsche, P. (1992). *A Nation of Fliers: German Aviation and the Popular Imagination*. Cambridge: Cambridge University Press.

Glete, J. (1983). *ASEA under hundra år: 1883 - 1983 : Ett studie i ett storföretags organisatoriska, tekniska och ekonomiska utveckling*. Västerås: ASEA.

Glete, J. (1984a). High technology and industrial networks: some notes on the cooperation between Swedish high technology industries and their customers. Paper presented at International Research Seminar on Industrial Marketing, organised by the Stockholm School of Economics, Stockholm, August 29 - 31.

Glete, J. (1984b). *Storföretag i starkström: Ett svenskt industriföretags omvärldsrelationer - en sammanfattning baserad på "ASEA under hundra år"*. Västerås: ASEA.

[Helén, M.] (1957). *Den tekniska utvecklingen: 1883 - 1948*, vol. 3 of *ASEA:s historia: 1883 - 1948*. Västerås.

Hunter, L. C. and Bryant, L. (1991) *The Transmission of Power*, vol. 3 of *A History of Industrial Power in the United States, 1780 - 1930*. Cambridge, Mass.: MIT Press.

Håkansson, N. G. (1983). *Historik för lågspända jonventiler, 1927 - 65*. Ludvika.

Konti-Skan (1962). *Konti-Skan: Untersuchungen über die Voraussetzungen für eine dem Verbundbetrieb zwischen der BR Deutschland und den Skandinavischen Ländern dienende Hochspannungsverbindung / Studier av förutsättningarna för upprättande av en samkörningsförbindelse mellan Västtyskland och de skandinaviska länderna*, 2 vols. Stockholm.

Lalander, S. and Gradin, R. (1962). Konti-Skan. *Teknisk Tidskrift*, 92, 1185 - 89.

Lamm, U. (1946). Report 133: Mercury arc converter stations for high voltage d. c. power transmission. *The International Conference on Large Electric Systems (C.I.G.R.E.), 11th Session,* vol. 1., Paris.

Lamm, U. (1947). Grundläggande problem vid högspänd likströmsöverföring. *Teknisk Tidskrift*, 77, 307 - 14.

Lamm, U. (1949). Kraftöverföring med högspänd likström – ett utvecklingsarbete. In *Företagande, ekonomi och teknik: Studier tillägnade Marcus Wallenberg*. Stockholm.

Lamm, U. (1950). Kraftöverföring med högspänd likström från fastlandet till Gotland. *Aseas Tidning*, 42, 145 - 6.

Lamm, U. (1956). Report 9: High voltage direct current transmission. *In Proceedings High Voltage Symposium*. Ottawa.

Lamm, U. (1976). Utvecklingen av högspänd likströmsöverföring. *In Eliasson–föreläsningarna 1976*. Göteborg: Chalmers Tekniska Högskola.

Lamm, U. (1983a). Kraftöverföring med högspänd likström: den tidiga utvecklingen : jonventilepoken. In P. Barnevik, et al., *Teknik i ASEA: 1883 - 1983*. Västerås: ASEA.

Lamm, U. (1983b). ASEAs lansering av kraftöverföring med högspänd likström (HVDC) i USA. In Barnevik et al.

Lindqvist, S. (1994). Spatial networks of technological change: social mobility between industry and university. *In The Curt R. Nicolin Seminar: Knowledge as Substitute for Natural Resources*. Stockholm: Royal Institute of Technology.

Lundholm, R. (1947). Likström genom jorden över långa avstånd. *Teknisk Tidskrift*, 77, 319 - 22.

Lundholm, R. (1952/53). Return current through the earth for high voltage d.c. transmission. *Direct Current*, 1, 79 - 86.

Maier, H. (1993). *Erwin Marx: (1883 - 1980) : Ingenieurwissenschaftler in Braunschweig, und die Forschung und Entwicklung auf dem Gebiet der elektrischen Energieübertragung auf weite Entfernung zwischen 1918 und 1950*. Stuttgart: GNT Verlag.

Rathsman, B. G. (1954a). Likströmsöverföringen till Gotland. *Teknisk Tidskrift*, 84, 1 - 8.

Rathsman, B. G. (1954b). Forskning bakom 380 kV systemet. *Teknisk Tidskrift*, 84, 301 - 5.

Rathsman, B. G. and Glimstedt, U. (1949a). *De tekniska och ekonomiska möjligheterna att överföra kraft från fastlandet till Gotland*. Stockholm: Vattenfallstyrelsen.

Rathsman, B. and Glimstedt, U. (1949b). Elkraftöverföring till Gotland. *ERA*, 22, 96 - 99.

Rathsman, B. G. and Sviden, S. (1956). *Likströmsöverföringen från fastlandet till Gotland*, Blå-vita serien 15. Stockholm: Vattenfallstyrelsen.

Rissik, H. (1935). *Mercury-Arc Current Converters: An Introduction to the Theory of Vapour-Arc Discharge Devices and to the Study of Rectification Phenomena*. London.

Rosenberg, N. (1982). *Inside the Black Box: Technology and Economics*. Cambridge.

Rosenberg, N. (1992). Science and technology in the twentieth century. In G. Dosi, R. Giannetti and P. A. Toninelli, (eds.), *Technology and Enterprise in a Historical Perspective*. Oxford: Clarendon Press.

Rusck, Å. (1945). Svensk kraftförsörjning i dag och i morgon. In *Morgondagens teknik: Aktuella problem och framtidsperspektiv inom teknik och naturvetenskap*. Stockholm: Teknisk Tidskrifts förlag.

Samuels, R.J. (1994). *"Rich Nation, Strong Army": National Security and the Technological Transformation of Japan*. Ithaca: Cornell University Press.

Svenska Dagbladet (1934). Högspänd likström även från Asea: von Platens system får svensk konkurrent, *Svenska Dagbladet*, 17 February 1934, 3.

Svenska Vattenkraftsföreningen (1921). *Svenska Vattenkraftsföreningens tolfte ordinarie årsmöte den 29 april 1921*, Svenska Vattenkraftsföreningens publikationer 130. Stockholm.

Sylwan, E. (1944). Storkraftöverföring: svensk försöksanläggning för kraftöverföring med högspänd likström. *ERA*, 17, 55 - 57.

Teknisk Tidskrift (1954). Elkabel under Engelska kanalen. *Teknisk Tidskrift*, 84, 556.

TMA (1975). "Intervju med Uno Lamm", 1 July 1975, *Archives of Tekniska Museet, Stockholm (TMA)*, 3771, IVA's Teknikhistoriska råds intervjuer, Uno Lamm.

TMA (1977). [N.-G. Håkansson], "Försökslikriktaren : jonventilepokens inledningsskede", LFA 151A. Ludvika: ASEA, *TMA*, 3771, IVA's Teknikhistoriska råds intervjuer, Uno Lamm.

Vi Aseater (1960). Asea håller tiden i kanalprojektet. *Vi Aseater*, 23 (8), 20.

Vi Aseater (1963). Konti-Skan: ett verkligt kraftprojekt. *Vi Aseater*, 28 (1963), No. 4, 4 - 6.

Vi i Vattenfall (1963). "Europaväg för elkraften". *Vi i Vattenfall*, 16 (3), 3.

Wollard, K. (1988). Uno Lamm: inventor and activist. *IEEE Spectrum*, 25 (March), 42 - 45.

VA (1940a). Letter Borgquist to Sylwan, 27 July 1940, *Vattenfall Archives, Råcksta (VA)*, KB, F1BA:8, ASEA:II.

VA (1940b). A. Olsson, "Några litteraturdata rörande kraftöverföring med högspänd likström", 18 October 1940, 1, 15, *VA*, EBB, Allmänt, F1AA:3204, Storkraftöverföring:I - III.

VA (1941). Letter Borgquist to Lamm, 1 April 1941, *VA*, KB, F1BA:8, ASEA:II.

VA (1942). Letter Borgquist to ÖD et al., 11 May 1942, *VA*, EBB, Allmänt, F1AA:3204, Storkraftöverföring:I - III.

VA (1943). Åke T. Vrethem, "PM beträffande försöksanläggning for högspänd likström", 8 October 1943, 1, *VA*, EBB, Allmänt, F1AA:3204, Storkraftöverföring:I - III.

VVA (1942). Letter Rusck to Elektrobyggnadskontoret, 11 June 1942, *Vattenfall Väst Archives, Trollhättan (VVA)*, HK, F1ABC:68, Försöksanl. för högspänd likström 1942 - 1947.

VVA (1943). "Avskrift: Avtal", 20 December 1943 & 28 December 1943, *VVA*, DA, F1CC:1, Högspänd likström Juli - Dec 1944.

VVA (1959a). "Sammanställning över gjorda investeringar och driftskostnader jämte pålägg och beräknad ränta", n.d. but probably 1959, *VVA*, HK, F1ABC:68, Kontrakt.

VVA (1959b). "PM. angående kostnader i samband med försöksanläggningen för högspänd likström", 31 December 1959, *VVA*, HK, F1ABC:68, Kontrakt.

4. A CASE STUDY OF THE SWEDISH PUBLIC TECHNOLOGY PROCUREMENT PROJECT "THE COMPUTER IN THE SCHOOL" (COMPIS), 1981-1988

T. Kaiserfeld

1. INTRODUCTION

The object of this study is to describe and analyse the Swedish public technology procurement project Compis (Computer in School) aiming to develop a Swedish school computer, from 1981, when the Swedish government decided to sponsor the project, until production of the computer was discontinued in 1988.[1] The Compis project was a public technology procurement project proper: a government agency, The Swedish Board for Technical Development (*Styrelsen för teknisk utveckling*), formulated a functional specification of requirements for a product, personal computers for use in schools, which did not exist at the time (1981). After this the Board formally invited bidders, placed an order and signed a contract with the bidder judged most favourably. Since additional or new technological development work was required to fulfil the demands of the buyer, the Compis project was an "ideal type" of public technology procurement project. Moreover, since new products were supposed to be created, it was a creation-oriented project and therefore a developmental public technology procurement.

The Compis was a technical product whose creation had two purposes: to provide the Swedish compulsory school and upper secondary school with modern and inexpensive computers, partly in order to introduce computer science, and to give Swedish industry the chance to develop new technology. A school computer was called for by the Ministry of Education and its agency the Swedish Board of Education (*Skolöverstyrelsen*), whose enquiries had established that there was no computer on the Swedish market meeting the necessary performance standards to ensure the success of the plans to introduce the new subject. At the same time as the Board of Education made its enquiries, the Swedish computer industry, and particularly the two largest manufacturers, Datasaab and Luxor, began to run into financial difficulties (Sjöström, 1996,

[1] The project has also been treated in Kaiserfeld (1996).

vol. 1: 104 - 106).[2] It seemed natural to co-ordinate the decision of the Ministry of Education on computer science in the schools with the Ministry of Industry, and through it with the manufacturers (Borgström, 1989).[3] This gave rise to the idea of a technology procurement project designed to produce a Swedish-made school computer as well as to help the Swedish computer industry through a period of difficulties. In this way the two problems to be solved by the Compis project were formulated on two different hands. The educational problem was formulated by the agency that represented the future buyers, the schools. The industrial problem was formulated by the contemplated sellers, the Swedish computer manufacturers.[4]

In 1981, the Swedish government instructed the Board for Technical Development to handle the school computer procurement project, "The Computer in the School", stipulating in its formal decision that the Board of Education should pay the Board for Technical Development the sum of four million Swedish kronor (approximately $723,000) (Swedish Government Office Administrative Section, 1981 b). In October 1981, a reference group had been formed for the Compis-project. This group consisted of people drawn from various sectors of Swedish society who were considered to have the expert knowledge needed for successful management of such a project. It included representatives of the Swedish Association of Local Authorities (*Svenska kommunförbundet*), who were the prospective purchasers, as educational aids for Swedish schools were purchased at the municipal level, and representatives of the Swedish Agency for Administrative Development (*Statskontoret*). Furthermore, an industrial consultant was included, together with two representatives of the Board for Technical Development. These three formed the project management group (NUTEK, 1981 a).

The first task of the reference group was to select a number of municipalities to take part in the project. Of Sweden's 280 municipalities, no less than 110 expressed an interest and from these the reference group chose four to send representatives to form the actual project group which was to begin its work by writing a specification of requirements (NUTEK, undated). The project group also included representatives of the Board for Technical Development and the Board of Education. The first meeting of the Compis-project group was held in December, 1981 (NUTEK, 1981 b). Those present were the teacher representatives, one for each of the four municipalities, and the project management group, and it was these people who would henceforth manage the project.[5] The reference group, whose task was to meet the schools' need for computer equipment, had largely discharged their duties by selecting suitable local authorities to take

[2] Luxor had severe financial problems in the beginning of 1979 and the loss was about 15 percent of the turnover in 1980 and 1981 (see Sjöström, 1996, vol. 2: 25 & 100).

[3] The industrial policy motive of the Swedish school computer to be is also plain from the NUTEK Letter from the Ministry of Budget to the Board for Technical Development (1982).

[4] The earliest evidence of a technology procurement project designed to produce a Swedish-made school computer is in Swedish Government Office Administrative Section (1981 a).

[5] The teacher representative of Stockholm, Ingvar Gratte, gave his views on computer education policy (in Gratte and Månsson, 1982).

part in the project. They were now replaced by teachers as local authority representatives of those who would be working with the planned school computer in the field.

In early 1982, the project group started to draft a set of specifications and the work seems to have made good progress throughout the spring – not that there was any reason for conflict between the teacher representatives and the management group. They were working towards a common goal – an effective Swedish school computer – and any conflicts of interest seem to have been easily resolved.[6] In addition to the meetings of the project group once or twice a month during the spring of 1982, the management group also arranged an industrial seminar in Stockholm. This was attended by a number of representatives of different computer companies, especially the major Swedish ones, as well as the members of the project group (NUTEK, 1982 a). Among these, a group consisting of Luxor, Digital and Dataindustrier AB was prominent, as were the educational aids companies Esselte Studium and Liber (NUTEK, 1982 b).

The final specification of requirements, which was presented in May, 1982, pointed out that the schools had hitherto been obliged to rely on commercially developed computers which had not met the schools' requirements for accessibility, durability, speed (instruction took place in forty-minute periods) and cost (TUDIS Kravspecifikation, 1982). All in all, the specification of requirements may be said to be a relatively loosely worded document, largely following earlier guidelines laid down by the Board of Education. The difference was that the Board's reports envisaged local purchase of computers by individual schools, whereas this specification had been drawn up at a national level to apply to the purchase of computer equipment for at least forty municipalities.

The specified requirements, considered individually, were audacious but not impossible. Even if no single computer could meet them in 1982, they contained no more elements of science fiction than other expressions of hope in the computer market at the time. In this sense, the requirements seem to have been well timed. It was the combination of the generally high technical and functional requirements and the request for "low price" that was the real test. In many ways the specification also left doors open, but one exception was the requirement for an interpretative program language which was to "allow structured programming." (TUDIS Kravspecifikation, 1982: 15) In practice, this requirement limited the choice of program language to Comal which would later prove problematic. However the most optimistic feature of the specification was the time-table requesting delivery of a prototype by June 1, 1983. Educational aids and the installation of the equipment in schools were desired by August 1, 1983, and the training of teachers was to be completed by August 30, 1983 (NUTEK, 1982 c: 4).

2. INTERACTION ENVIRONMENT

The specification of requirements was indeed a challenge to the Swedish computer industry, which had emerged out of military interests after the Second World War when the predecessors of the modern computer had been developed primarily for military

---

[6] From the minutes of the meetings of the project group during spring 1982 (NUTEK, 1981).

purposes in the United States. The Swedish military had then been quick to show an interest in building a "mathematics machine" (Annerstedt et al., 1970: 99 - 137).[7] Early efforts by the Swedish government and various interest groups to introduce the new technology in Sweden provide a good illustration of Swedish technology policy after the Second World War. The government set up commissions in the late 1940s to study the military and civil requirement for mathematics machines. Their findings resulted in a project to develop a Swedish-made computer in the 1950s, and the knowledge thus gained by Swedish engineers and researchers became in part, thanks to agreements between government authorities and industrial companies, also available to private industry. In short, the Swedish computer industry had its background in state initiatives in the early 1950s. These initiatives were later transformed into private enterprises relying on government as the most important customer.

By the beginning of the 1980s, the Swedish electronics industry in general was dominated by communication applications, representing half of the production of the electronics sector, while the computer industry stood for about a quarter (Jørgensen, 1988). The Swedish computer industry had a protected state market as well as a profitable customer in the military. This had resulted in a rather well developed national self-sufficiency regarding computers. On the other hand, the profitable military market had weakened the internationalisation of the Swedish computer industry and its ability to adapt to the civil market (Jørgensen, 1988: 73). This was the situation when the Compis-project was launched in 1981. By then contacts between representatives of the Swedish computer industry and the Board for Technical Development had long since been well established. The year before, for example, the Board, in its largest single investment ever, had supplied a total of 78 million Swedish kronor (approximately $14 million) over three years for the development of Swedish electronics manufacturing (Anonymous, 1980: 12).

When the Board for Technical Development had been formed in 1968, it was created by the government to assist industrial development in general (Weinberger, 1997). One way of doing this was the method of public technology procurement. (See Figure 1.) This became a standard procedure in which the Board for Technical Development compiled a specification of requirements in collaboration with the ordering authority.[8] The specification of requirements was then distributed to potential manufacturers and suppliers. Ideally, the one submitting the best tender then won the order. To support industrial development in this way, the new body was also granted increased financial resources.

This arrangement was supposed to yield benefits for all concerned. The advantage to the client was that it was able to buy the most state-of-the-art products the market could offer. The supplier gained access to a guaranteed outlet at a sensitive stage of product development. There was also a national advantage. For a country with a small population like Sweden's (approximately eight million people in the 1970s, unevenly spread over an area somewhat larger than California), it was a heavy financial burden

---

[7] A somewhat different perspective is supplied in De Geer (1992) and Johansson (1997).

[8] Concerning technology procurement, see Teknikupphandling (1976); Linder (1980); Edquist (1996).

to manufacture and maintain its own advanced technology in the form of, for example, fighter aircraft (Annerstedt, 1976). For such projects to be realised, it was important for dealings between manufacturer and customer to take place in a rather rational manner. It was in the field of the large Swedish fighter projects, intended to produce internationally competitive Swedish fighters for the post-war market, that the technology procurement procedure was best manifested, although in other sectors, too, where sophisticated technical equipment was essential (e.g., energy supply) such projects were very significant.

Figure 1. Diagram of a technical procurement project. Note that all the steps in the process are marked within the area of the arrow. External factors which might affect the project were not mentioned (En svensk skoldator, 1985: 25).

In the early 1970s, a government committee had reported on technology procurement and recommended the method also in cases of high-tech consumer goods such as computers. The secretary of that committee was Ulf Eklund who more than a decade later became manager of the Compis-project. With a bachelor's degree in political economy and with fifteen years of experience from administrative work, the last six years in the service of the Board for Technical Development, the forty-year old Eklund's role was important. The work of the committee had resulted in a report that proposed highly centralised technology procurement with long production runs and a single big customer (Teknikupphandling, 1976). Being in charge of the School Computer Project, Ulf Eklund decided how the work was to be organised and must therefore have had an excellent opportunity to put his views on technology procurement into practice.

Thus even though the School Computer Project reference group and also, later, the project group itself based their opinions expressly on the reports published by the Board of Education in the late 1970s, a more centralised approach to procurement than the one advocated there gradually came to characterise their activities. In this case it meant the production of a uniform school computer for use throughout Sweden. It is not improbable that the personal link which Ulf Eklund formed between the government committee which had examined technology procurement and the School Computer Project may have had a bearing on the situation.

3. COMPETENCE AND USER-PRODUCER INTERACTION

The potential users, i.e., teachers and pupils, were co-ordinated through four teacher representatives, one for each chosen municipality. These teachers, although still active in the classroom, had experience with fact-finding and organisational work for the government and local authorities and also with teaching with the aid of computers. Therefore, user competence was hardly any problem in the procurement process. In addition, the Board for Technical Development got help from the Agency for Administrative Development which had expertise in buying all different types of material for use within the governmental administration. The involvement of the Agency for Administrative Development and the experience of the Board for Technical Development meant that buyer competence hardly was any problem.

So far, only the user and buyer sides of the project have been outlined. On the producer side, the activities were not as co-ordinated, at least not at the beginning of the project. However, when the work began with the specification of requirements, this changed and one sign of this was the industrial seminar in Stockholm that has already been mentioned. Again it was the Board for Technical Development that oversaw the co-ordination of the actors. The seminar was attended by representatives of different computer companies, but more importantly the educational aids companies Esselte Studium and Liber.

By August 31, 1982, the Board for Technical Development had received sixteen tenders. After a screening of them there remained three to choose from, the rest having been rejected on the grounds of price or inadequate capability of upgrading a single-user station to a multi-user system (NUTEK, 1982 d). What is not apparent here is that the main rivals in the struggle to become the suppliers were not the computer companies but the educational aids companies Liber and Esselte Studium. Liber had submitted a tender based on the school computer ABC 800, manufactured by Luxor, who already had a large part of the market.[9] Liber's material thus enjoyed the advantage of being well established in many of the schools. Esselte Studium, for its part, had submitted three different tenders in collaboration with different hardware manufacturers. Two of these had reached the final round.

[9] A consortium of companies consisting of Luxor, Liber and Dataindustrier AB was at this time making computers, the ABC 80 and ABC 800, which accounted for 63 % the equipment which existed in 150 schools surveyed in the spring of 1981 (NUTEK, 1982 e: 2). The ABC 80 had been sold since 1978 and ABC 800 since April 1981, their development having been described in Johansson (1997: 137 - 172). By the end of 1980, altogether 17,000 personal computers had been sold in Sweden and of these 45 % were ABC 80 and 20 % Commodore PET (see Johansson, 1997: 145).

On August 29, 1982, two days before the tender period had expired, Bo Jansson, managing director of the small computer company Management Computer International (MCI), had happened to receive a copy of the specification of requirements (Jansson, undated).[10] The company was at that time in the process of developing a 16-bit computer for the Dutch electrical engineering company Philips; but Jansson, who had acquired experience from the computer market a few years earlier when he had managed the Computer Sales Department of Texas Instruments in Sweden, quickly realised that this prototype not only represented the general trend on the personal computer market but also met the specification. After obtaining Philips' permission, Bo Jansson and the owners of the small computer company AB Olesen & Lindgren Innovatörer drew up a tender the day before the period expired. The tender was submitted with Olesen & Lindgren as designers and MCI as program developer (NUTEK, 1982 f).

A 16-bit computer could simultaneously process quantities of information made up of 16 binary characters (i.e. ones and zeros) and should not be confused with an 8-bit computer with a 16-bit processor. In the first case the whole architecture of the computer was entirely based on the communication of 16 bits at a time, while the other only handled 16 bits in the central processing unit, the rest of the computer dealing with 8 bits at a time as did older computers. In the early 1980s, most computers used a "word length" of 8 bits, although some were beginning to operate with 16-bit processors, the best known being IBM's PC, which entered the market in August 1981. Increasing the word length enabled the computer to handle more information at the same time. One of the disadvantages was that the hardware was more difficult to configure. For example, the number of conductors (or buses) between memory and processor had to be increased.

As early as 1979, Jansson had voiced ideas concerning a very logical development in the personal computer field (Jansson, 1980). In his view a 16-bit computer was the natural "next stage" of development for the mid-1980s. He anticipated development in a series of steps, each step involving a doubling of the number of bits in the computer architecture. (See Figure 2.) His thinking was very consistent with an ideology of the rationalisation of technical development, and Jansson decided to submit a prospectus for the 16-bit computer on which he was working as a tender to the Board for Technical Development.[11]

As mentioned, the basic idea of central technology procurement in general and of the School Computer Project in particular was not only to purchase material but also to develop new technology. This desire to spur new technology was probably what led the procurement group to decide to give the tenderers the opportunity to submit supplementary tenders. Bo Jansson's proposal had turned out to be the only one built around a 16-bit processor. The others offered only 8-bit computers. The School Computer Project group therefore wished to give the others the chance to submit tenders for a 16-bit computer. However no such tenders at a reasonable price were received, and the

---

[10] The article, which was turned down by the editors of *Ny Teknik*, was written in response to another article in the paper (Borgström, 1989).

[11] All the original tenders are in NUTEK (1982 g).

project group now had to choose between an established computer with established educational aids (Liber's and Luxor's ABC 800, as Liber's supplementary offers contained only increased graphic capabilities and a program catalogue) and a computer which was only on the drawing board (Esselte Studium in collaboration with another designer) (NUTEK, 1982 h). The project group decided that the most far-sighted course would be to opt for a 16-bit computer, and they contacted Bo Jansson at MCI, who was willing to meet their demands.

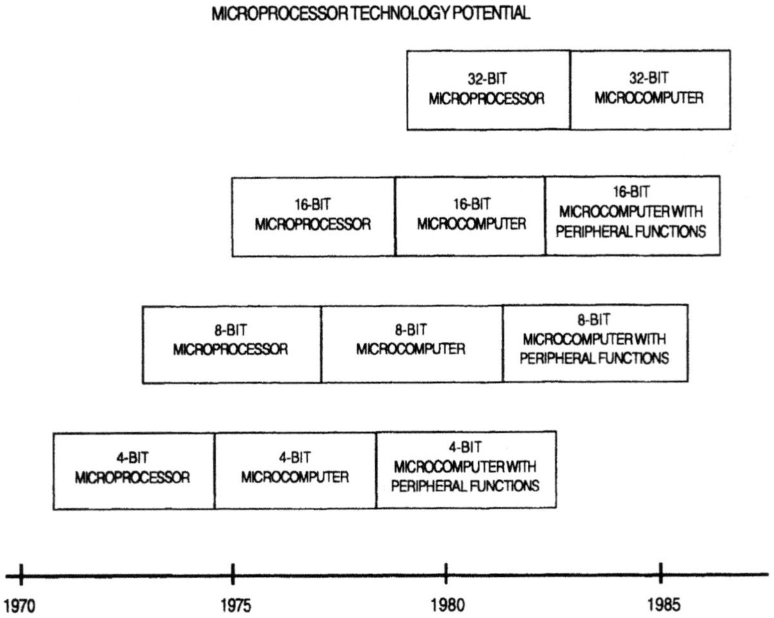

Figure 2. Bo Jansson's diagram of the anticipated logical development of the personal computer field (Jansson, 1980: 22).

The problem was, however, that his company did not have the capacity to produce or market educational aids for the Swede 16, as the computer was called in the tender. The project group recommended instead that Jansson contact established educational aids companies to explore the possibility of a joint venture. Esselte Studium was interested, and the discussions quickly led to a joint venture agreement between Esselte on the one hand and MCI and Olesen & Lindgren on the other. In December, 1982, an agreement was signed between the Board for Technical Development, representing the four municipalities, and the main supplier Esselte Studium (En svensk skoldator, 1985: 7). The agreement provided for the supply of computer equipment for seventy-six pupil work stations, together with software and educational aids, for testing in a total of fifteen upper secondary schools from October, 1983. By promoting a type of computer that was not yet at hand, this case of public technology procurement hardly resulted in a reduction of technological uncertainty for the manufacturers besides the testing of the equipment in different schools.

Earlier the same year the agreement was signed, in 1982, Bo Jansson and his collaborators had started to scour the market for suitable processors. They soon thought they had found a candidate in Intel's 80150. The tight schedule made this processor particularly attractive to Jansson as it contained the built-in operating system CP/M. This choice had to be revised when production of the processor was delayed and it turned out to function unsatisfactorily (Jansson, 1991). However, another of Intel's processors then caught their eye. From the smoothly running and widely used 8-bit processor 8086, Intel had developed two different types of 16-bit processors. One, the 80186, was integrated with various other functions to reduce costs. The other, the 80286, was more expensive and had to be wired with other components to make a complete central processing unit (Zingale, 1983; Wells, 1984). Neither of these had earlier been used in functioning computers. (The first computer found to carry the 80186 processor was the Pronto Series 16 presented in June, 1983; the 80286 was later to be used in the second generation of IBM's PC (Hansen, 1983).)[12] The Swedish development group therefore seems to have been groping in the dark.

Moreover, the choice of operating system was an open question in 1982 and 1983. There was a large number of different computers on the market, with various operating systems, and speculation concerning which would become standard was continuous.[13] The fact that IBM had opted for MS-DOS did not discourage other designers until later in 1984, when the variety of software applications for this operating system by comparison with others became clear. In the end, the severe cost constraints of the project probably made Jansson and his colleagues choose the 80186, which seemed cheap and simple in comparison to the 80286. At the same time, they also chose to retain the operating system CP/M that had come with the 80150 – two decisions that would later prove fateful.

The hardware problems of the 80186 processor in combination with the CP/M-86 operating system were not unique to the Swedish Compis project. Among American manufacturers the same solution was used by Onyx Systems, Monroe Systems for Business, and others. Perhaps the best known personal computer based on the 80186 processor was Radio Shack's Tandy Model 2000, which was sold at $4,500 (including hard-drive) at more than a thousand Radio Shack stores all over the United States in 1984 (Jennings, 1984). At the time of its introduction on the market, the Model 2000 was characterised as superior to IBM PC in speed, display options and disk-storage capacity, its drawbacks being the limited availability of software. Despite the hardware advantages, the Radio Shack's Tandy Model 2000 was described as built on the "powerful but infamous 80186 microprocessor chip" and what proved to be its limited life on the market compared to the IBM PC (which was the common computer for comparison in the mid-1980s) seems to have been attributable chiefly to the lack of software and the flaws of its processor (Malloy, 1984: 307).

---

12 Two more early computers using the 80186 are listed in Williams (1983).

13 The main forum for discussing which operating system should be chosen was the specialist press in the late 1970s and early 1980s.

Ever since MCI and Olesen & Lindgren had taken on the task of developing a 16-bit computer, it had been evident that the actual manufacture of the computer must be carried out by some appropriate industrial company. In May 1983, the management of Teli decided that the company, with about 4,000 employees and a turnover of 1 billion Swedish kronor, was now "strategically interested in alternative manufacture in view of the gradual abolition of the National Telecommunications Administration monopoly [*Televerket*]." (Jansson, undated).[14] Their reason for turning to the manufacture of computers indicated that the deregulation and privatisation of the Swedish telephone monopoly was, although only indirectly, a significant component in the Compis project. When Jansson's company ran into financial problems due to soaring development costs, an agreement was reached with Teli whereby other obligations and rights were assigned, and in this way Teli became a subcontractor to Esselte Studium (NUTEK, 1983; Anonymous, 1983 b).

While work proceeded on the hardware, the teachers of the project group in co-operation with the software supplier Esselte Studium concentrated on the production of educational aids. Gradually they had been encountering more and more problems with the fixed programming language Comal. As the Comal core is not big enough to meet all the demands made on a programming language, different dialects of Comal had to be developed with no possibility of checking the changes against a uniform standard (NUTEK, 1984 a). The difficulty of checking changes in software on the computer market in the early 1980s says something significant about this part of the development work. New software was often written by unorganised "amateurs" who sat at home and sent their programs to friends and acquaintances. If anything needed amending, this was done without reference to relevant standards.

The lack of a standard for Comal meant that Esselte Studium preferred to use other compiling languages than Comal for producing educational aids, a practice which simplified and expedited the work but was contrary to the stipulation in the specification that the programs for the school computer should be listable in the language that was being taught. Indeed, the notion of unorganised amateurs (i.e., the intended users themselves, being unpaid adolescents) writing the software to be used in the computer, was attractive not only to Esselte Studium, which established "Compis Clubs" with the aim of organising the unorganised for economic reasons, but also to the teachers involved, who embraced the idea of schoolwork being democratised.[15] (See Figure 3.) Surely they saw nothing wrong in having pupils developing teaching aids in the form of software.

Throughout the development work, the group at Teli kept in close touch with Intel in the USA, who supplied the 80186 processor for Compis. Nevertheless there were delays, mainly because the software for the CP/M-86 operating system in Compis was

[14] Although there never was a formal governmental monopoly of telecommunications in Sweden, all telephony had been run by the National Telecommunications Administration since 1918. By the end of the 1970s, the situation had started to change with the introduction of radio and satellite communications which made the Telecommunications Administration seek co-operation in new development projects. (See Karlsson, 1998: 92 - 124; Brattberg, 1983; Anonymous, 1983 a).

[15] The same strategy had been tried for ABC 80 in 1980 (see Johansson, 1997).

not delivered from the United States in time for the development of educational aids (Anonymous, 1983 b; NUTEK, 1984 b: 4). Through the summer and early autumn of 1983, the development group at Teli worked frantically on hardware problems, during which time they were under great pressure from the teachers of the project group as well as the mass media.[16] The first Compis computers were delivered on time to the test schools during the second half of October, but they did not work satisfactorily and proper trials could not be started until the spring term of 1984.

Figure 3. The name of the computer, Compis (which is short for "Computer in School," and also means "pal" in Swedish), may perhaps reveal the manufacturer's dreams concerning the relations between pupils and the school computer. The word "kompis" (for companion or comrade) is used mainly by younger school pupils and therefore also has a connotation of identification with the school. Swedish school policy has been marked by a striving for equality and perhaps the symbol for the computer, the rotund and cheerful little man wearing glasses and bearing a garland, also stands for the ideal computer science teacher in the Swedish comprehensive school.

Although testing in the schools had started only 4 - 5 months after the original deadline of October 15, 1983, it had been beset with serious problems; for example, the introduction of the multi-user system had taken considerably longer than anyone in the project group had foreseen.[17] While testing was taking place, moreover, operating systems other than CP/M-86 had begun to appear as standard-setting, particularly MS-DOS. For these systems there was a wide range of software (NUTEK, 1984 a). Soon after-

[16] In his article, for example, Bo Jansson states that "Liber-Luxor [...] chose [...] to launch a campaign of disinformation in true Dallas style designed by the advertising consultants Bergström & Nissen [Bergström & Nissen-Lie Information AB]." (Jansson, undated). It has not been possible to find evidence for the assertions although it is clear that Luxor was anxious to have good public relations (see Sjöström, 1996: 69).

[17] This is clear from minutes of the so-called control stations 8 - 11 during 1984 in NUTEK (1984 c).

wards, in March 1984, serial production of Compis began, with delivery taking place in May (Nilsson, 1984). During the summer, the Compis was even displayed in Intel's showcase at the National Computer Conference in Las Vegas as an example of the most promising 16-bit computer (Malmius, 1984). By the autumn, Esselte Studium had delivered some 1,400 computers, approximately 350 of them for the Swedish market (NUTEK, 1984 d). Testing of the Compis computer continued at the different schools throughout 1984. By September that year, there was at last a functioning system of interconnected computers for use in a class or a school (NUTEK, 1984 d: 15).

To sum up, the supplier competence proved to be a bigger problem than the buyer competence. The constructor MCI and Olesen & Lindgren got help from the American component manufacturer Intel and soon enough Teli stepped in with necessary financial resources. Although it is hard to judge the competence acquired at Teli, it seems probable that they got the help they needed from Intel and other companies. The problems of the software supplier Esselte Studium were more severe. This was partly a result of a lack of buyer competence since the requirements on the programming language had effectively ruled out any alternatives to Comal. But Esselte also lacked people with the necessary programming experiences and they more and more desperately sought help from future users, the teachers and even their pupils. Moreover, there was no sign of any technical research institute or university taking part in the development of competence or capability on either the user or producer side.

4. POLITICAL INFLUENCE AND LEVELS OF INTERVENTION

In the case of the Compis-project, the public technology procurement process was thoroughly organised along formal lines and it did not emerge "organically" in an evolutionary manner. The boundaries between the different groups of users and producers taking part in the process are most simply drawn on the basis of the organisational base of each representative. This reveals two main groups. On the one side were three people from the Association of Local Authorities, the Agency for Administrative Development and the Board of Education, whose task was to represent the customer. On the other side were the industrial consultant and the two representatives of the Board for Technical Development, whose concern was the industrial policy aspect of the project. It is clear that the first-mentioned group represented organisations which had a formal interest in the implementation of an education policy, whereas the second group stood primarily for the concept of a rational industrial policy.

The testing of equipment in schools shows the continued active participation of the teacher representatives. In addition, the boldness – some would say recklessness – of the designer confirmed that the Board for Technical Development had found what it had been looking for. Although the specification of requirements, being largely the work of the teacher representatives in the project group, was addressed primarily to one of the two large educational aids companies in collaboration with a computer manufacturer, the members of the Board for Technical Development attached such importance to ensuring that the funds actually did go to technical development and not merely to general industrial support, that they chose a small manufacturer with great visions. As mentioned, the agreement was signed between the Board for Technical Development, representing the four municipalities, and the main supplier, Esselte Studium, in December, 1982. The formal contract, providing the supply of computer equipment for seventy-six pupil work stations together with software and educational aids for testing in a total of fifteen upper secondary schools from October, 1983, can hardly be called innovative as such. However, framing the agreement had taken a good deal of time and ingenuity.

Early in 1983, after suppliers had been chosen and details of their bid had been revealed, the project was swamped with criticism in the trade press, both of the specification of requirements and of the selected tender. The criticism came from politicians of the right-wing opposition parties, from economists, and from representatives of larger but unsuccessful computer companies. Criticism concerned everything from the choice of programming language, Comal, to the choice of developer, a small company with little standing in the school computer sector.[18] Many thought that it would prove impossible to supply the computer at the promised price. The criticism also extended to the hurried planning. But there weren't any allegations of either corruption and nepotism, or protectionism.

Although criticised, the Swedish designers still held high hopes for Compis well into 1984. In only a couple of years, between 1982 and 1984, the project's manufacturers had succeeded in developing and had begun to produce a computer which was considered so promising that it was allowed to represent one of the world's largest electronics manufacturers at a computer fair in the United States. Bo Jansson's ideas on a foreseeable development of the personal computer field had also proven correct when the giant among manufacturers, IBM, decided to develop a 16-bit computer.

Despite the successes, however, development did not proceed so straightforwardly and rationally as had been assumed when the project was planned. Teli had taken on the role of supplier, not because of any acknowledged competence as a developer of personal computers, but rather for reasons of business strategy. The processor which was supplied from the USA turned out to suffer from various problems and never reached serial production. It was instead the next generation of 16-bit processors from Intel, the 80286, that became IBM's choice for its 16-bit personal computer. Moreover, the cost of those components of the Compis that came from the USA had risen steeply as the exchange rate for the dollar had soared (NUTEK, 1985 a).[19] The production price of the Compis at this stage was well above the option price to which Teli was tied by the agreement with the Board for Technical Development, and the option price did not cease to apply until the summer of 1987. Moreover, once the agreement had been signed between Esselte and the Board for Technical Development in December 1982, the Board by and large kept itself out of the actual development work.

---

[18] A number of articles were published, particularly in *Dagens industri* and in *Datavärlden* during 1983. Articles also appeared in *Skolvärlden, Kommunaktuellt, Datornytt* and other newspapers.

[19] At the time of the government decision of September 1981, the dollar was at 5.50 Swedish kronor. When the agreement was drawn up in December 1982 it had risen to 7.35 kronor and in March 1985 it reached a peak notation of 9.40. The increase during 1984 was 12.6 % and by March 1985 a further 5.0 %.

In 1984, the different phases of the procurement process as foreseen by the Board for Technical Development were still succeeding each other according to the diagram in Figure 1. After the initial negotiations between buyers and suppliers (study phase), a specification of requirements had been produced which had resulted in different bids (project phase). After a bidder had been chosen as a contractor and supplier, the project went into a stage of design and construction (design phase). The resulting computer was later tested in schools (production phase) (Edquist, 1996: 158 - 159). However, during the testing of the Compis, unforeseen institutional changes occurred that made the procurement project enter stages that had not been anticipated originally. Still, they proved to be of great importance to the outcome of the Compis project.

While the Compis project was in progress, the Ministry of Education drew up a plan for financing the purchase of computers for the schools. In April 1984, the Social Democratic government decided to allow the local authorities twenty million Swedish kronor a year for three years to buy computer equipment for the schools (Förordning om statsbidrag, 1984: 1).[20] The decision became known as the three-year campaign and meant that those municipalities that wished to do so could have half of the money they spent on computers for the compulsory school or the upper secondary school furnished by a grant from the National Board of Education, which also decided which computers qualified for grants.

To be able to provide this support, the Board of Education turned to other government and local authority departments which had previously been involved in the purchase of computers. The most important of these was again the Agency for Administrative Development, which had made a public purchase of computers for the government at the end of 1982. In the autumn of 1981 both the Board of Education and the Agency for Administrative Development had been involved in the appointment of the Compis project group. But once the project group had been appointed, the project had run without supervision from these authorities and they could hardly have felt any responsibility to the Board for Technical Development. Of the individuals from the Agency for Administrative Development, the Association of Local Authorities and the Board of Education who had originally been involved in the Compis project, only one remained when the guidelines for purchase of computer equipment for the schools were drawn up (*En svensk skoldator*, 1985; Förteckning över bidragsberättigad datorutrustning och programvara för grundskolan, 1985: 1).

Admittedly, the teachers who had been appointed were still members of the group, but their successors in the authorities who were now responsible for purchasing school computers no longer had any obligation to follow any particular line of technological change. The representatives of different agencies no longer tried to meet the requirements which had been made of Compis, with regard either to technical innovations or to the performance of Swedish school computers. With their re-entry into the procurement process, the educational policy base of Compis stressing uniform technical equipment throughout Swedish schools had been replaced by another stressing a more function-oriented public purchase. Their main concern now was not that the schools should

[20] The three-year campaign is described in Nissen and Riis (1985: 3 - 9 & 29 - 37).

buy Compis but an efficient computer at a reasonable price of any make (Förteckning över bidragsberättigad datorutrustning och programvara för gymnasieskolan, 1985: 1).

The 1982 specification to which the Compis group was working was expressed in general terms and from the point of view of the needs of the users (i.e. teachers and pupils). Bengt Bengtsson, the representative of the Municipality of Gothenburg in the project group, was one of those who foresaw the problems that Compis might encounter when the prices of competing computers were falling. The whole point of the project as far as Teli was concerned was that the computer should be guaranteed the ***whole*** school computer market. If outside competition were allowed, the whole project would be in jeopardy. In the autumn of 1984 Bengtsson wrote concerning the Board of Education assessment of school computer equipment:

> Unfortunately my fears were confirmed. The assessment was very equipment-oriented. The demand for programs and material adapted to the compulsory school was toned down. It is likely that a large number of makes will be approved, which will be a disaster (NUTEK, 1984 e).

The first list released from the Board of Education granted approval to nine computers including Compis, which was judged favourably since it had been developed particularly for use in schools (Förteckning över bidragsberättigad datorutrustning och programvara för grundskolan, 1984). The software, however, was not in all respects fully developed and the text on the screen was not stable. In addition, the network had proven slow in certain applications, and the start-up complicated.

The two policies on which Compis was based went together: one concerned industrial motives, the other educational ones. When the National Board of Education did not act on one of these ideas – that the subject of computer science in the compulsory school should for reasons of fairness and uniformity make use of the same computer all over Sweden – Compis suddenly found itself out on a limb. One of the two pillars of the project had been removed, and the whole project was wobbling.

5. THE DIMENSION OF TIME

The school trials of Compis were completed just as the three-year campaign started. (See Figure 4.) With the accomplishment of the trials, Esselte had discharged its formal obligations to the Board for Technical Development. In the evaluations which were submitted to that authority by the trial schools in the spring of 1985, the verdict on Compis was rather mixed.[21]

Although the formal procurement project was complete and final settlement had taken place, Teli would not be free from the option price for another two years. A multi-

[21] The schools' assessments were at first classified, but they were nevertheless quoted in a progress report from the Board for Technical Development, (see *En svensk skoldator*, 1985). The opinions quoted in this publication were selected so as to present the schools' opinions as largely positive. The assessments were made public later and reactions included headlines like "Classified school trial report says: Thumbs down for Compis!" (Junberger, 1985). Here the assessments were presented as wholly negative. The original copies of these assessments are in NUTEK (1985 b).

user system for Compis (eight work stations, network, disk memory, printer and software) cost 130,000 Swedish kronor (approximately $14,650) at the option price, while an equivalent system from other suppliers approved by the Board of Education cost at least 200,000 Swedish kronor ($22,220) (Anonymous, 1985: 34). During the spring of 1985, in the face of stiff competition, and presumably largely because of the low price, Compis won 60% of the school computer market (Swedish Patent Office, 1984/85). Compis was still a deficit-ridden company, but it had been established to compete outside the school market and internationally. Thus, the course of the project was once again to be influenced by external factors.

Figure 4. Single-user equipment (with users) of Compis make *(Tele*, 1984: 1).

Not only was the operating system CP/M-86 becoming non-standard by comparison with the IBM PC by mid-1985, but meanwhile different standards were being developed in the graphics field for software and for communication between the processor and the screen.[22] The Compis solution, involving a separate graphics processor, appeared

[22] Regarding IBM's choice of MS-DOS, see Freiberger and Swaine (1984: 269 - 275). Admittedly an "emulator" was developed for Compis, making the computer IBM-compatible in accordance with the specifications. In actual fact, however, there was always difficulty in using programs intended for computers other than the one they were used on—in Compis' case very considerable difficulty.

cumbersome (Jansson, undated). During the spring of 1985, the threat of competition was still not taken very seriously by Teli. Compis dominated the school market and the flow of orders was good. Esselte Studium was inclined to regard CP/M-86 almost as protection for its software.

During the summer of 1985 all the 5,000 computers so far supplied were recalled to Teli for upgrading, a new improved version of the hardware-implemented operating system being fitted (NUTEK, 1985 c; Gustafsson, 1991). With this upgrading, the Board for Technical Development regarded the technology procurement project as concluded. The project group rounded off its work with a final report (*En svensk skoldator: Slutrapport*, 1985). There, the project group and the project management boasted:

> The School Computer Project has shown that technology procurement is an effective method of developing products and systems in collaboration with a competent customer. These products have a good chance of succeeding on the market. [...] Good effects are achieved by involving the customer's organisations (in this case, the schools) at an early stage in the project [...] The project has been completed in four years, which is the intended time for a project of this type (En svensk skoldator: Slutrapport, 1985: 11).

No comment was made here, however, on the fact that the project was about a year behind time. Trials in the schools did not take place in 1983/84 as planned but in 1984/85. As the computer industry was so dynamic in the mid-1980s, this single year made a lot of difference to Compis, even after the project properly had been completed.

In 1986 and 1987, the prices of competing computer equipment fell dramatically. When Teli was finally freed from the option price in the summer of 1987, it was scarcely lower than that of competing computers. Sales of Compis slumped and, after the three-year campaign had been completed, Teli's subsidiary company, TeleNova, which manufactured Compis, was purchased in November, 1988, by the computer company, Svenska Victor AB, and manufacture of the Compis was discontinued. The total losses of the manufacturer, the Swedish state through its Telecommunications Administration, were between 125 and 250 million Swedish kronor (approximately between $22 million and $44 million) not counting the four million Swedish kronor (approximately $723,000) initially paid by the Board for Technical Development (Widmark, 1991; Borgström, 1989). When the project was abandoned, some 25,000 computers had been manufactured and sold, 11,000 of them in Sweden.[23]

6. CONCLUSIONS

The choice of processor, operating system and programming language for the Compis computer was in many ways made with a view to intended buyers and users, who were supposed to demand cheap but advanced technology that at the same time would be simple to use. In this way, the Compis-project was meant to be an innovation policy instrument, although it hardly came up to expectations as an instrument to speed up

[23] In addition about 4,000 were sold in Norway, 2,000 in Finland, and others in Denmark and Italy (Svanborg, 1991).

technical development.[24] This was not primarily because of its one-year delay, but rather due to the fact that other companies meanwhile had managed to design and manufacture MS-DOS-compatible computers at lower cost. So the Compis-project can hardly be said to have resulted in more rapid technical change, at least not as seen over a period of more than a few years. Nor did it result in technology developing in a different direction, thus changing trajectory. However, Compis meant an increased technical diversity since the schools got computers of yet another make to choose from when they equipped classrooms for computer science.

The market failure of Compis raises questions about the project as an industrial policy instrument and although Compis never became a market success, its importance to the Swedish computer industry should not be underestimated. Many of those who worked on the development of the computer and the software acquired a great deal of know-how and experience from the project. Many presumably continued to work in the computer industry and their competence spread when the project was abandoned and the various people involved looked for new jobs or started their own companies. In this way, the computer created the impetus for new computer projects. Therefore, Compis may, at least partly, be seen as a successful industrial and governmental venture after all.

The policy relevance of the Compis-project is mainly negative in the sense that it is more useful for understanding when a technology procurement project is a bad policy instrument, than when it is a good one. In the case of Compis, one initial problem was that the anticipated bidders did not submit tenders that were far-sighted enough for the technological policy motive of the Board for Technical Development. Since the tenders of the more established manufacturers weren't ambitious enough to further technical development, the Board finally felt itself forced to intervene by choosing the tender of a smaller and less established company. This problem might have been eliminated, had the Board for Technical Development earlier clarified the technology policy motive of the procurement project.

Another implication regarding political influence and levels of intervention, is the importance of isolating the policy motives behind a specific technology procurement project. The policy motives should be as few and as distinctly formulated as possible in order to secure the continuity of longer-term initiatives in innovation policy and minimise the risks of unforeseen institutional changes due to altered political priorities. In particular, the institutional change brought about by the introduction of the three-year campaign turned out to be an obstacle for the success of Compis. The situation worsened from the perspective of the manufacturers of Compis when one of the buyer-organisations originally involved in the project, the Board of Education, accordingly acted against the idea of a monopsonistic school computer market and allowed several different computers to be sold to Swedish schools. One important conclusion may therefore be to try to keep down the number of public organisations involved in a technology procurement project, especially when institutional changes can be expected.

[24] Compare the discussion of Sherry Turkle's study of computer users (1984) in Mackay and Gillespie (1992: 705 - 708).

Still, it is important to consult the different organisations needed to establish a solid user and producer competence in the project. The need to involve many different organisations in the project in order to secure competence and the need to keep down the number of public organisations in order to avoid unforeseen institutional changes seem to counteract each other. However, the problem can be solved by facilitating user and producer interaction. Although it is hard to be more specific on how to choose the participating organisations and foresee institutional changes, it can at least be concluded that interaction environment is a crucial factor when setting up technology procurement projects.

However, the most important policy implication of the Compis project is related to the dimension of time. The Compis project is namely an example of a technology procurement project in a technical field exposed to rapid technical change. This proved to make it of an especially high-risk nature in which any type of delay, no matter how short, risked the outcome of the whole project. This was also pointed out by the 1976 government committee investigating technology procurement with Ulf Eklund as secretary (Teknikupphandling, 1976). The implication becomes especially important since technology procurement projects seem to attract most attention as a policy instrument in exactly these types of technical fields.

REFERENCES

Annerstedt, J. (1976). Några drag i den svenska statliga forskningspolitikens framväxt. In V. Andersen & R. Kalleberg (Eds.), *Forskningen – Staten og Kapitalet* (pp. 151 - 188). Copenhagen: Nordisk sommaruniversitets skriftserie nr. 9.

Annerstedt, J. et al. (1970). *Datorer och politik, Studier i en ny tekniks politiska effekter på det svenska samhället*, Lund: Zenitserien 10.

Anonymous. (1980). Kraftig satsning för att ta igen de förlorade åren. *Teknik i tiden, 2:2*, 12.

Anonymous. (1983 a). Intervju med Kai Lervik, VD för Teli. *Skolan och datorn, 3:3*, 21.

Anonymous. (1983 b). Skoldatorn COMPIS. *Nordisk Datanytt med data, 13:17*, 9 - 10.

Anonymous. (1985).*Datornytt, 9:1*, 34.

Borgström, H. (1989). Compisen som inte fick några vänner. *Ny Teknik, 23:45*, 22 - 24.

Brattberg, A. (1983). En presentation av Teli – Televerkets industridivision. *Skolan och datorn, 3:3*, 20.

De Geer, H. (1992). *Vägen till datorsamhället: Datatekniken i politiken 1946 - 1963*, Stockholm.

Edquist, C. (1996). Government Technology Procurement as an Instrument of Technology Policy. In M. Teubal, D. Foray, M. Justman & E. Zuscovitch (Eds.), *Technological Infrastructure Policy: An International Perspective* (pp. 141 - 170). Dordrecht.

En svensk skoldator: Presentation av teknikupphandlingsprojektet Datorn i skolan. (1985). Stockholm: National Swedish Board for Technical Development, STU-information no. 455 - 1985.

En svensk skoldator: Slutrapport. (1985). Stockholm: National Swedish Board for Technical Development, STU-information no. 524 - 1985.

Freiberger, P. and Swaine, M. (1984). *Fire in the Valley: The Making of the Personal Computer*, Berkeley.

Förordning om statsbidrag till datorutrustning och programvara för dataundervisning på grundskolans högstadium. (1984). National Board of Education statute 1984:101, § 3, May 24.

Förteckning över bidragsberättigad datorutrustning och programvara för grundskolan. (1984). Stockholm: National Board of Education, December 10.

Förteckning över bidragsberättigad datorutrustning och programvara för grundskolan. (1985). Komplettering I, Stockholm: National Board of Education.

Förteckning över bidragsberättigad datorutrustning och programvara för gymnasieskolan. (1985). Rapporter – planering, uppföljning, utvärdering 85:34, Stockholm: National Board of Education.

Gratte, I. and Månsson, E. (1982) *Datorer? – att leva med datorer,* Lund.

Gustafsson, H. (1991). National Telecommunications Administration Radio. Personal communication to the author, July 25.

Hansen, S. (1983). The Pronto Series 16. *Byte Magazine, 8:6,* 168 - 184.

Jansson, B. (1980). The Computer is a Tool!. In B. Sundin (Ed.), *Is the Computer a Tool? Reports from a Symposium in Stockholm/Sigtuna, Sweden, June 1979* (pp. 17 - 24). Stockholm.

Jansson, B. (1991). Personal communication to the author, September 10.

Jansson, B. (undated). Compisen som inte fick några vänner – eller?. Manuscript in possession of author.

Jennings, M. S. (1984). The Tandy Model 2000. *Byte Magazine, 9:12,* 239 - 248.

Johansson, M. (1997). *Smart, Fast and Beautiful: On Rhetoric of Technology and Computing Discourse in Sweden 1955 - 1995,* Linköping: Linköping Studies in Arts and Science 164.

Junberger, L. (1985). Classified school trial report says: Thumbs down for Compis!. *Datornytt, 9:5,* 30 - 31.

Jørgensen, U. (1988). Elektronikproduktionen i Danmark, Norge och Sverige. In H. Glimell (Ed.), *Industriförnyelse i Norden: 80-talets programsatsningar på mikroelektronik* (pp. 49 - 85). Roskilde.

Kaiserfeld, T. (1996). Computerizing the Swedish Welfare State: The Middle Way of Technological Success and Failure. *Technology and Culture, 37,* 249 - 279.

Karlsson, M. (1998). *The Liberalisation of Telecommunications in Sweden: Technology and Regime Change from the 1960s to 1993,* Linköping: Linköping Studies in Arts and Science 172.

Linder, G. (1980) *Teknikupphandling,* Stockholm.

Mackay, H. and Gillespie, G. (1992). Extending the Social Shaping of Technology Approach: Ideology and Appropriation. *Social Studies of Science, 22,* 685 - 716.

Malloy, R. (1984). The Tandy TRS-80 Model 2000: A Powerful New MS-DOS Machine. *Byte Magazine, 9:3,* 306 - 317.

Malmius, L. (1984). Compis i Las Vegas. *Skolan och datorn, 4:3,* 22 - 23.

Nilsson, B. (1984). Skoldatorprojektet – en lägesrapport. *Skolan och datorn, 4:1,* 25 - 26.

Nissen, J. and Riis, U. (1985). Datalära på grundskolans högstadium, Stockholm: Departementstencil C 1985:15.

NUTEK [National Swedish Board for Industrial and Technical Development]. (1982). STU-Archive, Dnr 218-82-522, F 7:61 a, Letter from the Ministry of Budget to the Board for Technical Development, June 20, 1982.

NUTEK. (1981). STU-Archive, undocumented work file of Ulf Eklund.

NUTEK. (1981 a). STU-Archive, undocumented work file of Ulf Eklund, "Kallelse till sammanträde i referensgruppen 21 oktober 1981," October 13.

NUTEK. (1981 b). STU-Archive, undocumented work file of Ulf Eklund, "Kallelse TUDIS. Projektmöte," December 10.

NUTEK. (1982 a). STU-Archive, undocumented work file of Ulf Eklund, "Deltagare i seminarium om teknikupphandlingsprojektet 'Datorn i skolan'," February 22.

NUTEK. (1982 b). STU-Archive, undocumented work file of Ulf Eklund, "Sammanställning av minnesanteckningar från företagsvisa överläggningar i samband med TUDIS industriseminarium," February 22.

NUTEK. (1982 c). STU-Archive, undocumented work file of Ulf Eklund, "Anbud på datorutrustning, programvara och läromedel för skolan," May 12.

NUTEK. (1982 d). STU-Archive, F 7:53, Protocol A 241 - 82/83, Enclosure "Anbudsutvärdering".

NUTEK. (1982 e). STU-Archive, undocumented work file of Ulf Eklund, "Anteckningar TUDIS. Projektmöte," April 21.

NUTEK. (1982 f). STU-Archive, Dnr 218-82-362, F 7:60, "Anbud 11," August 31.

NUTEK. (1982 g). STU-Archive, Dnr 218-82-362, F 7:55 - F 7:60.

NUTEK. (1982 h) STU-Archive, F 7:53, Protocol A 241 - 82/83, December 15.

NUTEK. (1983). STU-Archive, Dnr 218-83-362, F: 7:53, Letter from Esselte Studium to the Board for Technical Development, September 6.

NUTEK. (1984 a). STU-Archive, Dnr 218-83-362, F 7:61 e, "Till Tudisgruppens medlemmar av Bengt Bengtsson," October 5.

NUTEK. (1984 b). STU-Archive, Dnr 218-82-362, F 7:61 e, Control station 5, September 1.

NUTEK. (1984 c). STU-Archive, Dnr 218-82-362, F 7:61 e.

NUTEK. (1984 d). STU-Archive, Dnr 218-82-362, F 7:61 e, Control station 11, September 24.

NUTEK. (1984 e). STU-Archive, Dnr 218-83-362, F 7:53, "TUDISRAPPORT NR 2 av Bengt Bengtsson," October 31.

NUTEK. (1985 a). STU-Archive, Dnr 218-82-362, F 7:53, "Beträffande indexjustering från 850315 av Kai Lervik," March 8.

NUTEK. (1985 b). STU-Archive, Dnr 218-81739, F 7:54.

NUTEK. (1985 c). STU-Archive, F 7:53, "Anteckningar vid TUDIS uppföljningsmöte," May 21.

NUTEK. (undated). STU-Archive, Dnr 218-81-739, F 7:53, "Fördelning av statsbidrag för anskaffning av datorutrustning för skolan budgetåret 1981/82".

Sjöström, R. (1996). *Positionering under strategisk osäkerhet: En studie av positionering i en ny bransch*, Linköping: Linköping Studies in Management and Economics, Dissertations No. 30, vol. 1 - 2.

Svanborg, G. (1991). Head of Victor DataNova. Personal communication to the author, March 1.

Swedish Government Office Administrative Section. (1981 a). Archives Unit, Ministry of Education, Government Decision no. 37, October 29, 1981, Ulf Keijer, "Skiss till planering av teknikupphandling inom projektet 'Dator i skolan'," June 11.

Swedish Government Office Administrative Section. (1981 b). Archives Unit, Ministry of Education, Government Decision no. 37, October 29.

Swedish Patent Office. (1984/85). Companies Department, TeleNova AB (556235-8738), Årsredovisning för 1984/85.

Teknikupphandling: Betänkande av teknikupphandlingskommittén. (1976). Stockholm: Swedish Government Official Reports 1976:69.

Tele. (1984). *Tele: Televerkets tekniska tidskrift, 90:2*.

TUDIS Kravspecifikation: Teknikupphandlingsprojektet Datorn i skolan. (1982). STU-information no. 279 - 1982, Stockholm: National Swedish Board for Technical Development.

Turkle, S. (1984). *The Second Self: Computers and the Human Spirit*, New York.

Weinberger, H. (1997). *Nätverksentreprenören: En historia om teknisk forskning och industriellt utvecklingsarbete från den Malmska utredningen till Styrelsen för teknisk utveckling*, Stockholm: Stockholm Papers in History and Philosophy of Technology 2030.

Wells, P. (1984). The 80286 Microprocessor. *Byte Magazine, 9:11*, 231 - 242.

Widmark, B. (1991). K-section, Teli AB. Personal communication to the author, July 24.

Williams, G. (1983). Benchmarking the Intel 8086 and 8088. *Byte Magazine, 8:7*, 147 - 162.

Zingale, T. (1983). Intel's 80186: A 16-Bit Computer on a Chip. *Byte Magazine, 8:4*, 132 - 150.

5. SWITCHING RELATIONS AND TRAJECTORIES: THE DEVELOPMENT PROCUREMENT OF THE AXE SWEDISH SWITCHING TECHNOLOGY

M. Fridlund

1. INTRODUCTION

The case study treats the Swedish Public Technology Procurement of the AXE telephone switch, 1954 - 1980. This electronic computerised switching system was developed by the semi-public company, Ellemtel, for the public telephone utility, Swedish Telecommunications Administration (STA), and the private manufacturer, LM Ericsson (LME). As an introduction to the institutional framework, a short outline is given of the Swedish telecommunications system, including key organisations and centrally important formal and informal relations. Following this, the main events preceding and following the AXE procurement are described.

In retrospect, the development of the AXE system can be divided into three stages. The first consisted of learning and search activities concerning the new technologies of electronic and computerised switching, while the second stage was devoted to a specific procurement project of developing the new switching technology. In the last stage, the technology was further developed, adapted and exported and started being used on a large scale in its country of origin. The study ends when the AXE had reached commercial maturity locally and globally, as well as having gone through its first process of Adaptive Public Technology Procurement (APTP). Even if this division into stages might give it an orderly and planned appearance, the development process was very much filled with contingencies and unforeseen technological events, as well as changes in opportunities and strategies concerning markets.

2. AN OUTLINE OF THE SWEDISH TELEPHONE SYSTEM, 1920–1980

The two main partners in the procurement of the AXE switching technology were the two main actors of the Swedish telecommunications system, the Swedish switching user, STA, and the switching producer, LME. Except for LME, the only other equipment producing firms in Sweden's telecommunications sector were the smaller manufacturer, Swedish Philips Inc., and American ITT's subsidiary, Standard Radio & Telephone. However, both of them were unimportant to the AXE development. The two joint 'bridging organisations' that were formed between STA and LME and became important in the AXE development were the R&D co-ordinator, the Electronics Council, and the joint developer, Ellemtel. In addition, two of LME's foreign subsidiaries, Northern Electric and EPA, played important roles. Both STA and LME were manufacturing switches and they were both procurers of AXE, although the specific project that will be most in focus here featured STA as the buyer. The company from which they procured the AXE development was their jointly owned development company, Ellemtel.

The regulatory framework of the Swedish telecommunications system that existed in the 1980s had remained more or less unchanged since the 1920s. Before the introduction of mobile telephony in the 1980s there had not been any formal monopoly on telephony, although STA held a practical monopoly. Telephony was rather un-regulated and there was no special Swedish telecommunications law until 1993. The reason was that STA's dominance made it possible to regulate the sector without legislation. STA, as a public agency, was directly subordinate to the Ministry of Communications, and indirectly to the Ministry of Finance (Ioannidis, 1998: 203 - 4). Concerning advanced technical education, Sweden before the 1960s had only two technical universities and the lack of engineers with knowledge in electronics and computer science became a major problem for the AXE development.

2.1 The User: The Swedish Telecommunications Administration (STA)

The Swedish Telecommunications Administration (today, the state-owned joint-stock company, Telia Inc.) was founded in 1853 and was the public agency responsible for the Swedish telecommunications network. It was a Public Telephone Administration (PTA) but not a PTT, as it was separate from the Swedish Post Office. From 1911 onwards, STA became formally institutionalised as a so-called state-owned public enterprise (*affärsverk*). Its tasks were to act as a state authority controlling and supervising all matters concerning telecommunications, and to sell telephony services and telephone equipment to Swedish subscribers and companies. STA was subordinated to the Ministry of Communications and had a legal status between those of a company and a government authority. STA's economic objective was to be self-supporting and it could decide by itself on how to use its operating income. To be in charge of STA the government appointed a board and a Director General (DG) who were responsible only to the government (Granstrand and Sigurdson, 1985: 153; Karlsson, 1998: 79 - 81).

STA's management was dominated by engineers and in the 1970s technological values were dominant within STA, whose "implicit technical objective" was high efficiency through continual technical development and rationalisation. (Hagström, 1997: 12). As a user, STA was highly competent and could both formulate demanding functional requirements and make independent evaluations of the resulting technologies. This was because STA had expertise in all important areas of telecommunications: It had experience of the problems of operating and maintaining telephony systems through its *operational department*; it was engaged in R&D on new telephone technologies in its *technical department*; and it had knowledge in industrial mass production through its *manufacturing branch*. In having this double competence of being both a user and producer STA was, together with the American giant AT&T, unique among Western utilities. In 1970, STA had factories in five Swedish cities and around 3,500 employees (Angerby, 1965: 48). In 1970, only 5 % of its manufacture was delivered outside STA but several engineers at STA had for several years desired to export its switching technology. ([Karlsson], 1963: 117). In 1994, STA's manufacturing branch was sold to LME.

2.2 The Producer: LM Ericsson Telephone Company (LME)

LM Ericsson Telephone Company (today, only Ericsson), founded in 1876, was the largest Swedish telephone equipment manufacturer. However, it had no captive market in Sweden, since STA manufactured most of its own equipment. LME's main product was electromechanical crossbar switches and in the 1960s it "was living on crossbar" (Meurling and Jeans, 1995: 2). LME depended on foreign markets and sold to around 130 international PTAs and private utilities. In 1970, the Swedish market accounted for 37 % of its turnover and the largest export countries were Brazil, France, Mexico, Italy, Norway, Denmark and Australia (Attman and Olsson, 1976: 262 - 64; Sunesson, 1970: 146). In 1980, the Swedish market was 22 %, and of PTA sales STA accounted for 3 %. (Heimbürger and Tahvanainen, 1989: 13 - 16; Granstrand and Sigurdson, 1985: 156). LME also owned shares in two companies important for the development of AXE and mobile telephony. The first was Ellemtel and the second was Swedish Radio Inc. (SRA), formed in 1919 and specialising in radio and mobile telephony technology. LME in 1965 increased its ownership in SRA from 57 to 71 % and as of 1983 SRA became wholly owned by LME and changed its name to Ericsson Radio Systems (Mölleryd, 1996: 81 - 83).

2.3 The Bridging Organisations: The Electronics Council and Ellemtel

In the 1950s, both LME and STA began developing new applications based on electronics and in the mid-50s decided to co-ordinate their efforts. In 1956, the two parties formed the joint Electronics Council, which was intended to supervise and co-ordinate their development projects and, in particular, study the possibility of developing electronic switching systems. Two supplementary agreements were signed in 1963 and

1965. The first concerned the development of two switching stations, A210 and AKE12, to be developed separately by the two parties. The contract can be seen as a recognition of the fact that the collaborative development had not been altogether successful. The second agreement concerned the development of advanced electronic telephones (Lernevall and Åkesson, 1997: 703). The Electronics Council meetings continued until Ellemtel was formed in 1970.

Ellemtel Development Inc. was a development company jointly owned by LME and STA. (The name reflected the shared ownership: the first part sounded like "LM" in LM Ericsson and the second (tel) comprised the first letters in STA's Swedish name, *Televerket*.) The company was founded to develop strategic telecommunications technologies in general, and the AXE electronic switch in particular. There had been resistance from some elements in LME's management to 'giving away' this very strategic development project to an outside and partly state-owned company, but after tough negotiations an agreement was reached in 1970. Ellemtel employed engineers from STA and LME and thereby acquired competence from the user- as well as the producer-side. (Lernevall and Åkesson, 1997: 706). In October 1995, all STA's shares in Ellemtel were sold to LME.

2.4 The Regulators: The Ministries of Communications and Finance

As a state-owned public enterprise, STA was formally regulated by the Swedish parliament and government and was the direct responsibility of the Ministry of Communications. However, the Ministry did not apply direct rule and its regulation before the mid-1980s was, instead, rather indirect, through the yearly negotiations regarding STA's annual estimates of expenditure. During this period STA, financed most of its investments on the basis of income from its subscribers and without additional capital from the public treasury (Lernevall and Åkesson, 1997: 72 - 73). Since the government's main financial responsibility belonged to the Ministry of Finance, this ministry was also instrumental in regulating STA. The Minister of Finance in charge throughout nearly all (1955 - 76) of the AXE procurement period, and the one who governed STA's investments, was the powerful Social Democrat, Gunnar Sträng. His wish was that STA should contribute to the public treasury while at the same time keeping down its investments, and he did not want STA to grow "unnecessarily". This meant keeping a "suitable" stock of subscribers waiting for telephones (Bjurel, 1997: 6). During the first part of the 1970s, the parliament and government agreed with STA on trying to reduce this queue. Subsequently, large amounts of STA's total investments were spent on switching stations for that purpose (Lernevall and Åkesson, 1997: 73).

2.5 Institutions: Formal and Informal Switching Relations

With respect to institutions, the most important for the AXE procurement were a number of long-term social relations: one was a national relation and the others were international. The most crucial was the national one: a long-term user-producer relation – the Development Pair – that had been established between STA and LME. The international relations included that which STA had cultivated with the dominant international telecommunications company, AT&T, and those between LME and two of its foreign subsidiaries.

2.5.1 National relations: the Development Pair, STA–LME

Probably the most important informal institution in Swedish telecom has been the long-term close development relation existing between STA and LME. Starting in the 19th century, LME and STA had conducted a plethora of more or less formalised joint construction and manufacturing projects in most product areas (Sterky, 1968: 27 - 56; Patricks and Jacobæus, 1972: 41 - 46; Bjurel, 1976: 1 - 9; Lundvall, 1977: 129 - 131). This is one of several Swedish examples of public technology procurements institutionalised through a *Development Pair* (DP), i.e., a tightly linked long-term user-producer relation formed around several joint development projects between a manufacturing firm and a government customer (Fridlund, 1993).

The DP between STA-LME went through two major institutionalisations in the 1950s and 1970s that progressively 'hardened' the relation. In the beginning of this joint collaboration it was mostly conducted through single engineers employed at STA, who came to LME to get support, resources, and money to develop their inventions into commercial technologies. There was also a lot of informal information exchange between the two organisations, because it was very common for engineers to change jobs back and forth between the two organisations. This was very widespread and existed all the way up to the highest executive levels. This relation became more formalised with the Electronics Council in the 1950s and 1960s, and even moreso with the development company Ellemtel from 1970 onwards. Concerning previous joint projects, a majority concerned automatic electromechanical switching. In the 1910s two of STA's engineers invented two automatic switching technologies and went into co-operation with LME to develop them into commercial products. STA later procured experimental switches, conducted field tests and, in the 1920s chose one of the switches for automating its major network. In 1943 - 45, LME and STA jointly developed an electromechanical crossbar switch that fit both STA's domestic needs and LME's export ambitions. In the 1950s this technology made LME into one of the major international telecommunications companies (Attman and Olsson, 1976: 191 - 92).

2.5.2 Foreign relations: AT&T, Northern Electric and EPA

The other important relation was between STA and LME and the large American Telephone & Telegraph Company (AT&T). This American giant owned the R&D organisation, Bell Telephone Laboratories (Bell Labs), and the manufacturer, Western Electric. This relation came to be an important channel for information exchange concerning technological developments. The first important exchange, and one that is supposed to have laid a basis for the relation between Bell Labs and STA, came when engineers from Bell Labs visited STA in the 1930s and were given crossbar technology to bring back home (Broberg, 1977: 53; Vedin, 1992: 39). Based on this, AT&T developed its own crossbar switch which from the 1940s became the dominant switching technology internationally. In 1947, STA's DG used his friendship with Bell Labs' chief engineer and sent one of STA's leading engineers to spend half a year at AT&T in the USA. The engineer developed relations with engineers at AT&T which became important when he became STA's next DG in 1966. In 1950, AT&T approached LME about a mutual patent exchange which included the patent on the transistor. In 1951, LME and Western Electric closed a similar exchange contract, which allowed LME's engineers to visit Bell Labs and Western Electric. After LME and STA had started their joint collaboration in the Electronics Council, a similar contract was signed between STA and Western Electric in 1957. Several key engineers in the AXE project had the opportunity to visit Bell Labs and discuss possible electronic switching solutions with AT&T's leading R&D engineers. Furthermore, they had access to operational results from AT&T's first transistorised and computerised switching station ESS 1, installed in Morris, a small town outside Chicago, and becoming operational in 1960. STA kept on renewing their exchange contract until 1993 (Jacobæus et al., 1976: 409, 411; Neumeyer and Rönnblom, 1965: 266; Fredriksson, 1990a: 3; Lernevall and Åkesson, 1997: 701 - 702).

To get a foothold on the US market, LME in 1951 bought the American company North Electric Co (NE). Through NE, LME in 1960 got its first commercial order – from the US Air Force – for an electronic switch. The project was finished in 1963 and the knowledge LME acquired was later important for the AXE development. When LME later also sold this switch to the Swedish national defence, it was the first electronic telephone station outside USA. NE was, however, not profitable and was sold in 1967. The sales agreement included future co-operation and technology transfer and some of LME's former engineers also stayed on at NE. The relation continued and became important in the AXE project when Ellemtel engineers visited NE to learn the new "wire technology". This process technology made it possible to make changes 10 times as fast as before and through this the first AXE switch could be delivered more or less on time in 1977 (Meurling and Jeans, 1995: 174; Vedin, 1982: 149 - 50).

Since 1950, LME had also had an Australian subsidiary, LM Ericsson Pty. Ltd. (EPA). Through EPA, LME got large orders from the Australian Post Office (APO), and in 1963 LME held a third of the Australian switching market. Since around 1970 EPA had also conducted a major part of LME's development of digital switching. One reason for this was the good possibility of recruiting engineers. It was also important for LME's relation to the large customer, APO, to show that it was performing advanced and independent construction work in Australia (Attman and Olsson, 1976: 247; Vedin, 1982: 187). This relation became very important in the 1970s for the first exported AXE digital switch and a large Australian order.

3. PROTO-PROCUREMENT: LEARNING SWITCHING TRAJECTORIES, 1952-69

Learning figures importantly in this study. Both LME and STA had by the 1950s developed an advanced competence in telephone switching by being actively involved in the established trajectory of electromechanical technologies and automatic switching. So, in several important ways, the new AXE switching technology included areas where the two parties already had extensive knowledge and experience. However, the two new technological trajectories of electronics and computerisation were in the early 1950s still unknown and untried by virtually all engineers within LME and STA.

STA realised in the 1950s that telephone switches using electronics and computer technologies could make possible new services for the subscribers and improve the efficiency and capacity of old services. LME also saw the possibilities but was not equally enthusiastic. It considered its electromechanical switches sufficient, but invested in electronics R&D as an insurance against sudden technological breakthroughs. LME and STA began development of electronic switching along two different development paths. In the mid-50s, attempts were made to co-ordinate these projects through the formation of the Electronics Council and, in 1963, LME and STA decided to co-ordinate their development by initiating projects on two complementary switches, one small switch for rural towns by LME, and a larger one for major cities by STA. The switches became operational in 1968 and 1969 but were far from commercially viable technologies and would demand several years of further development. To reach a functional solution as early as possible, and to minimise the duplication of development work, the two parties decided in 1969 to pool their resources with an agreement to form a joint development company.

3.1 Established Trajectories of Electromechanics and Automatic Switching

STA's and LME's learning experience began in operating, manufacturing and developing various automatic electromechanical switching technologies. The *electromechanics trajectory* arose in the 1800s as the fusing of mechanical technologies of wood and metal with electricity to power and to operate the mechanical components, which became the basis for all telecommunications technology. Another major trajectory was *automatic switching* – to connect telephone calls from one subscriber to another automatically by machines instead of by human switchboard operators. STA was among the leading utilities in automating. Its first automatic station was inaugurated in 1924 with switches from LME, and Sweden became one of the first countries in automating its national network in 1972. Simultaneous with the last stages of this automation process, STA began another similar large process of switching-over to a new telecommunications trajectory when in 1968 they also began computerising their automatic stations. Similarly, LME, after being a laggard in the 1920s, established itself in the lead of the trajectories of electromechanics and automatic switching. This was much thanks to STA and their crossbar switch, which LME started to successfully manufacture for exports in the 1950s. By the end of the decade it had made LME into a major international switching manufacturer.

3.2 Entering the Electronics and Computerisation Trajectories, 1952-57

Electromechanics was firmly entrenched in the 1950s and many telephone engineers were very sceptical of electronics and its possibilities. Before the invention in 1948 of the semiconducting transistor at Bell Labs, electronics technology consisted of valve technologies such as electron tubes and radio valves. The great advantage of electronics to switching was its high speed, 10,000 times as fast as electromechanical relays. This also made it possible to send more calls in the same telephone line through the principle of "time-division multiplex" (TDM) instead of traditional "space-division multiplex" (SDM). TDM meant that when 100 telephone calls were transmitted at the same time in the same telephone line, each call alternated in using the common telephone line for a micro-second at a time. Because the pauses between the micro-second parts was small (milliseconds), the human ear heard it as a continuous telephone call.

In the 1950s, LME and STA separately studied the possibilities of upgrading electromechanical switching by exchanging electromechanical components with electronics (Heimbürger and Tahvanainen, 1989: 19). In 1954, LME finished its first small prototype electronic switch. It was a "pure laboratory product" using old valve technology and LME planned to go on with a larger practical switch (Jacobæus et al., 1976: 137). However, in 1955, STA approached LME regarding a collaboration in electronics. STA's director of development had spent half a year at Bell Labs, thus becoming a believer in electronics. He had convinced STA's DG to start using the transistor in electromechanical switching and to do it in collaboration with LME (Vedin, 1982: 85 - 86). STA's DG argued that electronics was going to be very important and to develop it would demand large resources. LME was not eager to start a large development project on the new and unknown transistor technology. Its electromechanical crossbar switches were successful exports, and LME "neither believed in electronics nor wanted it". LME however saw the R&D investments as an insurance against unwanted technological surprises and after some deliberation agreed to a limited collaboration (Vedin, 1982: 31). In 1956, LME and STA signed an agreement "to collaborate and co-ordinate" their R&D activities in electronic switching and formed the Electronics Council to co-ordinate the various development projects. (Patricks and Jacobæus, 1972: 44). The same year, the breakthrough of the electronics trajectory also came into focus when the Nobel Prize for Physics was awarded to three engineers at Bell Labs for development of the transistor.

The next breakthrough of Bell Labs and the electronic telephone station came the following year with the introduction of the *computerisation* trajectory into telecommunications. LME and STA discovered this at the Morris-symposium arranged by Bell Labs to present the first computerised telephone switching station which applied the computer technology of "stored program control" (SPC) to telephony. The idea was to control all switching operations of the electronic and electromechanical elements with a programme stored in a computer memory inside the station. This was a completely new technology, with which neither LME nor STA had much experience, and several senior LME engineers were very sceptical of SPC (Vedin, 1982: 34; Lernevall and Åkesson, 1997: 114; Fredriksson, 1991: 17). Among the Swedish engineers at the symposium, only one – who in the early 1950s had worked with the development of the two first Swedish computers – really understood the new technology. After the symposium, LME and STA began a period of learning in electronics and computerisation. What they had been shown and soon discovered themselves were the advantages the SPC technology gave to telephony. These were high flexibility that made it easy to change the functions of a switching station by changing the programmes stored in its memory, the possibility of automatically localising errors in the station's functions, and the potential to introduce new kinds of automatic telephone services such as "call-waiting signal", "automatic wake-up call", et cetera. But in addition to advantages they also learned about obstacles, technical as well as social, associated with developing the new technologies. As LME's and STA's ignorance at the Morris-symposium had shown, it soon became clear that a critical problem would be to acquire enough engineers knowledgeable in computer technology. To solve this, LME and STA started a joint advanced education programme in computer science for new graduate engineers (Vedin, 1982: 62). This became centrally important when STA and LME began their first more practical development projects.

3.3 Switching to Separate Paths, 1957-63

After the Morris-symposium, STA initiated their first electronic switching project by replacing electromechanical components of a traditional crossbar switch with electronics (Lernevall and Åkesson, 1997: 114). However, STA soon decided the switch was not computerised enough and an unnecessary detour. The project was terminated in 1958 and instead the new TEST 1 project was started. In this project, switching would be operated by SPC and memories built into the switch. This yielded a flexible but expensive switch, since it used many electronics components. TEST 1 was contrary to the development path of LME, which kept its old electromechanical switch, but sped it up by substituting electromechanical components with electronics and by using a separate computer for regulating the switching operations. To STA this was too conventional and rigid a system, without the cost advantages of a more computerised solution and without flexibility for further development (Norinder, 1987: 33). The two parties also took separate paths regarding the choice between the computer-oriented TDM or the traditional SDM technologies. But here the roles were switched, with LME using TDM, while STA went ahead with SDM.

Even when TEST 1 had gathered momentum, the joint collaboration between STA and LME was derailed for a couple of years while LME put almost all of its time and resources into a new switching project. At the end of 1959, LME had planned to continue its first project with a more commercial version of a large electronic switching station (Vedin, 1982: 69). In 1960, however, LME through its US subsidiary, North

Electric, suddenly got a commercial order to construct and deliver a small electronic switching station to the US Air Force. This order was important in generating practical expertise in electronic switching. Although the switch was relatively small – a few hundred lines – and did not use SPC, it was a very advanced project. It used TDM, and had high demands for reliability, speed and flexibility. Thanks to this, LME built up a large group of engineers that was to become the core of the future development groups formed around computerised switches. When the project started, the group numbered around 20 persons and after two years they were 120 (Jacobæus et al., 1976: 140; Fredriksson, 1991: 18). Meanwhile, STA's TEST 1 project continued, even if it had not become fully operational. Although there were only about 20 subscribers connected to the experimental switching station, STA had learnt a lot. LME would also learn from STA's experience gained in this project when the partners restarted their collaboration in the Electronics Council for their next electronic switching project.

3.4 Switching into Parallel Paths: A210 & AKE12, 1963-69

In 1963, LME and STA signed a supplementary agreement to the Electronics Council agreement regarding their collaboration on the development of two computerised switching stations for STA. The joint development strategy had changed for this second generation of electronic switches. In 1961, parallel to the US switch, LME had begun developing a new switch and STA in 1963 terminated TEST 1, also in order to start a new project (Vedin, 1982: 65 - 66; Lernevall and Åkesson, 1997: 115). Although LME's and STA's top managements had tried to induce co-operation through the Electronics Council, it had not spread to the engineers in the projects where "the lack of co-operation, sharing and joint development had become embarrassing" (Meurling and Jeans, 1995: 28). It was therefore decided that they should concentrate on their own separate development but that the projects should complement each other in that the partners would switch traditional markets with each other. STA should concentrate on developing a switching station, A210, for cities – LME's traditional product market – while LME should focus on a larger switch, AKE12, for smaller communities and rural areas. Switches for the latter type of market made up the bulk of STA's switching manufacture. In this way, the joint collaboration took off, since the two parties could now benefit from exchange of each other's traditional expertise (Norinder, 1987: 34). But even though it was collaboration around information exchange, it was still aiming for two separate technologies – with the exception of joint development of computer memories. Another thing was that the Electronics Council in 1964 presented a joint proposal for new automatic telephone subscriber services such as "call waiting", "automatic wake-up call", etc., indicating that these should be tried out on the subscribers in the two new telephone stations (Heimbürger and Tahvanainen, 1989: 21).

A serious critical problem, common to both projects and seriously underestimated previously, was constituted by the large computer programmes that operated the switching stations (Fredriksson, 1991: 18 - 19). Critical to the development of the two switching systems was the need for access to full-scale and realistic field testing with subscribers connected to the switching stations. This was because of the specific characteristics of computer software construction. Large computer programmes, in contrast to electronics hardware, can not be tested out in dry-runs. As described by STA's project manager:

> "Most of the time a hardware constructor can sit with his gadgets around him on top of a table with measurement equipment [...] and can verify that the thing he has been assigned the task of doing is properly done. Those who shall create software are forced to perform full-scale tests; several subscribers have to get in and test it. And this we can't work out on paper. It is out there in the field, not before, that it is decided" (Norinder, 1987: 36).

Both of the two new switches suffered from serious problems concerning programming errors. LME's AKE12 was the first switch to go into subscriber field tests in 1967. These showed that a small single error in the programming could cause a break-down of the switching station. In addition, it was difficult to localise the error in the large programme. These first installations showed that much more work on programming was needed to make computerised stations operational than with the previous electromechanical stations. After long delays because of these problems, the first AKE12 station was inaugurated in 1968. The critical problems with the program errors raised serious doubts that AKE12 could ever become a functional technology for local stations, but it appeared possible to use a modified switch, AKE13, for international transit stations. LME had got an order for such a switch from a telephone operator in the Netherlands. When working on AKE13 in 1967, an engineer came up with a solution to the critical programming problem. This solution was contrary to the traditional system of having one large computer with one large programme that controlled all the switch's operations; instead, it consisted of constructing a switch with "modular programs". Thus, the large programme were divided into smaller programmes that controlled only a specific function, each of which was isolated together with its own data. Also, the switch should be divided into different isolated blocks, depending on what functions it was performing. This would make it easy to localise errors in programming or function, since the errors could not spread outside the module in which they occurred. Furthermore, it would make modification of the faulty programmes easy, since it would only be necessary to correct the small programme within the module instead of the whole programme (Vedin, 1982: 104).

3.5 Growing Concerns about Losing Time

During 1968 and 1969, the STA's DG and LME's President discussed further co-ordination and even a merger of their switching projects. It had become obvious to them that the development work would be of a completely different magnitude than that of the previous electromechanical switching technologies: several hundreds compared to tens of man-years. And the number of engineers was a critical problem for this development work. Between 1965 and 1970, LME's costs for its R&D almost doubled, from 122 to 226 million SEK, and a large part of this was due to the projects concerning electronic switching (Attman and Olsson, 1976: 265). It was clear to both LME and

STA that they had started "to run out of time" because neither had a "competitive technical solution in sight" for the largest share of the switching market, which consisted of *local* switching stations for cities or densely populated areas. Internationally, LME sold two semi-electronic switching systems: for local stations it sold an old electromechanical switch with some "electronic make-up applied", and for the smaller volume of trunk stations, the AKE13, which used SPC but was expensive. LME had begun to panic. Getting to a new switching generation was a matter of destiny to LME, and its existence as an independent manufacturer was in peril (Meurling and Jeans, 1995: 91 - 92; Bjurel, 1997: 8; Vedin, 1982: 57).

At this time, the secrecy among international manufacturers was pronounced, and everyone suspected each other of having "reached far in their development" (Forss, 1970: 13; Bjurel, 1997: 8). The American companies, Western Electric and ITT, had in ESS1 and Metaconta electronic switches using SPC. The British Post Office was, together with five British manufacturers, developing switches in the Joint Electronics Research Committee; the German Bundespost collaborated on switching systems with electronic control with Siemens; and in France the government agency, CNET, collaborated with domestic manufacturers to develop both a computerised switch and a small digital switch. Furthermore, IBM experimented with a fully electronic and computerised switch, and in Japan there were the experimental computerised DEX-1 and the DEX-T1 digital switches, developed in collaboration with NTT (Forss, 1970: 11 - 13; Meurling and Jeans, 1995: 24 - 26). LME experienced this foreign competition first-hand when LME's traditional good customer, the Australian Post Office (APO), invited tenders for a large exchange. The project suited AKE13 perfectly and would mean a good reference project for developing a new computerised switch. LME had "high hopes" about getting it, and it came as a "serious blow" when APO in 1969 chose ITT's Metaconta switch (Meurling and Jeans, 1995: 29).

As it is evident from the international projects, it was not unusual for the public agencies to collaborate with national manufacturers in different kinds of joint organisations such as the Electronics Council. But the next step that LME and STA took together was more unusual. In 1969, STA's DG and LME's President met with the Chairman of the Board of LME to discuss a new phase in the joint development and it was then decided to form a jointly owned development company. STA got an approval from the Ministry of Communications that was positive to the sharing of development costs (Fredriksson, 1990a:4 - 5; Meurling and Jeans, 1995: 31 - 32; Vedin, 1992: 95 - 97; Åkermark, 1996: B:5; Bjurel, 1997: 8) This meant that the old development relation had switched to a more procurement-directed relation. Thus, the procurement phase began in 1970 with the institutionalisation of the previous joint development in the company Ellemtel.

4. PROCUREMENT: DEVELOPING TRAJECTORIES, 1970-77

The signing in 1970 of the Ellemtel contract was the start of the formal procurement phase. Ellemtel was a pure development company without any manufacturing and its main mission was to construct and develop equipment for electronic switching. In 1972, LME and

STA formally procured the development of the computerised electronic switch that came to be known as AXE from this subsidiary. The process, until the first AXE telephone station was brought into operation in a small Swedish community in 1977, was fraught with lot of socio-technical development concerning crucial decisions and critical problems. Of these, the most central are discussed below. AXE came to include several new technologies such as electronic switching, computer control, functional and service modularisation. It also allowed for change-over to digital switching and transmission.

4.1 The Ellemtel Development Contract

In April, 1970, STA and LME signed a contract "concerning certain development and production collaboration in the area of telecommunications". The contract stated that the new telecommunication technology would put stronger demands on development resources and personnel and that its advantages (the societal needs it would meet) were that the new switches could be "adapted to the varying conditions of the time and the customers can be given new kinds of [telephony] services". The new switches were also seen as providing increased operational reliability and lowering maintenance costs. Ellemtel was owned on a 50-50 basis and regarding its costs LME was accountable for about 65 % and STA for 35 % (Sunesson, 1970: 148 - 49).

The two largest contractual problems had concerned patent rights and the division of the profits from the sale of innovations. For LME's part, the largest obstacle concerned the intellectual property law agreement. Who should own the future innovations? STA had the ambition of starting to export their products and LME did not want to fund a potential competitor on its export market. The outcome was that all Swedish patents were to be owned by Ellemtel, with LME and STA having full rights to use them, while all foreign patents were to belong to LME.

The uneven situation was balanced through an agreement concerning collaboration in manufacturing. This was to ensure 'equal growth' of the two parties in the manufacture of products developed by Ellemtel. This meant that STA would sell and produce Ellemtel's products on the Swedish market, while LME would produce for the export market, but only in the same quantities as STA's domestic production. Since the export market was expected to grow more than the Swedish market, any amount of export above that level was to be shared between LME and STA and sold internationally by LME. This was seen as implying a large future expansion of STA's manufacturing branch, and it was estimated that by 1980 STA's exports would be 25% of its total manufacture. It looked as if the export dreams of STA's manufacturing branch would come true (Sunesson, 1970: 147 - 149; Vedin, 1992: 98 - 99). However, this agreement had to be removed when the contract was changed in 1977. The reason was that when the collaboration on production according to the contract was to come into effect in 1975, LME's growth in switching had became negative and LME had threatened to close a Swedish factory. The positive assumptions about the growth of LME exports, which were the basis for the joint production agreement, had changed and the contract had to be changed accordingly (Vedin 1992: 100, 98 - 99).

4.2 Organisation of the Development

Ellemtel's development work was organised in such a way that, at almost all levels where decisions were taken, the experience and opinions of both LME and STA influenced it. On the top level was the board with six ordinary members, three from each owner. The board among themselves appointed a co-ordinative committee (CC) consisting of five members: two from STA and LME each, and Ellemtel's Director. The CC decided which proposed development projects Ellemtel should pursue and handled questions regarding allocation of time and resources. On the next level, functions committees and product committees (PKs) for each main product or technology area prepared and discussed all issues to be decided in the CC. In these, STA, LME and Ellemtel were represented equally. In the product committee for telephone stations (PK-TS) all formal decisions regarding the technological choices of the AXE project were taken. The PK-TS met only about every sixth week and, therefore, internally appointed members to a co-ordinative group (TS-KG) that handled practical administrative and technical questions that had to be handled quickly. This group also consisted of representatives of the three parties. Connected to these committees was a number of other co-ordination groups, expert groups, steering committees, etc. All these different committees functioned as arenas for problem solving on different levels. These ranged from the practical micro-level questions regarding the actual development work dealt with in the TS-KG, to co-ordinative issues in the PK-TS, and questions of the project's funding and resource allocation in the CC and the board. Another important purpose, aside from the "purely" technical problem solving activities of these various groups, "was to create consensus regarding different decisions about changes during the construction work." It was in these different collaborative institutions "that the natural state of tension between the three co-operating companies would be expressed, the opinions broken against each other, and a common strategy and a handling finally be worked out to solve both problems on the level of requirement specifications as well as technical problems that arose during the construction work" (Lernevall and Åkesson, 1997: 707 - 8).

To begin a project like AXE, the purchaser filed a procurement order for the development of a technology. This was then given to the CC, which decided if it should be approved depending on estimates of whether the product would earn its cost, if there were sufficient demand for it in Sweden or on the export market, etc. When a procurement order had been approved, it became the responsibility of its PK to handle the project. The PK could not increase the cost ceiling but could change the level of ambition and shift the allocations of resources. After this it became Ellemtel's task to develop the product (Fredriksson, 1978:4, 2 - 3). The following section deals with the critical decisions and problems before and during the development of the AXE procurement order. With the acceptance of this procurement order from STA and LME to develop a computerised electronic switching station, the formal procurement of the AXE technology started.

4.3 Reconstructing Supply and Demand, 1970-72

A few weeks before the signing of the Ellemtel contract, STA's A210 switch was inaugurated. This was a big failure. The station did not work because of an error in the computer program and it took several days to make it functional. When Ellemtel began, STA decided to terminate the project and most of its personnel transferred to Ellemtel. Thus, the first and only A210 station was shut down after only seven months of operation (Fredriksson, 1990b: 56 - 57).

Ellemtel started up its operations in July, 1970. One of its first tasks was to evaluate STA's and LME's previous three proposals for computerised switching systems. By the end of 1970 Ellemtel had finished a study that proposed a switching technology in two versions: one for local switches of up to 20,000 subscribers, and one for larger switches. The initial problem of the AXE project was to come up with a proposal that would fit the demands of both owners. These were in many ways conflicting. STA was mainly interested in a switch that would fit the specific Swedish telephone system, while the export-directed LME wanted a switch that fitted as many systems as possible, preferably the whole world. Therefore, Ellemtel began working on a proposal for a new computerised switching system from a technical supply perspective. Parallel to this, LME worked on a requirement specification from a demand perspective regarding what the system should be able to do, how much it could cost, etc. The purpose was to describe the "ideal system" according to the demands of LME's foreign markets (Meurling and Jeans, 1995: 36). STA also prepared a similar specification that reflected the specific demands of the Swedish system. Drawing on its operational knowledge, STA made an especially concentrated effort to specify various functions in the operation and maintenance functions. These were later developed into a range of "feature packages of great sophistication" that became important when the AXE-system was marketed to other telephone operators (Meurling and Jeans, 1995: 37).

LME's requirements were presented in January, 1971, and a dialogue began between the three parties. By mid-1971, some common central conclusions regarding the requirements' specifications had been reached: it should be a computerised SPC switch with large flexibility in switching functions; it should be adaptable to all kinds of telephone networks; it should be constructed modularly to make it possible to add or remove applications and functions modules, depending on the various demands of different markets; it should be "flexible in time" and allow for new undeveloped services and technical innovations to be included, such as digitalisation and more efficient memory circuits; it should be easy and fast to manufacture, install and maintain; it should localise any faults and stop them from spreading in the system; and, finally, it should have a capacity of 40,000 subscribers (Vedin, 1982: 138; Meurling and Jeans, 1995: 38).

Ellemtel's systems specifications were finished in mid-1971. A central aspect of the AXE system was that it was going to have a functional modularisation. This was an extension of the ideas developed by LME in connection with the development of the AKE13 switch. Now a special signalling system was introduced that isolated the different functional modules from each other. This was the solution to the critical problem of the error-sensitivity of large software programmes. The majority of other switching manufacturers had more or less accepted this difficulty as a 'fact of life'. This problem had led them to the critical solution of 'modular software'. In AXE, this solution was now also extended to include the hardware that belonged to the functions. That is, the hardware would also be incorporated in the modules. The AXE proposal therefore included the construction of a special new processor to support the specific characteristics of the modular software. However, Ellemtel's proposal was not concrete enough and the company got orders to produce a more detailed proposal, which was presented early in 1972.

In the spring of 1972, an evaluation group from LME presented a new report that focused on three main issues: the high cost for smaller exchanges below 4,000 subscriber lines; the fact that Ellemtel still couldn't reach the capacity of 40,000 subscribers; and the problem of low reliability. LME had analysed previous installations for electromechanical switches and had found that their customers for their first order had ordered a switch for 2,000 subscribers. LME saw a risk that such a customer would not buy the AXE if the prices for small switches were too high; instead this potential customer would buy itself into another system. But since the prices for electronics components were going down and another market looked more promising, the Ellemtel proposal was nevertheless acceptable from the point of view of costs. The 'other market' referred to in this connection was the replacement market for very large city stations between 20,000 and 40,000 lines. It was a market which made the capacity requirements important. Since these stations contained the oldest electromechanical switches on the market, they should be among the first to replaced. Ellemtel's specification only reached 32,000 but managed to achieve the same traffic capacity as that of 40,000 subscriber requirements by speeding up its processor – a solution which was accepted by LME (Vedin, 1982: 141: Meurling and Jeans, 1995: 141 - 2). The requirement for reliability was related to the regional processors that handled simpler receptive switching functions and were different from a central processor such as APZ 210, which handled more complex and less frequent switching operations. In contrast to the central processor, the regional processors had not been duplicated. This meant that if a processor failed around 100 subscriber lines would be disconnected. The solution found was to duplicate these processors as well.

One central technological feature of the Ellemtel proposal was the development of a new central processor, the APZ 120. This caused a major crisis within LME and forced it to make a crucial decision. This was because LME had not, like STA, stopped their other electronic switching projects but continued with the development of their AKE switch. LME had made an inquiry concerning a local station switch, based on the AKE 13, that would use the APZ 150 processor developed for the AKE switch. During the autumn, LME's development department showed that it would be possible to construct an AKE switch less expensive than the one described in Ellemtel's AXE proposal. Among the major differences that came to be pitched against each other was the fact that the AKE proposal did not demand the development of any new, specially designed processor. This made for a difference in cost of about 100 M SEK, which was the price of developing Ellemtel's processor (Andersson, 1987: 7 - 8). If LME decided to go ahead with the proposed AXE project, it meant that they had to stop with further development of AKE switches. Even though the AXE project, with the new processor, seemed to promise larger gains in capacity, there were several arguments in favour of developing a switch based on the AKE concept. It was based on known and proven technology and would be ready before the AXE project, which was not expected to be finished until 1976. Meanwhile, LME's foreign competitors had released new models of computerised switches, and if LME opted for AXE it would mean 4 years without a competitive product to offer its customers. This crisis within LME continued until May, 1972. Two persons from LME and two from Ellemtel were given the task of comparing the two processors. Even though STA supported Ellemtel's AXE proposal, the decision had to be taken by LME, since it was the latter organisation that had an alternative to give up in APZ 150. The evaluation concluded that APZ 150 was going to be cheaper and faster, but would have less capacity potential. The choice LME had to make was between capacity or speed of development. At the time of the decision in May, 1972, LME decided to go ahead with Ellemtel's proposal. The rationale was that it was Ellemtel's task to develop the new switch, and it would be best if they did something that "they really believed in" instead of using a processor that they already had renounced. This meant that LME stopped the further development of AKE and bet all their money on the AXE card (Vedin, 1982: 144 - 46; Andersson, 1987: 8).

This meant that the procurement order for the development of the AXE system which had been presented May 1st now could be approved by Ellemtel's Board. At about the same time, a procurement order concerning a new way of building the switching racks modularly was also approved. Through this, the AXE technology had now got its major design determined and could go into practical development (Vedin, 1982: 200).

4.4 Critical Development Problems, 1972-76

After the AXE procurement had been approved, there followed an intense period of practical design of the system's various software and hardware. Some of the most important innovations were the construction practice of modular building and a new programming language adapted to telephony (Meurling and Jeans, 1995: 78). During early 1974, all new improvements were frozen as it became crucial to finish the system on time. Later improvements were collected and saved for adaptations in future orders. Thanks to the specially designed central processor, APZ 210, it became possible to introduce so-called high-level language for programming the station. One of Ellemtel's engineers had, while working at LME, developed a special telephony language that was now further developed into the special high-level telephony programming language, Plex, which made programming three times as fast as in the normal machine language. Around 20 - 30 % of the total development cost of the AXE project, amounting to about 700 M SEK, was saved through this (Andersson, 1987: 5). However, not all of the development related to Plex reduced time. In autumn, 1974, a method for faster checking of the programming codes was introduced; it had to be abolished in 1975, after a lot of misspent development work (Vedin, 1982: 204).

One of the problems for Ellemtel and LME during 1974 was to avoid getting involved in another project, AGAX, that STA proposed. The idea was to develop a computer system to control STA's first generation of electromechanical switches. The inquiries about AGAX were drawn out in time and LME had problems finding personnel for the joint committees. This procurement order was never completed and the project did not come into effect (Vedin, 1982: 203). But the AXE project did, and during 1974 the date of the inauguration of STA's first AXE station was discussed and changed several times.

4.5 Entering Operation: Delivery of the First AXE-Station, 1976-77

In 1973 it had been decided that the first practical installation of the AXE technology would be in April, 1976, in an STA switching station in Södertälje, a small community close to Stockholm. This was a rather small station, of only 3,000 subscriber lines and about the tenth of the size for a normal city station. Accordingly, this was a suitable first project for the new technology. The electronic and computerised AXE technology would replace the station's previous switches of the first electromechanical generation. The date stipulated by LME was earlier than Ellemtel's 1977 estimate. This was because it was supposed to be in time for the centenary birthday jubilee of the telephone and the LME company. But as the estimates of employee-years increased, the opening date was pushed ahead several times. In January, 1974, it became October, 1976, and in October, 1974, it became December, 1976. Finally, in October, 1976, the date was changed to March 1st, 1977. Despite this, it was decided to demonstrate a working station in 1976, and to start the trial operations in December, 1976. Issues regarding the process of handing over the AXE technology to LME were accentuated during 1976. Ellemtel during 1975 - 76 borrowed several LME engineers to work on the AXE technology and learn the new technology first-hand, through "hands-on" work. At the same time, parts of the switching equipment for the Södertälje station were manufactured by LME in their new electronics work-shop and through this manufacturing knowledge was transferred from Ellemtel to LME (Vedin, 1982: 206, 176).

The first large public presentation of the AXE technology was in the spring of 1976, in connection with LME's centenary jubilee. It was considered very important that LME should be seen as the manufacturer of the AXE technology. Thus, at presentations to foreign customers Ellemtel was not mentioned and their engineers were presented as coming from LME. Finally, the station went on trial operation during late 1976. All subscribers connected to the switch were employees of STA who lived in Södertälje. Through this, several smaller errors in the electronics set-up as well as in the programmes were found and could be corrected before the official inauguration. The actual installation came during the night of March 1st, 1977, when all of the station's 3,000 subscribers were switched from the old crossbar switches to the new AXE technology. The switch-over took place without any major problems. The first telephone call from the new station went to a LME manager who the following day was going to demonstrate the AXE station to engineers from the Australian Post Office, which was considering ordering the new technology. The call came through and he got the message that LME now could demonstrate an operating computerised switching technology to their international buyers. With this, the AXE development switched fully into a path that it had entered a couple of years before, but which had now become more accentuated: to export AXE.

5. POST-PROCUREMENT: ADAPTING TRAJECTORIES, 1973-81

The post-procurement phase overlapped with the end of the procurement phase. In this period, LME went out and tried to sell its new technology on the export market, and STA filed its second order for an AXE station. The first export order came in 1973, from Finland, and was followed in 1976 and 1977 with new orders from France and Saudi-Arabia. Since most of these new buyers were public agencies, these were also cases of public technology procurement. Many of the new orders were for AXE switching stations that were not fully developed. Others included adaptation to the major new technology trajectories of digitalisation and mobile telephony. In this way, these trajectories assisted, and were assisted by, the breakthroughs of the AXE technology.

5.1 Switching Expansion: Adaptation to Foreign Markets 1973-77

In 1973, LME started to market its new technology abroad. Its first strategy was to introduce gradually small AXE stations in markets with established customer relations, such as Denmark, Finland, Mexico, Brazil and Australia and, after this, expand to other regions. However, this only worked for the first export order, which came 1973 from a Finnish utility with which LME had a long customer relation. LME managed to make the utility change a previous order for an AKE switch to an AXE subscriber switch of 4,000 lines. This was also the first AXE adapted with the new digital technology developed by LME's Australian subsidiary, EPA, and Ellemtel. This first digital AXE switch was installed in Turku, Finland, in 1978 (Vedin, 1982: 187, 203 - 5; Meurling and Jeans, 1995: 111 - 12).

In 1974 the effects of the oil crisis were felt in the telecommunications sector and most utilities stopped buying switching stations. An exception was APO in Australia, which was to make a long-term choice regarding switches for local stations. LME wanted to offer AXE, while APO insisted that it could not consider an unproven technology. Nevertheless, LME and EPA managed to convince APO to let AXE be included in the invitation to tender. LME submitted its AXE proposal and in 1976 APO announced that the final choice fell between ITT's Metaconta and AXE (Meurling and Jeans, 1995: 113 - 14). Meanwhile, another AXE order had suddenly become very promising, when France decided to modernise its old network. In 1975, the Ministry of Telecommunications invited tenders for computerised local switching stations and AXE was chosen – but on the condition that it should be manufactured under license by the French firm, Thomson-CSF. To get a large export order under conditions of tough international competition was seen as very important. Therefore, LME accepted this condition and got the order in 1976. This also meant that LME had to sell its French subsidiary to Thomson. The irony was that this first large export order on AXE was not given to LME but to another company (Vedin, 1982: 222 - 23; Meurling and Jeans, 1995: 120 - 21).

Concerning the Swedish market, although the first AXE-station was working without problems it took more than three years until AXE began to be introduced on a larger scale by STA. This was pushed through in 1977, in connection with the Australian order. In 1977, LME because of the recession, had got a promise from the government to support AXE purchases. But when APO's decision was to be taken in 1977, no orders had yet been placed by STA. APO then told LME that if not even STA had bought any more AXE exchanges, how could APO dare buy any? LME's CEO got into contact with STA's DG, who got a quick approval for a purchase from the Ministry of Communications and the Ministry of Finance. One day before the decision was to be taken by APO, STA's DG called and told them that STA had ordered 10 new digital AXE stations and an extension of the Södertälje station – orders totalling 240,000 subscriber lines (Vedin, 1982: 223; Lernevall and Åkesson, 1997: 122; Lervik, 1977: 57). In 1977, APO announced that it would chose AXE as the standard for local switching stations. The Battle of Australia was won (Meurling and Jeans, 1995: 114).

The major international breakthrough came in 1977, when LME, in a joint bid with Philips and Bell Canada, got the order to modernise Saudi Arabia's telephone network with SPC exchanges for half a million telephone lines. This was world record and "the telecommunications affair of the century". The AXE delivered in this APTP incorporated digital switching technology and the contract was signed in 1978. It stipulated an operational network within a year. something which LME managed, thanks to the construction practice of modular building (Vedin, 1982: 224 - 27; Meurling and Jeans, 1995: 117 - 20). The breakthrough in Saudi Arabia gave AXE the lead in digital switching, but the first digital AXE switch had been inaugurated in May, 1978, in Turku, Finland. This was about the time when the digitalisation trajectory had its breakthrough among the international telephone utilities. LME was, together with the Canadian Northern Telecom, among very the first companies to offer full digital switching systems. Through AXE, LME could offer functioning digital switching systems in 1978, while most of its competitors were still busy developing theirs. This happened in 1980, when two Swedish AXE stations were inaugurated that included digital signal systems. STA's further installation of digital AXE switches led to the fact that Sweden in the late 1980s became the first country in the world to offer national digital coverage.

5.2 Switching to a New Trajectory: Introducing Mobile Telephony, 1977-81

STA, together with LME and its subsidiary, SRA, had been pursuing research and development in mobile telephony since the end of the 1940s. The mobile telephony trajectory went into new phase in 1969, when STA informed the other Nordic PTAs that it was intending to work on the next generation of mobile telephony. This led to a plan for a common Nordic Mobile Telephony network (NMT). Several technical NMT standards were decided in co-operation between LME and STA. In 1977, a test system was installed in the Stockholm region and, until 1978, LME and STA, in co-operation, conducted extensive tests on the prototypes of this switch (Ny Teknik, 1994: 27; IVA, 1992: 84). STA, around 1977, convinced LME that they should construct a modified version of their AXE-switch for mobile telephony. STA wanted a modern, computerised switch for their new mobile telephony system, but LME considered that a computerised electromechanical switch would suffice. But since STA insisted on an AXE switch and threatened to buy from a Japanese firm, LME changed its mind and offered a modified AXE-switch (Meurling and Jeans, 1995: 198). LME got the NMT order which included switches to Sweden, Norway, Finland and Denmark. In 1978, STA signed the contract for the new switches and the NMT-network was opened in 1981.

6. RESULTS: SWITCHING ADVANTAGES

The AXE procurement has been very profitable for both STA and LME. STA has benefited from the AXE-system when it comes to the quality of its network. According to its own evaluation, STA with the help of AXE had built "one of the most modern telephone networks in Europe and could show that the telephony fees still were among the lowest in the world" (Lernevall and Åkesson, 1997: 80). If STA had not bought a Swedish switch, it would probably not have been able to handle a large part of the manufacture itself. Furthermore, it most likely would have been more difficult for STA to have one single system for the whole country if it was going to buy new switches gradually from abroad. Thanks to the extensive build out of AXE stations Sweden became the first country in the world to offer national digital coverage (Meurling and Jeans, 1995: 140). Last but not least the Swedish end consumers got a modern telephone system with a large array of new telephone services.

To LME, the AXE procurement meant three major results. First, LME managed to switch from the old electromechanical trajectory to the new, higher value-added electronics trajectory. Ellemtel's development cost for the AXE project was around 500 M SEK, with LME's part – around 75 % – coming from profits from sales of the older electromechanical switches (Eliasson and Granstrand, 1981: 243). LME had in AXE one of the market's most advanced switches and doubled its share of the world switching market. AXE has been the most widely sold automatic switching system in the world. In 1990, LME had around 13 % of the market and in 1992 had sold AXE to 87 countries. Thereby, it was the fourth largest manufacturer after Northern Telecom (15 %), Alcatel (15 %), and AT&T (14 %). Moreover, in the late 1980s LME managed to enter the large US market thanks to the AXE system. And the last and maybe the most important result was that the flexible adaptability to mobile telephony of the AXE switch laid a foundation for LME's early entry into mobile telephony in the 1980s. In 1992, LME had 40 % of the world market, with orders for almost 400 AXE switching stations for mobile telephony networks. According to LME's former President, Björn Svedberg, it would have been a "catastrophe" for Ericsson to be without AXE. LME would most likely not have been able to remain an independent telecommunications company (Vedin, 1982: 249, 266, 251, 16).

REFERENCES

Andersson, G. (1987). Egna lösningar bäddade för AXE-framgång. *Tele*, 93 (2), 3 - 8.

Angerby, A. (1965). Manufacturing activities of the Swedish Telecommunications Administration during the past 25 years. In *Tele: Information from the Swedish Telecommunications Administration : Retrospective Survey of the Last Twenty-Five Years*. Stockholm, 41 - 48.

Attman, A. and Olsson, U. (1976). *Räddning återuppbyggnad världsföretag 1932 - 1976*, vol.2, *LM Ericsson 100 år*. Örebro.

Bjurel, B. (1976). 100 år med Telefonaktiebolaget LM Ericsson som leverantör och samarbetspartner. *Tele*, 82, 1 - 9.

Bjurel, B. (1997). Elektronikens barn AXE-systemet och mobiltelefonen föds: Televerket får ny organisation. In Lernevall and Åkesson.

Broberg, W. (1977). En historisk anläggning tas ner: länkkopplat system i Sundsvall. *Tele*, 83, 51 - 54 & 78.

Eliasson, G. and Granstrand, O. (1981). The financing of technological investments. In O. Granstrand and J. Sigurdson (eds.), *Technological and Industrial Policy in China and Europe: Proceedings from the First Joint TIPCE Conference 1981* Lund: Research Policy Institute.

Forss, B. (1970). Elektronisk kopplingsteknik just nu. *Tele*, 76, 11 - 16.

Fredriksson, G. (1978). Så fungerar ELLEMTEL. *Tele*, 84 (4), 1 - 4.

Fredriksson, G. (1990a). AXE – en lång historia. *Tele*, 96 (4), 3 - 5.

Fredriksson, G. (1990b). Pålitlig övervakning i 20 års tid. *Tele*, 96 (4), 56 - 57.

Fredriksson, G. (1991). Etikettsbrott bäddade för RACE-medverkan. *Tele*, 97 (2), 15 - 20.

Fridlund, M. (1993). The 'development pair' as a link between systems growth and industrial innovation: co-operation between the Swedish State Power Board and the ASEA company. Working papers from the Department of History of Science and Technology 93/9. Stockholm: Royal Institute of Technology.

Granstrand, O. and Sigurdson, J. (1985). The role of public procurement in technological innovation and industrial development in the telecommunication sector: the case of Sweden. In O. Granstrand and J. Sigurdson (eds.), *Technological Innovation and Industrial Development in Telecommunications: The Role of Public Buying in the Telecommunication Sector in the Nordic Countries*. Gothenburg.

Hagström, T. (1997). En brytningstid. In Lernevall and Åkesson.

Heimbürger, H. and Tahvanainen, K. V. (1989). *Telefon, telegraf och radio 1946 - 1965*, vol. 6 of *Svenska telegrafverket: Historisk framställning* Stockholm.

Ioannidis, D. (1998). *I nationens tjänst? Strategisk handling i politisk miljö : En nationell teleoperatörs interorganisatoriska strategiska utveckling*. Stockholm: Stockholm School of Economics.

IVA (1992). *Framgångsrik utveckling av energiteknik*. IVA-M 278. Stockholm: IVA.

Jacobæus, C. et al. (1976). *Teletekniskt skapande 1876 - 1976*, vol.3, *LM Ericsson 100 år*. Örebro: LM Ericsson.

[Karlsson, L.] (1963). *Automatik för millioner: En presentation av televerkets verkstadsrörelse*. Stockholm.

Karlsson, M. (1998). *The Liberalisation of Telecommunications in Sweden: Technology and Regime Change from the 1960s to 1993*. Linköping: Dept. of Technology and Social Change, Linköping University.

Lernevall, S. and Åkesson, B. (1997). *Från myndighet till bolag 1966 - 1993*, vol. 7 of *Svenska telegrafverket: Historisk framställning*. Stockholm.

Lervik, K. (1977). Satsning på Teli. *Tele*, 83, 55 - 65.

Lundvall, B. (1977). LME's samarbete med televerket. *Tele*, 83, 129 - 31.

Meurling, J. and Jeans, R. (1995). *A Switch in Time: AXE – Creating a Foundation for the Information Age*. London: Communications Week International.

Mölleryd, B. G. (1996). *Så byggdes en världsindustri: Entreprenörskapets betydelse för svensk mobiltelefoni.* Stockholm: Stockholm School of Economics.

Neumeyer, F. and Rönnblom, N. (1965). Uppfinningar och patent hos televerket. *Tele*, 71, Special issue, 266 - 69.

Norinder, B. (1987). Nu satsar AXE:s pappa på framtidens telesystem. *Tele*, 93 (2), 32 - 37.

Ny Teknik (1994). De baxade Ericsson mot toppen. *Ny Teknik*, (45), 27.

Patricks, M., and Jacobæus, C. (1972). LME's samarbete med Televerket och medverkan under automatiseringen. *Tele*, 78, 41 - 46.

Sterky, H. (1968). Ett högfrekvent halvsekel. In *En del om el: Minnesskrift utgiven till Sveriges Elektroindustriförenings 50–årsjubileum 1968.* Stockholm.

Sunesson, B. (1970). Televerkets utvecklings- och produktionssamarbete med LME. *Tele*, 76 145 - 51.

Vedin, B.-A. (1982). *Teknisk revolt: Det svenska AXE-systemets brokiga framgångshistoria.* Stockholm: Atlantis.

Åkermark, B. E. (1996). Drivkraften bakom nallen. *Dagens Nyheter*, 14 October 1996, part B, 5.

6. INDUSTRIAL TRANSFORMATION THROUGH PUBLIC TECHNOLOGY PROCUREMENT? THE CASE OF NOKIA AND THE FINNISH TELECOMMUNICATIONS INDUSTRY

C. Palmberg

1. TECHNOLOGY POLICY AND PROCUREMENT IN FINLAND

In Finland it seems fair to say that the practice of technology policy has been dominated by a science and technology push-oriented approach whereby the primary mode of support to industry has been R&D subsidies and loans granted by such agencies as the Ministry of Trade and Industry, the Technology Development Centre of Finland (Tekes), and the Finnish National Fund for Research and Development (Sitra). Following a period of a quantitative increase of funding in 1970s it was proposed, in the early 1980s, that major projects should be initiated in key areas of technological development. This proposal led to the introduction of a technology programme procedure, commissioned by Tekes. In the 1990s the emphasis has been on regional initiatives, such as centres of excellence and technology parks, as well as Finnish participation in the EU's research policy. (Lemola, 1994). Despite increasing diversity in approaches to the practice of technology policy, however, demand-oriented policies, such as public technology procurement, have received less attention.

In the early 1980s Sitra initiated a fact-finding study of the possibilities for initiating public technology procurement schemes or policies (see Mäkinen, 1984). In retrospect, though, it is clear that the proposals of the fact-finding study have not been followed – there is no Finnish tradition of public technology procurement as defined in this book. Perhaps this is related to the lack of a significant defence sector, as well as a sizeable home market, whereby public procuring agencies have not been able to offset risks associated with major infrastructural R&D projects.

The main policy rationale of public procurement, including public procurement of new technologies, has been to increase the domestic content of procurement orders as substitutes to imports, and thus contribute to job creation and the fostering of infant industries. An interesting feature of procurement has been the establishment of a set informal institutions and monitoring committees with the purpose of assuring that bigger orders are given to domestic producers. Such arrangements have been based on regulations which stipulated that *"...decisions concerning public procurement should be guided by national interests, and procurement orders should be placed domestically..."* (Valtioneuvoston päätös 136/1930). On the other hand, there has also been an official tendency to adhere strictly to various international free trade and competition agreements, such as the GATT and EEC agreements. These contradictions between infant-industry arguments and high standards of competition have prevailed up until closer integration with the EEC/EU in the late 1980s and early 1990s, when the remaining norms and regulations favouring infant industry arguments were dismantled. The only innovation-enhancing clauses which are identifiable are paragraphs concerning preferential procurement in the case of especially technically sophisticated products or systems which demand close negotiations between users and producers. These have been derived from similar clauses in, e.g., the GATT agreement.

The lack of a Finnish tradition does not mean that public technology procurement has not occurred. Rather, it has occurred in a decentralised, uncoordinated, fashion, mainly in the transport and telecommunications sectors wherein public agencies typically have been providers and guardians of services regarded as public. In this chapter, I describe public technology procurement practices in a major Finnish high-technology industry – the telecommunications industry. The role of technology procurement is exemplified here through case studies of the development of Nokia's platform technology in the cellular switching business, the DX 200 switching system, and network equipment for the NMT. While these cases are interesting in their own right, they have also been of crucial importance for Nokia's diversification into mobile telecommunications. First, however, I will highlight certain characteristics of the telecommunications sector in Finland – the framework conditions for technology procurement.

2. THE FINNISH TELECOMMUNICATIONS SECTOR

2.1 The Regulatory Structure of Telecommunications Operators as the Procurers and Users of Telecommunications Technologies

As opposed to the public monopoly situation that has prevailed in most other European countries, Finland's telecommunications operators are organised on a decentralised basis. The Imperial Telephone Decree, which was enacted as late as in 1886, at the time when Finland was a Russian Grand Duchy, authorised the Finnish senate to grant licences for private sector local telephone operators (telcos). The first Helsinki telco was established in 1882.

In 1917, when Finland became independent, the state entered telecommunications operations through the founding of Posts and Telecommunications of Finland (the PTT, since 1998 Sonera), in order to build trunk networks and local networks in certain scarcely populated rural areas. Meanwhile, the number of telcos started to grow rapidly. In 1921 the Association for Private Telephone Companies (the Finnet group) was founded in order to co-ordinate the activities of the telcos in certain administrative matters, and in matters related to procurement and modernising the networks. (Mäenpää & Luukkainen, 1994; Telecommunications Statistics 1989/96). Today the PTT is the biggest single operator. Of the remaining telcos, the Helsinki telco, the Tampere telco and the Turku telco are operating in the biggest cities. The remaining telcos are smaller and typically operate small local networks with a limited number of subscribers. In terms of the market structure of the users, this decentralised set-up has fragmented the market for telecommunications equipment for the fixed networks, while the PTT has exercised a monopsonist's market power in the field mobile telecommunications up until the early 1990s, when Finnish telephony was further liberalised. (Ministry of Transport and Communication 1996; Telecommunications statistics 1989/96).

Subordinated to the Council of State, the main regulatory bodies are the Ministry of Transport and Communications and the Telecommunications Administrations Centre, while the Ministry of Finance and the Ministry of Trade and Industry (since 1983 Tekes) have affected developments through more general industrial- and technology policy measures (e.g., subsidies, loans, technology programmes).

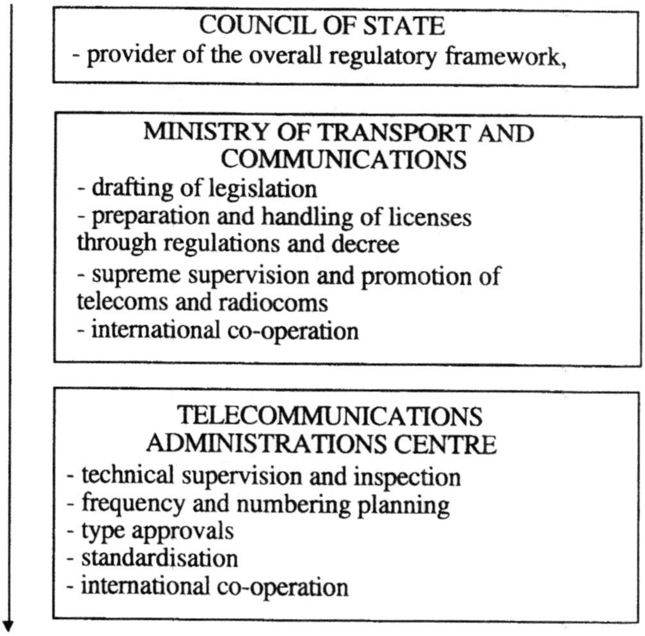

Figure 1. The structure of the telecommunications regulatory system. (Telecommunications statistics 1996: 8).

Figure 1 illustrates the structure of the telecommunications regulatory system. The tendency in recent years has been to decentralise and liberalise telecommunications further by means of regulations and decrees emanating from the Council of State. Through the *Telecommunications Act* in 1987, the Ministry of Transport and Communications received regulatory powers over technical supervision and inspection of telecommunications services, allocation of radio-frequency and number planning, type approval and national standardisation. In 1988, the Telecommunication Administration Centre was established for these purposes. Thus, until 1987, the PTT had had the double role of a telecommunications regulator and operator and had handled the above mentioned tasks. In 1990, the PTT was transformed from a state-owned monopoly into a state-controlled commercial enterprise, bound by the same rules as the telcos. In 1994, it was converted into the limited company, and, in 1997, the Parliament authorised the Council of State to sell 49% of their share of the company. Hence, since 1998, when state-ownership decreased significantly, the PTT has been in fact largely privatised.[1] Meanwhile, competition on the networks has been encouraged. Licenses have been granted for long-distance and international calls, in effect ending the monopoly of the PTT in these markets. Within the field of mobile telecommunications, the monopoly position of the PTT prevailed up until 1990, when the telcos received a license to operate the GSM network through the founding the joint-venture, Radiolinja Ltd. The PTT's monopoly over mobile networks until 1990 has been a logical extension of the fact that the PTT has controlled the radio-frequencies, along with the state-owned broadcasting association, YLE (Telecommunications statistics 1989/96).

2.2 The Evolution, Structure and Competencies of the Finnish Telecommunications Equipment Industry – The Evolution of Nokia[2]

The origins of Finland's telecommunications industry can be traced back to the early 1960s, despite the fact that some roots extend even further. Prior to the 1960s, an oligopoly of foreign companies, consisting of CIT-Alcatel, Ericsson and Siemens, shared the market – even though state-owned R&D units had established small-scale production in closely related technology fields. (Mäenpää & Luukkainen, 1994; Koivusalo, 1995). While the foreign companies, and especially Ericsson and Siemens, have had a strong position on the Finnish market throughout, it is interesting to note the gradual emergence and market take-overs of the domestic telecommunication equipment firms. These trends are very much interrelated with the growth of Nokia and an associated network of sub-contractors and component producers. Other smaller Finnish telecommunications equipment firms have typically entered the market much later, in the 1980s or 1990s.

---

[1] For the sake of clarity I refer to the abbreviation PTT throughout this paper even though the name changed to Sonera in 1998, and the share of state-ownership decreased significantly.

[2] The evolution of Nokia's telecommunications divisions is illustrated in the appendix.

Initially, production stemmed from the defence sector. During early developments of the industry, the defence sector had established an R&D unit to cater to defence sector demands for electronics and telecommunications equipment (especially switching systems) as well as maintenance and repair. In 1948, this became part of the PTT's activities. In 1962, the name Televa was adopted. In the private sector, telecommunications equipment was produced at Suomen Kaapelitehdas (established in 1912) and Salora (established in 1928). Suomen Kaapelitehdas produced various cable transmission systems for fixed telephone networks, but had also established a small electronics department. Later on, in the 1960s and early 1970s, the electronics department diversified into computer and data processing services. Salora (established in 1928), on the other hand, was preoccupied with early consumer electronics (television sets and transistor radios) and radiotelephones. (Koivusalo, 1995).

Valtion Sähköpaja, Suomen Kaapelitehdas and Salora were the main domestic telecommunications equipment suppliers at the time. From the early 1960s onwards, the telecommunications industry evolved in response to a combination of factors: successful strategic choices by firms, tough competition, extensive visionary R&D, public support and technology push on the part of the main telecommunications equipment producers, as well as demand pull, often involving close interaction with various private and public sector organisations, including the PTT.

One milestone was the merger in 1966 of Suomen Kaaplitehdas, Suomen Kumitehdas (rubber works) and Nokia (forestry and energy) into the Nokia Group, or Nokia for short. Another milestone was the establishment of Telefenno in 1977. Telefenno was a 50-50 joint-venture between Televa and Nokia in the field of switching systems. The establishment of Telefenno, and Nokia's seizure of Televa, coincided with a failed socialisation plan which envisioned that the Finnish electronics industry should be reorganised as a state-owned conglomerate. In 1981, Telefenno was absorbed completely by Nokia, through the establishment of the telecommunications division Telenokia, and now became completely de-coupled from the state. Meanwhile the remains of Televa were sold to other domestic electronics firms. (Lovio, 1993).

If the Finnish telecommunications equipment industry of the 1960s and the 1970s was characterised by small-scale pioneer production and the involvement of state-owned R&D units and firms, the early 1980s marked the start of the growth of the Nokia, in terms of both market shares and accumulation of competencies and other resources. During the same period, Nokia diversified more forcefully into electronics, and especially into telecommunications, as it divested itself from maturing forestry, cable- and rubber-works industries. This diversification process is illustrated in Figure 2 below.

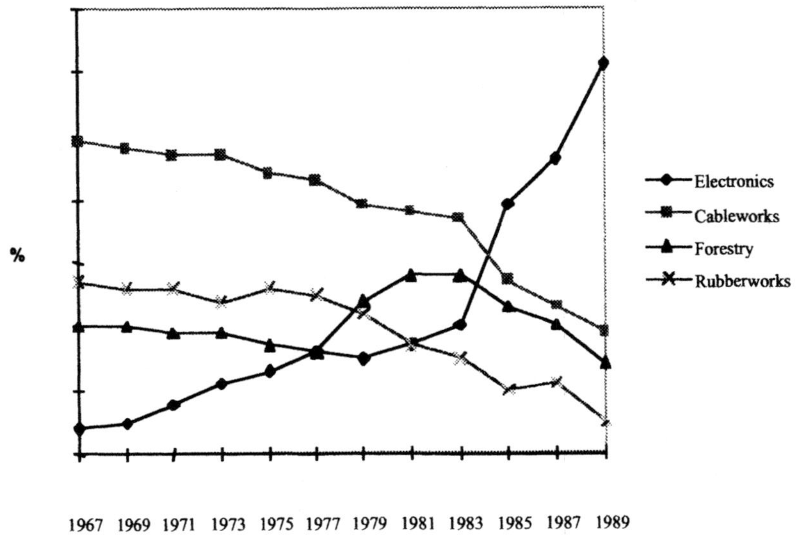

Figure 2. Distribution of net sales 1967 - 1989. (Lemola & Lovio, 1996: 207).

By 1980, Nokia had become the largest domestic electronics and telecommunications firm, while the foreign firms had lost market shares. Within the field of switching systems, resources at Telenokia were shifted towards developing early versions of the DX 200 digital switching system. Meanwhile, in 1979, a decision had been made to formalise collaboration between Nokia and Salora by the founding of a development company, Mobira, to meet more satisfactorily the anticipated future demand for mobile telephony. This was a third important milestone in the development of the industry. During the 1980s Nokia also acquired key component technologies and know-how, through firm acquisition and outsourcing. At the same time, the Nordic PTTs were engaged in Nordic Mobile Telephone (NMT) standardisation collaboration, and networks for the first generation of mobile telephones and network equipment were under construction. (Lovio, 1993; Koivusalo, 1995).

The NMT standard was crucially important for Nokia's diversification from fixed networks equipment into mobile telecommunications. It also paved the way for Nokia's aggressive internationalisation strategy, from the mid 1980s onwards. The final push for Nokia's breakthrough was the GSM standard during the early 1990s. In 1988, the Nokia Cellular Systems unit was formed and Telenokia was absorbed into a larger business unit named the Nokia Telecommunications division. Mobira was likewise absorbed through the founding of the Nokia Mobile Telephone division. By the early 1990s, Nokia had become an important player in the mobile telecommunications industry, and R&D funds have subsequently increased significantly. Nokia Telecommunications accounts for a lion's share (40%) of the funds. (Lemola, 1996; Koivusalo, 1995).

In comparison with other major telecommunications equipment suppliers – for example Ericsson – the scale and scope of Nokia's R&D activities seem modest. In the

Finnish telecommunications industry, Nokia has a very dominant position, accounting for some 90% of exports in 1996. The remaining firms are medium-sized, or small. Many of these firms have been established as Nokia spin-offs or sub-contractors. Nokia also plays a central role in the Finnish system of innovation, accounting for some 30 - 40% of total Finnish high-technology exports, and some 50% of total R&D performed in industry (excluding R&D undertaken abroad), according to figures in 1996 (Ylä-Anttila, 1996).

Next, I turn to the case studies of the development of the DX 200 switching system and network equipment for the NMT, examined from the perspective of public technology procurement.

3. THE DX 200 SWITCHING SYSTEM – NOKIA'S PLATFORM TECHNOLOGY

3.1 Entering Digital Technology – Converging Interests and Needs

The history of digital switching in Finland commenced in the early 1970s. At the time, all the major firms active in telecommunications – Salora, Nokia and Televa – followed different paths of entry to the digital paradigm in telecommunications.

The sole domestic supplier and carrier of switching exchange know-how was state-owned Televa, and Televa was the only domestic competitor alongside Ericsson, CIT-Alcatel and Siemens when the fixed networks were automated. Televa's first electromechanical switches were based on relay technology; they were small (25 - 50 subscriber lines) and especially suited for less-densely populated rural areas. By the mid 1960s, Televa had delivered some 200 switches to networks maintained by the PTT as the fixed networks were automated. Next, Televa developed the so-called crossbar switch, which had the capacity to handle a significantly larger number of subscriber lines. (Olkkola, 1986; Mäkinen, 1995).

During the late 1960s, Televa started to develop the first generation of electronic switches. Meanwhile the technological frontier had moved further, through the introduction of computer controlled switches, and through the advent of digital technology. In 1971 at Televa, the first official initiative towards developing a fully integrated, computerised, digital switching system was taken. This system was named the ADS system. An R&D collaboration scheme was initiated together with a telephone laboratory at the Technical University of Helsinki. Alongside this R&D project, Televa had also initiated a second R&D project for the development of the next generation computerised switching system, named the KKY, which was based on analogue technology. This switch was designed for catering to the PTT's demand as the final stage of the automation of the networks was under way. However, even though the ADS system applied digital technology, whereas the KKY system was based on analogue technology, the ADS and the KKY projects were very much competitors for funds. Initially, the KKY project was favoured, while the ADS project was merely regarded as an exotic experiment. (Olkkola, 1986; Interviews, 1996).

At the time, the general understanding was that computer controlled switching systems would provide significant cost-efficiency advantages over the previous generation of electromechanical and electronic switches. The more disputable questions were related to timing and the specific technological design of a digital system. In most quarters it was believed that digital switching systems, although promising, would not penetrate markets for decades to come. The PTTs had invested heavily in electromechanical and electronic switches analogue switches and hence analogue computer controlled systems, such as the KKY, could be used to upgrade older switches rather than to replace them completely (Olkkola, 1986; Mäkinen, 1995; Interviews, 1996).

In the early 1970s, the telcos had already automated close to 100% of the networks while the PTT still maintained a few manually operated networks. As the PTT was drafting the last set of specifications and procurement orders for the remaining local rural networks there was increasing awareness that computer controlled switches would indeed be more cost efficient. In part, this conclusion was derived from in-house R&D and more general trends in the field, but it was also realised that the KKY and ADS projects at Televa indicated that domestic production could be viable. So far, foreign companies, especially Ericsson and Siemens, had been the main suppliers and the PTT often had to pay excessive prices for larger switches which were not optimally designed for the typical sparsely populated rural areas in Finland. If Televa could present a small SPC system which would be especially suited for the typical rural network conditions, the PTT had every reason to support development work. Against this background, the PTT placed a first prototype order for the ADS system in 1972. This first order also marked the start of more focused user-producer collaboration in digital switching technology (Tele, 1993; Interviews, 1996).

3.2 The ADS-Project as a Collaborative R&D Venture

The PTT's prototype order was for a small switching system with a limited number of subscribers in the south-western part of the Finnish archipelago, covering a couple of islands. The procurement order was based on the procurement regulations covering preferential procurement at the time, and the formal contract was formulated as a collaborative R&D venture between Televa and the PTT. Noteworthy is the fact that the system was to be based on a combination of an electromechanical switch and the KKY system, but included an option whereby Televa had the right to construct parts of the network using digital ADS technology. Hence, the PTT also came to favour the KKY project while the ADS system was still regarded with suspicion. (Mäkinen, 1995; Interviews, 1996).

This first prototype order was crucial for the continued support for the ADS project within Televa, since the ADS and KKY projects still competed for funds; the management could not decide on the priority of the projects. At the same time, the PTT advanced very cautiously on the digital front, monitoring Televa's and other foreign producers' technological advances. Essentially the first prototype order was a field-trial order, and the site was soon transformed into a pilot laboratory where various competing technical solutions and technologies of the different projects were tested and developed. Apparently, development work was organised on the basis of a division of labour between the PTT and Televa, whereby Televa focused on the purely technical side of the hybrid system while the main role of the PTT was to conduct extensive quality and reliability tests. The PTT also played an important role in adapting specific technical solutions to emerging international CCITT standards – a necessity for future exports of the systems under development – as well as choosing reliable components for the system. (Olkkola, 1986; Interviews, 1996).

Meanwhile, it was already becoming evident that the next generation switching systems would indeed be digital systems and that transmission and switching technologies would become integrated along the lines prescribed by the IDN and ISDN standards. In the south-western Finnish archipelago, the PTT also started to shift their interest from the KKY projects in favour of the digital ADS system, and Televa came under increasing pressure to finalise the ADS prototype. At the same time, CIT-Alcatel, Siemens and Ericsson, were introducing digital prototype switching systems to the market. The next significant milestone for the ADS project was the founding in 1977 of Telefenno, the 50-50 joint venture between Televa and Nokia. (Mäkinen, 1995; Interviews, 1996).

3.3 Reorganisation and Market Deployment – Telefenno and the DX 200 Project

By the mid 1970s, Nokia's electronics department had already accumulated competencies in the field of digital so called PCM transmission and signalling technology. The electronics department had delivered the first ever PCM systems for networks operated by the Helsinki telco, finalised with 30 channels according the prevailing CCITT standard. Next, there were important breakthroughs on the Soviet market. In addition Nokia had decided to gain quick access to additional digital switching technology through a licensing agreement with CIT-Alcatel. The licensing agreement covered the E 10 switching system. The E 10 was a large switching system especially suited for densely populated areas, and Nokia envisioned that the E 10 could complement both their own projects in digital technology, as well as the KKY and ADS projects under development in the archipelago pilot laboratory. However, as the ADS project gained followers while at the same time development costs were rising, there was also increasing pressure within Nokia to consolidate resources and competencies further.

Interestingly enough, there now existed two major and complementary digital switching R&D projects in Finland: the ADS and E10 projects. There was also the KKY project. These developments coincided with the above mentioned socialisation initiative, which came to play an important role for the subsequent history of digital switching in Finland. (Mäkinen, 1995; Interviews, 1996).

State-owned firms have traditionally played a visible role in the process of industrial renewal in Finland, especially in the metal and forestry industries. The plans, during the mid 1970s, to reorganise the Finnish electronics industry into a large state-owned conglomerate, was part of a broader motivation for industrial policy, wherein the electronics industry was seen as the major growth industry in the future. An additional argument was the need to diminish the negative trade balance in key electronic components. On the basis of these plans the position of Televa was strengthened, and in 1976 Televa was transformed into a state-owned stock firm. The same year, the state-owned firm, Valco was founded for the production of vacuum-tube components for application in the consumer electronics industry (Lovio, 1989; Turpeinen, 1996). The next sequence of events was crucial and had direct repercussions for the ADS collaborative procurement and R&D venture between Televa and the PTT.

In the late 1970s, both Televa and Valco ran into financial difficulties. In part, the reasons were poor planning and commitment on the part of the government, but the socialisation plans also met opposition from some political parties as well as from Nokia – the biggest single electronics and telecommunications firm in Finland at the time. During 1976, preparations were made to embark on a course alternative to that which had originally been planned; a course which to a large extent was designed by Nokia. On the basis of extensive negotiations involving both Televa and Nokia, as well as the Ministry of Transport and Communications and the Ministry of Trade and Commerce, the decision was taken to alleviate the Televa 'problem' through the establishment of Telefenno, the joint venture between state-owned Televa and Nokia. Hence, starting from 1977, Nokia had effectively gained access to all the most important technologies and know-how in Finland that were needed for the construction of digital switching systems. Nokia had also gained access to the ADS prototype project in the south-western part of the Finnish archipelago. The main task of Telefenno was to rationalise production in the field and to develop and market a digital switching system, while the remaining departments at Televa (radiotelephones, base stations and miscellaneous electronic equipment) were sold to other firms. (Koivusalo, 1995; Interviews, 1996).

The joint-venture Telefenno now possessed three competing R&D switching projects: Nokia's license-based E 10, and Televa's analogue KKY and digital ADS systems. Even though there were overlaps and cross-fertilisation between the different systems, they were still essentially competing projects. The E 10 system was marketed as the DX 100, especially suited for densely populated urban areas, while the ADS prototype, named the DX 200, was marketed as a system especially designed for sparsely populated rural areas.

The first prototype procurement order by the PTT was crucial for the continuation of the ADS project, but it is difficult to conclude whether procurement by the PTT or the telcos was more important for future developments of digital switching in Finland. Collaboration between the PTT and Telefenno continued, and in 1980 the first switching system based solely on Televa's original ADS prototype came into operation in the south-western archipelago area. The original contract was reformulated, whereby the DX 200 system was also extended to cover the parts of the network covered by the competing KKY system. On the basis of extensive field-trials, the PTT was now ready to embrace fully the DX 200 system. Meanwhile, the PTT had realised that the DX 200 would out-compete the DX 100 system, and in the end Telefenno only delivered 5 DX 100 systems to trunk networks maintained by the PTT. These procurement decisions contributed to a chain of events which gradually led to the market selection of the DX 200 systems. (Mäkinen, 1995; Interviews, 1996).

3.4 Market Selection and the Founding of Nokia Telecommunications

Amongst the local private operators the Tampere and Helsinki telcos were the most important buyers and users. Despite the fact that the telcos lacked the necessary centralised resources and competencies that characterised the PTT, close personal ties and a business oriented mind-set paved the way for collaboration. Like the PTT, the Tampere and Helsinki telcos also ended up favouring the DX 200 exchange as opposed to the DX 100 system. Apparently, the telcos were also strongly influenced by the successful cross-over of the ADS project and prototype procurement by the PTT, and they now followed suit. Collaboration between Telefenno and the above mentioned telcos resulted in the next-generation DX 200 system, which was a hybrid system combining Nokia's PCM technology and ADS technology. This next-generation version of the DX 200 included the first fully digitised subscriber switch and Intel's 8086 processor. In 1982, the first of its kind was taken into operation in networks maintained by the PTT. After extensive user-producer collaboration, involving the PTT and the Tampere and Helsinki telcos as demanding users, Telefenno now possessed a system which had not only outcompeted the domestic KKY and DX 100 system is, but also outcompeted the main foreign systems.

In hindsight, it appears that the main reason for the success of the DX 200 was the flexibility of the system owing to the decentralised architecture of the control processor. It had been proven possible to upgrade it to support increasing numbers of subscribers due to rapid technological advances in microprocessor technology (the processing speed of Intel's microprocessors was continuously upgraded during the 1970s and 1980s). Despite the fact that the DX 200 was originally designed for catering to demand for small switching systems in rural areas operated by the telcos, it could also be adapted to support densely populated areas or trunk switching functions. Hence, even though the telcos procured more systems, the volume of the procurement orders of the PTT was bigger and the orders tended to be technologically more demanding, as the systems had to be deployed in the trunk network. A second major reason was the fact that Nokia had developed a whole range of different signalling modes in order to adjust individual DX 200 systems to the various technologically different local networks in Finland. This was a major advantage when Nokia started to market the system abroad. In 1981, when Telenokia was founded and Nokia had purchased Televa's remaining share of the switching business, the DX 200 project was their main R&D project. At the same time, the DX 100 project was scrapped and had not led to any significant technological spill-overs. (Olkkola, 1986; Koivusalo, 1996; Interviews, 1996).

Certain elements of national champion policies and infant-industry arguments might explain the first prototype order and subsequent user-producer interaction during the 1970s, since Televa was state-owned and subordinated to the PTT. However, it is important to observe that the DX 200 was developed in a competitive environment. During the 1980s, Televa was gradually de-coupled from state ownership, through the establishment of Telefenno, Telenokia and later Nokia Telecommunications. At the same time, the PTT and the telcos also procured digital switching systems from foreign producers. Table 1 illustrates the distribution of orders over the networks in 1980.

Table 1. Suppliers for the telephone networks 1980 (subscriber connections). (Ekberg, 1985: 87).

	Networks upheld by the PTT	Networks upheld by the telcos
Televa / Telefenno	34%	2%
Siemens	28%	30%
ITT-Alcatel	19%	4%
LM Ericsson	18%	60%
Others	1%	4%

By 1980 Telefenno had the biggest slice of all telecommunications equipment for networks maintained by the PTT, followed by Siemens. Siemens was also a dominant supplier for private local networks. These figures clearly illustrate the presence of foreign competition. The table also illustrates the degree of technological variety, and competition, which characterised the telecommunications networks in Finland. (Ekberg 1985).

Following the founding of Telenokia in 1981, Nokia's position on the domestic market was further consolidated, as Telenokia received bigger orders from both the PTT and the telcos. In 1984, there were some 50 DX 200 systems in operation in Finland. The same year, the first systems were exported to the Soviet Union. Starting in 1985, export volumes increased significantly and at the end of the 1980s Scandinavia and the Far East had become the most significant markets. In the meantime, the DX 200 was upgraded and finalised to support the ISDN function and Intel 386 processor. During 1986, 40 additional fully digitised systems were delivered. The PTT has also played a significant role in transmitting information about international standards in general, as well as in specific procurement cases, up until the late 1980s and early 1990s. During this period, the telecommunications equipment producers increasingly participated in standardisation activities. In hindsight, it seems fair to say that the history of the DX 200 system has been not only a success in its own right, but also an absolutely essential innovation for the further growth and internationalisation of the Nokia group during the 1990s. (Koivusalo, 1995; Lemola, 1996; Interviews, 1996). And while the DX 200 system has been a platform technology for Nokia's move into the mobile telecommunications industry, it has probably also been on of the largest R&D projects in contemporary Finnish history.

4. STANDARDISATION AND PROCUREMENT – DEVELOPING NETWORK EQUIPMENT FOR THE NMT

4.1 Identifying the Need and Main Principles of a Nordic Mobile Standard

The Nordic NMT standard provided the next big technological opportunity for those telecommunications firms which had the competence needed. The official kick-off for the

NMT venture was the NordTel conference of the Nordic PTTs in Kabelvg, Norway, in 1969. At this meeting the Swedish PTT presented a report on the future development of mobile telephony, based on the experience gained with the local MTA and MTB mobile network systems in Sweden. It also included a survey of all existing mobile systems around the world. On the basis of the report it was proposed that mobile telephony could be an additional area for fruitful Nordic collaboration. (Toivola, 1992; Mölleryd, 1996).

Apart from the strong tradition of Nordic collaboration, there appear to have been three or four primary factors explaining the common goal to which the Nordic PTTs committed themselves. During the late 1960s the Nordic PTTs were operating local or regional manual mobile networks, such as the MTA and MTB systems, which were incompatible with each other. In Finland, the PTT initiated plans to extend the ARP network – the first nation-wide network – to cover the whole of the country as part of the plans which had been drawn up by a working group subordinated to the Military General Staff. On the basis of several market analyses in the individual countries it was anticipated that there was future demand for mobile telecommunication, and that a common Nordic standard could yield profit for the PTTs as the number of subscribers was expected to increase. Implicit in these considerations was the intention to provide the necessary economies of scale for the Nordic telecommunications equipment producers. This would drive equipment prices down while at the same time providing a bigger market. Closely related to this business-minded attitude on the part of the PTTs was the decisive fact that the people who controlled and assigned radio-frequencies in the Nordic countries (in Finland this was the task of the PTT) had a more liberal attitude towards harnessing these limited resources than was the case, for example, with the FCC in the US. The final factor was related to technical advances that had been made in the Nordic countries, but also and especially at the Bell laboratory in the US. (Mäkinen, 1995; Mölleryd, 1996; Interviews, 1996).

The NMT group was convened for the first time at the Swedish PTT's premises in Stockholm in 1970, and subsequently gathered 4 times in order to produce a first report. The first report was drafted in the form of a request for a mandate to pursue plans for creating a common Nordic standard, and was directed to the governing boards of the Nordic PTTs. In 1971, in Reykjavik, the governing board of the respective Nordic PTTs approved the mandate. In Reykjavik the very important decision was also taken to allow subscribers to the PTTs' mobile networks to purchase mobile stations, or mobile phones, directly from the manufacturer. Essentially, this meant that the demand for mobile phones would be regulated by consumer demand, while demand for other mobile telecommunications equipment would be regulated by the PTTs' procurement orders. The next meeting in Copenhagen in 1973 laid down some broad guidelines for the technical specifications, and in 1974 development work for drafting the specifications was kicked-off officially. (Toivola, 1992; Meurling & Jeans, 1994).

4.2 Drafting the Specifications the NMT 450 Standard

The cellular concept originated from Bell laboratories in the USA. Apart from local collaboration with the various Nordic equipment suppliers, the NMT group was acquainted with the AMPS system – the first cellular standard in the US – and it also closely monitored research at Bell's laboratories – the pioneer in the field. Relatively early on during the drafting of the specifications, it was concluded that a cellular architecture would also function as a basis for the Nordic system, in order to add flexibility and capacity to the system, but the 'state of the art' would be supplemented with certain innovative features. Hence, the NMT group contributed to the diffusion of cellular technologies in the Nordic markets, but also played a major role for further technical advances in cellular technologies. (Meurling & Jeans, 1994; Interviews, 1996).

The main components of the proposed Nordic cellular system essentially consist of *base stations, mobile stations (mobile phones)* and *mobile telephone switching systems (MTX)*. The first phase of drafting the specifications for these components was finished in 1975, and the NMT group submitted a second major report to the NordTel meeting that same year. Apart from including specifications both for the cellular system as a whole, as well as for the MTX switching system, the base stations and the mobile phones, it also included suggestions for technical solutions and guidelines for further action. In order to maximise the efficiency of the system the NMT group had decided to locate the crucial data-processing power of the air interface signalling in the mobile phones and the MTX system. The MTX system was specified as a computerised digital switching system capable of handling digital signalling of large amounts of information up to a speed of 1200 bits/second. It was also envisioned that the system should be capable of handling such features as the 'hand-off' and trunking mechanism, as well as the innovative roaming feature enabling automatic tracking of subscribers anywhere in the Nordic coverage area. Apart from the fact that digital signalling and the inclusion of roaming implied that the original plans for a manually operated system in effect were scrapped, it had also led to especially demanding specifications for the mobile phones and the MTX system. In hindsight this was a very risky decision, since microprocessors and digital switching systems had only recently entered the market. Furthermore, the report concluded that the system would support 180 channels and operate within the 453 – 467.5 MHz radio frequency band. It also recommended that a common test-field network should be constructed in order to confirm that the specifications were necessary and adequate. (Toivola, 1992; Meurling & Jeans, 1994).

While the drafting of the specifications had been a very collective effort – a kind of large Nordic technology procurement project – the procurement and market deployment stages of the NMT project was delegated to the respective PTTs. Hence, the various national framework conditions came to play their role. In Finland, the manually operated ARP network had been in successful operation for several years, and had proven the business potential of mobile telephony. The PTT was now in a position to embrace fully the specifications and recommendations of the NMT group. (Turpeinen, 1996; Interviews, 1996). Meanwhile, higher-level government authorities had also started to realise that mobile telecommunications services and subscriber tariffs could bring more income to the federal budget, and the Ministry of Finance was being persuaded by the PTT to set aside the necessary funds for procurement and construction. Gradually, it seems, the Ministry of Finance also became willing to endorse the plan. In 1978, the budget of the radio department was increased significantly as procurement and construction plans were being formulated. (Toivola, 1992; Interviews, 1996).

Compared to the small-scale prototype procurement of the first version of the DX 200 system described previously, the NMT 450 procurement project was significantly bigger and more complex. Hence, it also required more careful preparations and risk calculations on the part of the PTT. The peculiarities of the subsequent procurement process are better understood once certain background factors are accounted for.

First of all, it should be remembered that the PTT had a natural monopoly in the field of mobile telecommunications since the PTT also controlled the radio-frequencies. The radio department, and the associated radio-laboratory, was the main carrier of cellular technology competencies. This in turn implied that the PTT had every reason to assume that its procurement decisions would become the driving force in consolidating networks of relevant actors and competencies amongst the domestic producers. Hence, it seems evident that the PTT was indeed strongly biased to procure from domestic producers. Relatedly, there were the more long-term infant-industry considerations as to what degree the PTT felt that domestic production should be stimulated in order to avoid excessive dependence on foreign suppliers, such as Ericsson. Finally, there was the question of how the complex specifications should be transformed into functional user requirements accessible to the producers, as well as that of how the quality of compatibility of the procured equipment should be tested. (Toivola, 1992; Interviews, 1996).

In 1975, the PTT had organised a seminar with the purpose of informing domestic producers of telecommunications equipment about the various technical specifications. In 1977, the official procurement bid for tenders for the NMT system was announced to the public on the basis of the unaltered specifications drafted by the NMT group (Toivola, 1992). Turning to the main producers in Finland at the time, the industry was still very much characterised by small-scale production and the presence of the foreign firms. During the late 1970s, the Nokia division, Telefenno, was entering the early prototype refinement and market deployment stages of the DX 200 switching system for fixed networks. It had neither the vision nor the resources to initiate an MTX project. Telefenno surprised the PTT by not submitting a bid, even though the engineers at Telefenno had made some efforts to initiate a MTX project alongside the DX 200 project in 1975. To that end, a budgetary bid had been submitted to the NMT group. At Salora, the radiotelephone division had been developing and supplying base stations and radiotelephones for the various local mobile networks, the ARP and other mobile networks abroad, but was primarily concerned with export activities of existing generations of mobile phones. Hence, despite the fact that both Salora and Nokia had closely monitored, and even participated in, the drafting of the NMT 450 standard, there were no domestic bids. Consequently, the PTT was forced to rely on foreign producers, especially Ericsson.

The subsequent period, 1977 - 1979, was probably very important for the future of the Finnish mobile telecommunications industry. In 1978, the NMT group had been discussing different ways to formulate the complex specifications into more easily approachable functional requirements, as well as how to test various equipment prior to construction of the actual network. (Toivola, 1992; Interviews, 1996). Around the same time, Mobira, as well as the sub-contractors LK-Products (duplex filter components for mobile phones) and Micronas (semiconductors, AISC chips), were founded in order to consolidate competencies and resources in the field of mobile phones and base stations. (Koivusalo, 1995). In 1979, Mobira purchased an NMT simulator from the PTT – an important piece of equipment developed by a small engineering firm in Finland – and some time later on, in August 1979, Mobira eventually presented their bid for tenders for base stations. The PTT did not hesitate to award Mobira the contract, even though there were also competitive bidders from abroad. In connection with the base station delivery, Mobira also delivered some 50 mobile phones in accordance with the NMT 450 specifications, with special permission from the PTT. Mobira and the PTT also extended the original procurement contract to include collaboration and testing using the NMT simulators. During the period 1979 – 1981, the NMT 450 network was constructed, and in 1982 the network was opened to the public. (Toivola, 1992).

The first NMT 450 network only covered the Helsinki metropolitan area, and the PTT immediately initiated new plans to procure more equipment and extend the coverage of the network to include other population centres as well as rural areas. Both Mobira and LK-Products and Micronas gradually came to focus on developing NMT equipment and components. The ongoing NMT 450 project, as well as the aggressive procurement strategy, certainly played an important role for the strategies of the firms involved. However, even though the above mentioned firms had successfully moved into NMT technologies and were penetrating some newly opened NMT markets in continental Europe, development and delivery of MTX switching systems was still solely in the hands of Ericsson. Ericsson was about to become the main beneficiary of the NMT initiative, and was free to pursue a monopolistic pricing policy, contrary to what had been the original intention of the NMT group. While this fact irritated the Finnish PTT, the management of Mobira, rather than Nokia, also started to realise the strategic importance of being capable of delivering complete turn-key NMT networks. (Toivola, 1992; Interviews, 1996)

4.4 The Mobile Telephony Boom and the Drafting of the Specifications for the NMT 900 Standard

A salient feature of the NMT initiative was the fact that the PTTs severely underestimated the popularity of mobile telephone services. Around the year 1983, only some years after the first version of the NMT 450 network had been erected in the Nordic countries, there were already signs that the radio frequencies reserved within the 453 – 467.5 MHz bandwidth would not suffice. At the time the penetration rate of mobile telephone subscribers in the Nordic countries was the highest in the world, and the NMT 450 cellular system was by far the largest single mobile network in operation. In Finland the situation was very much the same as in the other Nordic countries.

Table 2. Number of mobile telephone subscribers in Finland 1982-1986. (Telecommunications statistics 1996: 33).

Year	ARP	NMT 450
1980	23,482	
1981	28,278	
1982	31,232	2,648
1983	33,571	8,655
1984	34,145	17,865
1985	35,330	32,309
1986	35,560	49,672

Table 2 clearly shows what could be called 'the Finnish mobile telephone boom' from 1984 onwards, as the capacity of both the manual ARP system and the new automatically operated NMT 450 reached their highs. The practical consequence of this boom was that networks, and especially the NMT 450 network in the Helsinki metropolitan area, became congested despite the fact that the cellular architecture of the network as well as advancements in semiconductor technologies allowed for flexible capacity expansion and efficient harnessing of the radio frequencies. (Toivola, 1992; Turpeinen, 1996).

During the period 1977 - 1983, the NMT group still gathered for meetings in order to finalise technical details and monitor construction work, even though there had been some changes in personnel. On the national level the various PTTs were confronted with different choices for coping with the unanticipated popularity of the service. In 1983, the NMT group reached the conclusion that work would commence on the drafting of the second Nordic standard – the NMT 900 standard. (Toivola, 1992; Mölleryd, 1996).

During the period 1983 - 1985, when the major part of the work was conducted, special attention was given to ways and means to assure that the new system would not reach capacity constraints as quickly as the NMT 450. Since the NMT 900 had been assigned a broader band width (890 - 960 MHz) covering more kHz of channel width, the number of channels could be increased from 180 to 1000, which in turn would automatically increase capacity. Furthermore, the network was specified as a small-cell system, whereby cells in each transmission area were to be further subdivided into so-called micro cells. These were to be supported by a larger number of base stations, which would enable the construction of smaller mobile phones. Finally, the signalling interfaces of the NMT 900 system were specified as having to be more complex, in order to provide some additional services and features related to code security and interference hindrance. From the point of view of the equipment producers, the NMT 900 specifications did include innovative features, but these were mostly related to the software requirements of the equipment that had already been developed for the NMT 450 network. Again, the specifications were the most challenging and demanding vis-à-vis the

MTX switching system, which was the most important and sophisticated component in the network. (Toivola, 1992; Interviews, 1996).

4.5 Initiating the Procurement Process – Prototype Trials and Preparations

The NMT group presented the finalised NMT 900 specifications at a meeting in Copenhagen in 1985, and the procurement stage was delegated anew to the national level. In Finland it is possible to identify a change in strategy, on the parts of both the PTT and the industry. Prior to what was above called 'the Finnish mobile telephone boom', when the PTT started to sense the imminence of mobile telephony and the associated expected capacity problems, actions had already been taken at the operational level of the radio department in order to initiate the NMT 900 project. It seems that representatives from the Finnish PTT, in collaboration with Mobira, played quite an active role in pursuing the NMT 900 agenda at the NMT group meetings.

At Mobira, the first generation mobile phones and base stations for the NMT 450 had been finalised, and the product line was now upgraded and developed in anticipation of further initiatives from the NMT group. In large part due to a so called Experimental Mobile Cellular Radio network, which Mobira had developed in collaboration with the PTT, R&D projects were also initiated for the development of mobile phones and base stations for the NMT 900 network. Hence, in 1985, when the NMT group presented the specifications and the procurement process was initiated, Mobira was already capable of producing and marketing 'state-of-the-art' mobile telecommunications equipment.

During the mid 1980s, the market structure amongst the producers could be characterised as oligopolistic. Ericsson was the main supplier of MTX switching systems, but had also managed to move into NMT mobile phone and base station technologies and thus had a significant competitive advantage as the main turn-key network supplier. Apart from Mobira, the other main mobile phone and base station equipment suppliers were Danish Storno and KO-MA. Gradually such foreign firms as Motorola, Siemens and the Japanese multinationals NEC, Mitsubishi, and Panasonic were also catching up. (Mäkinen, 1995; Koivusalo, 1995).

Despite the fact that both the NMT 450 and NMT 900 were open standards, and thus allowed for the purchase of different network equipment from different suppliers, the management at Mobira was very quick to realise that the only viable way to compete with such bigger suppliers as Ericsson and Siemens, or Japanese multinationals, would be to become turn-key suppliers, like Ericsson. This implied that Mobira also had incentives to gain access to MTX switching systems and other transmission components as well. There was increasing pressure for this kind of expansion of activities, since NMT 450 systems had already been constructed in several other European and Asian countries, while the competing TACS and AMPS, as well as other cellular standards such as the German Netz C, and the French R 2000 standard, were also being developed. It was also realised that turn-key deliveries would pave the way for special customer-tailored solutions, which could provide de facto monopoly positions over certain markets.

Starting in 1982, discussions were held between the Nokia subsidiary, Telenokia, and Mobira about the possibilities of co-ordinating R&D and production in the field of turn-key cellular systems. While the founding of Mobira had brought the Salora and Nokia teams closer together, Mobira continued to exert considerable pressure in order to initiate more extensive collaboration. This would also cover the development of an MTX system – the key to Mobira's further expansion. Lateral support was given by the Finnish PTT, and in 1984 Telenokia reluctantly agreed. In December of the same year, Telenokia was awarded a very crudely drafted contract for a single MTX system which contained essentially only information about delivery in 1987, as well as the price of the delivery. Among other things, Telenokia was left with the decision of whether to develop an MTX system in accordance with NMT 450 or NMT 900 specifications. Evidently, the contract was awarded in the same kind of infant-industry spirit that had prevailed previously – but Telenokia did manage to promise delivery at a lower price and on shorter notice than Ericsson's prior deliveries, even though there were no other bidders for that specific order. Another important reason for supporting Telenokia was related to the anticipated superiority of the hardware solutions, i.e., the expected technological adaptability of the system. (Mäkinen, 1995; Interviews, 1996).

4.6 The DX 200 MTX Project and Nokia Cellular Systems

In part 3 the history of the DX 200 switching system was accounted for, from the point of view of procurement by the PTT, up until 1986 when the ISDN function was fitted to the system and the DX 200, in a sense, was finalised (incremental upgrading of the system has continued ever since). From the point of view of the design of the NMT small-cell networks, the decentralised architecture of the control processor also implied that the hardware construction of the DX 200 system was especially well-suited as a platform for the software extensions which were required for the various signalling features which had been specified in the NMT 450 and NMT 900 standards. When the PTT placed the first order for the MTX system, these technical considerations paved the way for close collaboration between Telenokia and the PTT during the period 1984 - 1987, leading up to the first delivery of the so-called DX 200 MTX switching system. (Interviews, 1996).

The DX 200 MTX project might actually be traced back to the mid-1970s, when Telefenno had submitted a budgetary bid to the NMT group. While only a budgetary bid, it had acquainted the engineers with the technical details of an MTX system. Later on, as all resources were concentrated on market deployment of the DX 200 system, any further investigations into MTX software requirements were put on ice. On the basis of collaboration between Mobira and Nokia, and outright persuasion by managers at Mobira, Telenokia gradually started to set aside whatever spare resources were available for developing an MTX system. At this point, in 1984, procurement by the PTT was the crucial factor which committed Telenokia to initiate development work. Development work was still organised alongside the DX 200 project, although the DX 200 project had now passed the critical market deployment stages. More R&D personnel were hired and Mobira had agreed to share the R&D costs of the project. Meanwhile, the PTT provided important information concerning the functional requirements of the system insofar as they had been altered compared to the original specifications of the NMT 450 standard. In this way, the PTT also fed Telenokia with information about specific technical solutions and alternatives which Ericsson had been forced to rely on in order to deliver their first generation MTX system (the AXE 10 system). (Toivola, 1992; Interviews, 1996).

Taking into account the rather poor technical prerequisites, as well as the very crudely specified procurement contract, it is interesting to note how quickly Telenokia and Mobira managed to adapt the DX 200 system to an NMT software environment and enter the turn-key network business. In a sense, collaboration between the PTT and Telenokia paved the way for the required technical competencies, while mangers at Mobira very aggressively pursued exports. Following a predetermined strategy, it was decided to construct the system for the more advanced NMT 900 specifications, since it was anticipated that demand for NMT 900 equipment would grow faster than demand for NMT 450 equipment. Meanwhile, Telenokia and Mobira had already agreed to deliver the first turn-key network to Turkey in 1986. A year later, in accordance with the procurement contract, Telenokia delivered their second system to the NMT 900 network maintained by the PTT.

Following these very important reference orders, MTX systems were also delivered to France (through collaboration with Alcatel) and China. The final steps towards consolidating competencies in the field were taken in 1987, when collaboration between Mobira and Telenokia was internalised even further through the founding of Nokia Cellular Systems. The next step was to focus on the emerging GSM and DCS 1800 standards. Parallel to these developments, the second significant Nokia business division – Nokia Mobile Phones – had been founded. By the early 1990s, Nokia had emerged as a major supplier of both mobile phones and turn-key networks, alongside its main competitor, Ericsson. (Koivusalo, 1995; Interviews, 1996). Meanwhile the popularity of the NMT and GSM service had grown tremendously. Table 3 illustrates the continued growth of mobile telephony in Finland.

Table 3. Number of subscriptions to the mobile telephone networks 1987 - 1999. (Telecommunications statistics 1996: 33).

Year	ARP	NMT 450	NMT 900	GSM The PTT	GSM Telcos	Subs. / 100 inh.
1987	34,262	69,560	2,038			2.14
1988	33,589	89,442	15,129			2.79
1989	32,062	112,046	45,923			3.82
1990	29,372	133,686	92,297			5.16
1991	27,896	149,573	133,478			6.35
1992	26,260	159,490	191,423	1,060	2,248	7.64
1993	23,240	172,656	267,307	10,168	8,943	9.63
1994	19,171	186,780	352,228	61,543	48,612	13.24
1995	15,223	197,849	439,044	253,203	127,500	20.37
1.7.1996	13,967	194,842	443,388	410,324	198,544	24.76

5. STRUCTURAL TENSIONS IN THE FINNISH TELECOMMUNICATIONS SECTOR – THE HISTORY REVISITED

This article has examined critical periods of development of the Finnish telecommunications industry during the 1970s and 1980s, through two case studies of public technology procurement by the PTT. During these periods the industry underwent a transformation, from small-scale production under a foreign firms oligopoly towards a diversified industry with Nokia as the dominant firm, and further towards internationalisation through global growth and joint-venturing. On a very general level, the cases presented above provide examples of the rather successful coupling of the needs of public sector users and private sector producers. From a theoretical point of view, the cases clearly illustrate the limitations of strict linear science and technology-push, as well as demand-pull, interpretations of the innovation process (Kline & Rosenberg, 1986; Freeman, 1994). Rather, they are examples of technological development at the intersections of supply and demand factors, involving a whole range of different actors and institutions. As such, they are particularly compatible with evolutionary theory, and a systems of innovation approach to studying technological change.

Despite the fact that the chosen cases only represent snapshots of developments in a broader field, the story that unfolds is one wherein the Finnish telecommunications equipment industry has evolved in a sequence of something closely resembling Dahmén's and Carlsson & Stankiewicz's notions of Schumpeterian transformation processes, or structural tensions (Dahmén, 1950; Carlsson & Stankiewicz, 1991). Even though Nokia's managerial decisions and strategies (for example, concerning strategic alliances) and other supply-side explanations should not be underestimated, especially during the 1980s and 1990s, the cases suggest that the technology procurement by the PTT has played a more important role in this transformation process than has so far been accounted for (see for example Mäkinen (1995), Koivusalo (1995), Rouvinen (1996), Lemola & Lovio (1996), Pulkkinen, 1997). This observation seems especially relevant, given that the developing competencies in digital switching and cellular technologies have been absolutely essential for Nokia's diversification into mobile telecommunications – the DX 200 has functioned as an important platform technology for subsequent MTX modifications for different mobile standards. Similarly, competencies in the field of NMT-technologies was pivotal for further organic growth in the direction of turn-key deliveries to subsequent standards and markets.

It is outside the scope of this chapter to give a complete interpretation of the evolution of the Finnish telecommunications sector. However, in theoretical terms of the three constituent elements of development blocs, or technological systems – the institutional infrastructure, knowledge/competence networks and economic competencies – the following observations seem relevant (compare in particular to Carlsson & Stankiewicz, 1991):

1) The most important *institutional infrastructures*[3] shaping the development of the industry have been the decentralised system of operators as procurers and users of telecommunications technologies, the set of conflicting procurement regulations and norms, and the early presence of foreign competitors on the home market. The regulatory regime supporting the decentralised system of operators in Finland has frequently been identified as a crucial factor for explaining favourable developments in the industry (see, e.g., Mäenpää & Luukkainen, 1994; Mäkinen, 1995; Rouvinen, 1996). This is primarily due to the fact that several buyers have created a competitive market for the producer firms as they have had to engage themselves in continued technology upgrading and renewal, as well as price competition, to secure customers. Through this competitive set-up the market has supported a relatively large number of domestic and foreign competitors compared to what has been the case in many other European countries, where the PTT often has been the sole buyer. This has especially enhanced Nokia's prerequisites to compete later on in foreign markets as well.

From the perspective of evolutionary theory, the decentralised system of operators and the presence of foreign competitors are also significant, since these conditions have contributed to diversity in the system through the generation of technological variety. Producer firms have been forced to develop differing technical solutions, organisational and other competencies to satisfy different users' functional requirements. The home markets have functioned as a large telecommunications test-trial laboratory prior to internationalisation – a breeding ground for progressive and daring technological and organisational innovations which have enhanced firms' reactivity to rapid technological change.

---

[3] By institutional infrastructures Carlsson & Stankiewicz mean "a set of institutional arrangements (both regulations and organisations) which, directly or indirectly, support, stimulate and regulate the process of innovation and diffusion of technology" (Carlsson & Stankiewicz, 1991: 23).

The most evident example of the importance of variety is the DX 200 case, where the system was refined and upgraded continuously in line with the functional requirements of the PTT and the telcos. There was also cross-fertilisation and competition between the various parallel projects until Telefenno came to focus on the DX 200, which outcompeted the rest and was 'selected' owing to superior flexibility of the system. Later the DX 200 emerged as Nokia's platform technology, when Nokia entered the field of mobile telecommunications as well. In the case of the NMT similar variety generating mechanisms were activated through standard setting. The NMT standard provided a larger user base upon which new varieties of technologies could be introduced and tested. Later, competition between different standards came to favour the Nordic NMT standard, and paved the way for the strong growth of Nokia. From this perspective it could be argued that Nordic collaboration functioned as an extended 'test-trial market' prior to tougher competition (selection) on international markets.

In Finland the competitive situation brought about by the decentralised system of operators and the presence of foreign firms has not only been restricted to the producers side of the market. It is equally evident from the point of view of the PTT's activities. Although the PTT and the telcos have not competed for market shares on the same networks, the PTT has had to adjust to competitive pressure stemming from the rapid upgrading of other parts of the network. This competitive pressure has functioned as one major incentive for the PTT to develop competencies actively, and to promote competitive technology procurement in order to drive prices down, thus securing similar service levels on the national trunk network and the PTT's local networks.

The procurement regulations prevailing in Finland since closer integration with GATT and the EU have added additional elements of competition and variety, even though a clear mismatch between the infant industry-favouring stance of governmental authorities (the government, the ministries and parliament) and the more business-minded attitude of the PTT is evident.

Procurement contracts have primarily been awarded through some sort of formal competition amongst suppliers, although it is difficult to make any clear-cut distinctions between competitive procurement and infant-industry arguments in the two cases. Publicised cases of nepotism have occurred in connection with awarding specific procurement contracts, but these primarily concerned Siemens rather than domestic firms and have not been discussed here. However, the primary reason that it is safe to say that the PTT has not adhered to national champion policies relates to the above mentioned institutional factors. These have prompted competition on the markets and thus provided limited room for 'national champion' policies.

Hence, on a general level, the industry has not grown out of successful 'national champion' policies, even though elements of infant industry policies have existed at the governmental level. It should be emphasised that procurement strategies have stemmed primarily from the PTT's desire to develop a modern and functional telecommunications infrastructure to meet socio-economic needs, rather than very explicit industrial or technology policy objectives. No clear links to governmental procurement schemes are identifiable, and it is only the fact that the PTT has been a public agency which justifies labelling the PTT as a technology procurement policy maker. Rather, the PTT has had to fend off infant-industry arguments emanating from governmental authorities, as these might have endangered price and technological competition and thus slowed down the upgrading of the networks. In this respect, the Finnish 'model' of public technology procurement in the telecommunications industry has been a decentralised, uncoordinated one, wherein the PTT primarily has acted for selfish reasons rather than for the common good of industrialisation. This observation is also compatible with the fact that public technology procurement has not been harnessed as an integrated part of either Finnish technology policy or industrial policy. However, in the very specific cases, at specific points in time, there have existed nationalistic viewpoints concerning how the PTT should allocate procurement contracts. This has been best exemplified in the case of procurement of the MTX system for the NMT 900 standard. Nonetheless, these specific cases are more related to internal consensus and vested interests of the actors involved, rather than active governmental policies. Evidently, this kind of 'sauna-nationalism' has been a necessary ingredient, and underlines the very close personal ties and mobility from one organisation to the another which have characterised the network of engineers and industrialists in the field – a good blend of competition and collaboration.

A further observation is the lack of a clearly articulated telecommunications policy until the 1980s. Running through the DX 200 and NMT cases, in particular, is the poor commitment that governmental authorities initially showed towards the development of telecommunications. It was only through the more widespread diffusion of mobile telecommunications (and information technology more generally) from the mid-1980s onwards that the sector became prioritised from the point of view of technology policy. The primary function of the government has been to issue the various regulations governing further liberalisation of the markets as well as the gradual de-coupling of the PTT from state ownership. Rather it has been the PTT which has acted as a de facto policy maker through the control over granting radio frequencies and licences until 1987. Hence the passive and reactive role of government telecommunications policies has also strengthened the PTT's position as a major actor in the field. On the other hand, the strong position of the PTT has frequently given rise to quarrels over licenses between the PTT and the telcos. These episodes have not been documented here.

Turning to the various organisations (as opposed to the above discussed regulations, and regimes) in the telecommunications system of production and distribution of knowledge, the role of basic research at the technical universities and the Technical Research Centre of Finland has been important. This has been especially the case in pre-paradigmatic stages of development such as those prevailing before the wider diffusion of digital switching and mobile telecommunications in the case of the DX 200 and NMT equipment. R&D labs, both in industry and at the PTT and the larger telcos (especially the Helsinki and Tampere telcos), have shared the responsibility for more applied research. Especially important R&D, testing and market analysis have been undertaken in the R&D laboratory of the PTT's radio department. Prior to the wider diffusion of mobile telecommunications, the PTT's radio department was an important consolidator of cellular-technology know-how in Finland. Later these competencies diffused to the industry, through the role that the PTT played in disseminating information on ongoing standardisation during the drafting of the specifications for the NMT 450 and 900 standards.

2) When the perspective is broadened to cover in addition the significance of interactions between the various firms and other organisations, it seems obvious that *competence networks* of entrepreneurs and innovators have played an important role. Owing to the transparency and density of the networks, rather than their size, the networks have enhanced variety through knowledge flows and technology diffusion.

Due to Nokia's dominant position, the telecommunications equipment industry forms the core of the Finnish system of innovation. The role of Tekes, as a bridging institution between firms, universities and public research organisations, has been especially important since the mid 1980s, when telecommunications and information technology became prioritised areas of the technology programme procedure. However, from the point of view of the cases, the most significant networks are the procurer/user-producer interactions, as well as vertically and horizontally integrated relationships between the various telecommunications firms, subcontractors, Nokia predecessors and business units. Later on the networks have covered collaboration with several research teams at the technical universities and the Technical Research Centre of Finland, especially during the development of competencies anticipating the pan-European GSM standard (see Lemola 1996).

The cases presented in this chapter illustrate the blending of visions – technological expectations and reactions to opportunities – and vested interests of the various actors. This is manifested in the form of joint experimental R&D collaboration, investments and joint marketing efforts. The cases illustrate processes of networking around specific technological problems and opportunities – a set of complementarities – occurring at different points in time. The diffusion of digital switching technology is a good example of joint effort directed towards understanding the fundamentals of the emerging digital paradigm, and the IDN/ISDN standard. Following the pre-competitive research consortia, consisting of representatives from industry (among others Televa, Nokia), the technical universities, and the operators (the PTT, the Helsinki and Tampere telcos), a close mutually beneficial R&D-venture evolved between Televa/Telefenno and the PTT. This R&D-venture eventually paved the way for the digital switching trajectory in Finland and Nokia's important DX 200 project. The general impression is that awareness concerning the opportunities related to digital technology was high in Finland (this would also explain the rapid digitalisation of the networks).

In the case of mobile telecommunications, additional collaborative relationships have been those between the defence sector, the state railways and the PTT, especially during the construction of the first local radio-telephone networks in Finland (see also Palmberg 1997). Later on, the Nordic NMT initiative strongly influenced the consolidation of competencies of the prime carriers of mobile telecommunications know-how. Evidently awareness of the opportunities opened up by the NMT-initiatives was highest at Mobira, while Nokia only became committed through the establishment of the joint-venture Mobira/Mobira-Nokia (R&D and marketing collaboration between Nokia and Salora). In addition to Mobira and the subcontractors LK-Products (duplex filter components for mobile phones) and Micronas (semiconductors and AISC chips), an extensive network of sub-contractors was established in close relation to Nokia's business units. The role of the PTT was primarily to disseminate information of the technical details of the emerging NMT 450 and 900 standards which were developed through collaboration with the other Nordic PTT's. The crucial issues were related to the systematic nature of the NMT technologies – the signalling and software compatibility of base stations, mobile phones and the MTX systems – which in turn explains the need for joint-ventures, alliances and subcontracting around cellular network technologies.

Common to both cases is thus the crucial role that the PTT's procurement projects have played in the clustering of resources around the specific technologies. From this perspective, procurement projects undertaken by the PTT have had an apparent system-, or network-, creating effect. The PTT has acted upon opportunities created by new technology during periods of rapid technological change, and communicated an awareness of these opportunities throughout the networks. This interpretation is closely in line with Carlsson's & Stankiewicz's (1991) emphasis on entrepreneurial activity during industrial transformation.

3) Carlsson & Stankewicz (1991: 18) point to the importance *of "something or someone to get the process started...".* They suggest that *"...the role of the entrepreneur is to provide the spark of the vision that turns a network into a development block...",* and speak of the entrepreneur who *"...has to perceive the (future) need, identify the necessary ingredients, secure the resources that may be missing initially, and communicate his vision to the relevant agents..."(ibid.).* In the case of the evolution of Finnish telecommunications, entrepreneurial activity and economic competencies are primarily identifiable at the level of the tightly-knit networks of engineers employed by Televa, Telefenno/Telenokia, Mobira, other Nokia affiliates, and the universities – the producers of telecommunications equipment. However, the main point is that similar kinds of entrepreneurial activity and economic competencies are also identifiable on the user side of the market, especially at the different departments and hierarchies of the PTT, despite the fact that the PTT has been a public sector agency. These structural tensions, which have developed through intensive interactions between the various producers and the PTT as the user, have been one major force driving the evolution of the Finnish telecommunications equipment industry. This observation seems valid until the mid-1980s, when Nokia entered a process of internationalisation and global growth in the field of mobile telecommunications.

In the DX 200 case, the initiative for the development of digital switching stemmed from Televa, notwithstanding the fact that Televa was formally subordinated to the PTT. The ADS project (the forerunner to the DX 200) was endorsed by the PTT, through the first experimental preferential procurement order in 1972. It was part of a more long term aspiration to establish domestic production of digital switching systems to avoid over-dependence on Ericsson's strong presence in the field. The order was crucial to counterbalance the scrapping of electromechanical switching systems, which might eventually have driven Televa out of the business and hindered the further advancement of the digital technology trajectory in Finland. Meanwhile, a combination of selective capabilities (correct component choices, crucial follow-up orders and market openings, cross-fertilisation of the different switching technologies etc.), organisational capabilities (the consolidation of competencies in switching technologies through the foundation of Telefenno/Telenokia), and other technical and learning abilities on the part of the public and private sector actors involved, paved the way for Nokia's diversification into the fields of digital switching, MTX, and other cellular technologies for turn-key deliveries.

Turning to the NMT initiative, prior accumulation of competencies in mobile telecommunications laid the foundations for a set of positive transformation processes which were set in motion at the national level by the aggressive procurement of the PTT. The case study shows that Nokia at first was reluctant to become engaged in the MTX and cellular network turn-key business, as all available resources were focused on finalising and upgrading the DX 200 system. Mobira and, above all, the PTT (through collaboration with the other Nordic PTTs) sensed the potential of mobile telecommunications, and it was really the PTT's procurement order for base stations and MTX switching systems which 'forced' Nokia to enter the field. In this respect Nokia acted as the "reluctant entrepreneur" which possessed the necessary competencies but lacked elements of visionary strategic management at that specific, critical, point in time (compare to Jacobsson, 1989). As in the DX 200 case, the prime background motive of the PTT was to stimulate domestic production as the PTT feared becoming overly dependent on Ericsson's monopolistic pricing policy, even though there was competition amongst the suppliers. The NMT initiative created a new generation of telecommunications technologies (upgraded cellular technologies – for example, the pan-Nordic roaming function) and export markets (through the wider diffusion and 'selection' of the NMT standard), as well as new methods of production and marketing (miniaturisation, the cellular network turn-key-deliveries and value-added services for end-users of telecommunications). Again a combination of selective capabilities (production of components and duplex filters, NMT simulators, exports to Turkey etc.), organisational capabilities (the establishment of Mobira/Mobira-Nokia, LK-Products and Micronas, the ARPA 1 group etc.), and technical and learning abilities (the MTX project, joint engineering and marketing etc.) set the stage for Nokia's breakthrough. The PTT has also benefited as a monopoly provider of mobile telecommunications services up until the early 1990s, when the telcos entered the field. The effects of increasing competition on the mobile telecommunications market are primarily reflected in lower tariffs as well as the supply of VANS for the benefit of consumers.

6. CONCLUSIONS

The Finnish telecommunications industry is characterised by the dominance of one major telecommunications equipment producer – Nokia. Even though Nokia is now a multinational firm, the roots are strongly Finnish, and can be traced back to the post-war cable-works and electronics industry. In this article it is suggested that public technology procurement, involving Nokia and the Finnish PTT (today Sonera), has played a more important role for the strong growth and internationalisation of Nokia from the mid 1980s than has been accounted for so far.

This important role of the PTT stems not only from the fact that the PTT has been an important procurer and network operator in Finland. The PTT has also acted as an entrepreneur and competent user related to the above mentioned two key technologies for Nokia's growth. Similar kinds of entrepreneurship and economic competencies to those which are identifiable on the producer side of the market are also identifiable on the user side of the market, at the different departments and hierarchies of the PTT. Structural tensions, which have developed through intensive interaction between Nokia and the PTT, have been an important factor driving the transformation of the industry up until the mid-1980s. These characteristics of the PTT's activities originate primarily from an entrepreneurial mind-set, as well as from the decentralised system of operators, and the early presence of foreign competition, which has enhanced competition and technological diversity on the market. The PTT's procurement and collaborative strategies have not been an integrated part of a broader technology or industry policy, and hence Nokia's success has not depended on an explicit 'national champion' policy.

REFERENCES

Literature

Carlsson, B. & Stankiewicz, R. (1991). On the nature, function and composition of technological systems. *Journal of Evolutionary Economics, vol. 1.*

Dahmén, E. (1950). *Entrepreneurial Activity in Swedish Industry 1909 - 1939.* Stockholm: IUI.

Ekberg, J. (1985). The case of Finland. In Granstrand, O. & Sigurdson, J. (Eds.), *Technological Innovation and Industrial Development in Telecommunications: The Role of Public Buying in the Telecommunications Sector in the Nordic Countries.* Lund: Tryckericentralen AVC.

Freeman, C. (1994). The economies of technical change. *Cambridge Journal of Economics, no. 18*, 463 - 514.

Kline, S. & Rosenberg, N. (1986). An overview of the process of innovation. In R. Landau & N. Rosenberg (Eds.), *The Positive Sum Strategy: Harnessing Technology for Economic Growth.* Washington DC: National Academy Press.

Koivusalo, M. (1995). *Kipinästä tuli syttyy: Suomalaisen radiopuhelinteollisuuden kehitys ja tulevaisuuden haasteet.* Jyväskylä: Gummerus Kirjapaino.

Lemola, T. (1994). Characteristics of Technology Policy in Finland. In S. Vuori & P. Vuorinen (Eds.), *Explaining Technical Change in a Small Country.* Heidelberg: Physica-Verlag.

Lemola, T. (1996) Riittääkö kolme miljardia markkaa? In T. Lemola & R. Lovio (Eds.), *Miksi Suomi, Nokia?* Juva: WSOY.

Lemola, T & Lovio, R (Eds.). 1996. *Miksi Nokia, Finland.* Juva: WSOY.

Lovio, R. (1989) *Suomalainen menestystarina – tietoteollisen verkostotalouden lâpimurto.* Helsinki: HANKI ja *Jff*.

Lovio, R. (1993). *Evolution of firm communities.* PhD thesis. The Helsinki School of Economics and Business Administration, Series A:92.

Meurling, J & Jeans, R. (1994) *The mobile phone book – The invention of the mobile phone industry.* London: Communications Week International.

Mäenpää, K & Luukkainen, S. (1994). *Teletekniikasta monimuotoiseen viestintään.* Tampere: Tammer-Paino Oy.

Mäkinen, M. (1995). *Nokia Saga*. Jyväskylä: Gummerus Kirjapaino Oy.

Mölleryd, B.G. (1996) Så byggdes en världsindustri – entrepenörskapets betydelse för svensk mobiltelefoni. *Teldok, no. 28.*

Olkkola, K (1986) *DX 200 vaiheita I ja II.* Unpublished memo.

Palmberg, C. (1997) *Public technology procurement as a policy instrument – selected cases from the Finnish telecommunications industry.* (Working Paper Series, No. 28/97). Helsinki, Finland: VTT Group for Technology Studies.

Rouvinen, P. (Ed.). (1996). *Advantage Finland – The Future of Finnish Industries.* Helsinki: Etla Taloustieto Oy.

Tele. (1993). *Poimintoja teletoimen historiasta. Osa 6- Elämäntyö telealalla.* Helsinki: Painatuskeskus Oy.

Toivola, K. (1992). *Poimintoja teletoimen historiasta. Osa 4: Kertomus Suomen Matkaviestinnästä.* Karjaan Kirjapaino.

Toivola, K. (1989). *Virkamiehen Sattumukset.* Valtion painatuskeskus.

Turpeinen, O. (1996). *Yhdistämme 2. 200 vuotta historiaa – haasteena ttulevaisuus.* Helsinki: Oy Edita Ab.

Ylä-Anttila, P. (1996): Teollisuuspolitiikan ikuisuusongelman ratkaisu? In T. Lemola & R. Lovio (Eds.), *Miksi Nokia, Finland?* Juva: WSOY.

Statistical material

Telecommunications statistics 1989

Telecommunications statistics 1996

Legal documents

Valtioneuvoston päätös 136/1930

Personal interviews in 1996

Halme, Lauri (PTT)

Johansson, Veikko (PTT)

Koski, Aimo (PTT)

Makkonen, Matti (PTT)

Melamies, Lauri (Nokia Telecommunications)

Nevanpää, Tapani (PTT)

Olkkola, Keijo (Televa, Nokia)

Palonen, Vesa (Ministry of Transport and Communications)

Pasanen, Matti (PTT)

Rajamäki, Timo (PTT)

Räsänen, Erik (PTT)

Salminen, Sakari (Nokia Electronics/Telefenno / Telenokia, Nokia Telecommunications)

Toivola, Keijo (State Railways, PTT)

Toivonen, Seppo (Televa, PTT, Finnet)
Valle, Jorma (PTT)
Vallström, Eero (Mobira, Nokia)
Koivusalo, Mikko (Nokia)

Telephone interviews in 1996

Kirves, Liisa (PTT, freelance reporter)
Kangasluoma, Matti (PTT)
Pasanen, Juha-Heikki (Nokia Electronics, PTT, State Railways)

Appendix

The evolution of Nokia's telecommunications divisions Nokia Telecommunications and Nokia Mobile Phones (adapted from Koivusalo, 1995: 193).

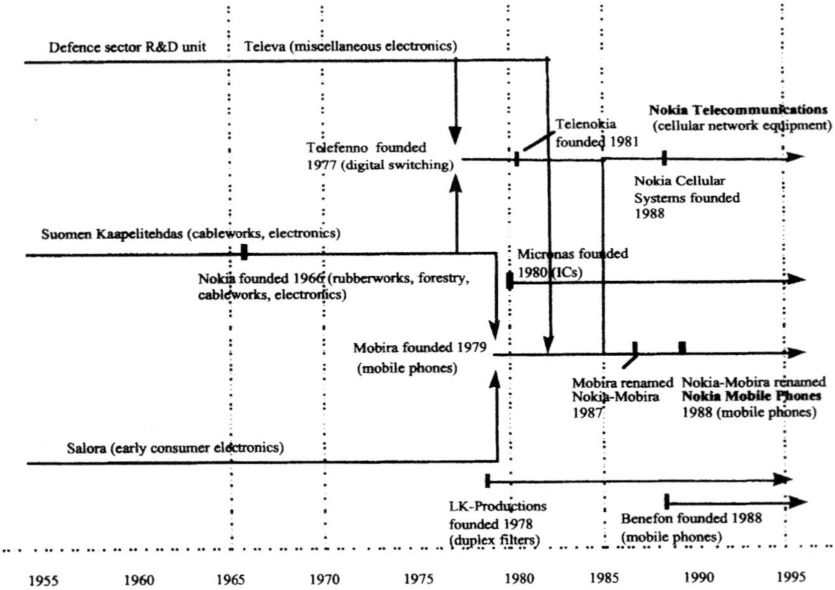

7. PUBLIC TECHNOLOGY PROCUREMENT: THE CASE OF DIGITAL SWITCHING SYSTEMS IN FRANCE

P. Llerena, M. Matt and S. Trenti

INTRODUCTION

In this chapter we will study the role of Public Technology Procurement (PTP) in the telecommunication industry in France. The case study outlined is the development of the E10 system. The period covered is from the 50s to the 80s. In section 1, the French national system of innovation will be briefly described and the institutional and legislative framework for procurement, as well as the industrial structure of the telecommunication sector, will be presented. Section 2 will introduce the evolution of the government policy interventions in this industry and section 3 will focus more precisely on the case of the evolution of the digital switching system, studying the specific role of PTP and of other policy instruments. In the last section, we will try to assess some of the societal and industrial results (section 4).

1. THE FRENCH TELECOMMUNICATION SECTOR

1.1 The French National System of Innovation

In this section, we will give a brief description of the French national system of innovation (NSI). A wide and comprehensive analysis of the characteristics of the French NSI is beyond the scope of this study. We will limit the analysis to some of the characteristics of NSI that are of major interest for the telecommunications case.

The French national system of innovation presents some peculiar characteristics (Chesnais, 1993):

1. The greatest part of fundamental research is funded and organised through a special institution, the CNRS;
2. The higher education sector is organised in a dual manner with, on one side, universities and, on the other side, "Grandes Ecoles" forming an elite of technical experts;

3. The state is pervasively involved in the production not just of general scientific and technical knowledge but often also of technology *per se* in the form of patentable and/or immediately useable products or production processes.

This last point is particularly noteworthy for the analysis of the telecommunication industry. The telecommunication industry has been one of the sectoral subsystems forming the vertically structured French National System (Chesnais, 1993). These subsystems have involved an alliance between the state and public and/or private business enterprises. Characteristic of the alliance between the state and firms, usually belonging to the French oligopolistic core, has been the building of an R&D capacity by the state and the search for industrial partner to whom to transfer the technology and knowledge produced. This transfer to the main French firms ("national champions") took place especially after the Second World War, in two phases of intensive Science and Technology institution-building (Chesnais, 1993).[1] The French NSI was characterised by the willingness of the French state to build an autonomous technological capability *versus* (mainly) the USA, which was seen both as a model and as an enemy. This "struggle for technological independence" was clearly linked with the desire to play a renewed important role in the military and political arena. The importance of the strategic and military content in the building of NSI is reflected by the choice of industries to be sustained. Besides the weapon industry, aeronautics, telecommunications, computers and the energy industry were chosen as main axes to build the autonomous French power.[2]

1.2 The French National System of Telecommunication

The brief description of the national system of innovation in France has introduced some of the main actors playing in the telecommunication industry. In this section, we will describe in more detail the main actors of the telecommunications sector introducing the institutional and regulatory framework for the public procurement and the industrial structure of the sector at the beginning of the period.

1.2.1 The regulatory framework in France: The role of the state

Since 1889, when the first service provider company was nationalised and put under the control of the PTT, telecommunications in France has been considered as a public service and as an object of government policy (Libois, 1983). The chart below (Fig-

---

[1] The first phase, immediately after the War, included the creation under the Ministry of PTT of CNET, which would play the key role in the development of telecommunication technology. During the second phase, started with the setting up of the Fifth Republic, innovation continued to be driven strongly by the state but a process of technology transfer towards nationalised or publicly owned firms took place.

[2] Other high-tech industries, such as the chemicals and pharmaceuticals industries, seemed to be judged less important from a strategic point of view and were thus partially excluded from the "technology transfer" from the state to industry. See Llerena, Matt and Trenti (1997).

ure 1) shows the main actors of the French telecommunication system. The telecommunication service has been the object of many reforms changing the links among the different actors and adapting its regulation to the new technological and commercial needs of the industry—though always confirming the central role of the state and the "public service" nature of telecommunications. Generally, the main interactions among the different actors were the following. The Prime Ministry and the Planning Committee influenced the other Ministries through budgets and the development of plans deciding the amount of funds available to them. The Ministry of PTT acted both as a regulating body and as a direct provider of telecommunication services. In the course of time, the operational branch was given a wider autonomy to separate the exploitation from the regulation and control functions. The Ministry of Defence was initially involved in the technological part of telecommunications through a direct link with CNET, but this link was dissolved during the late 50s. The Ministry of Industry played an important role for the restructuring of the national industry during the 70s and 80s, both directly and indirectly, through the Ministry of PTT plans and the DGT procurement. Finally, CNET, as part of the Ministry of PTT, has been the main actor of the strategy of technological transfer for the national industry, across the whole span of the electronic "filiere" (telecommunications, computers and semiconductors).

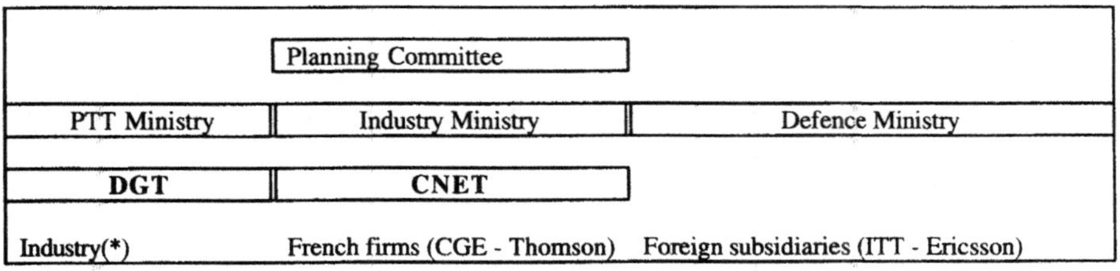

Figure 1. The main actors in the French National System of Telecommunication

(*) The structure and evolution of the French telecommunication industry will be presented below

As shown by the figure, there are two main players acting at the interface between the state and the industry: the DGT (Direction Generale des Telecommunications), now France Telecom and the CNET (Centre National d'Etude sur les Telecommunications) the national research centre. Both organisations were used, during the 30-year period analysed here, as major policy instruments for the telecommunication sector.

The DGT (now France Telecom) is the operation branch of the PTT, providing the telecommunication services, and was established at the end of the past century. The structure and the internal organisation of DGT has been modified in the course of its history. An important reform, which influenced the evolution of the digital switch, was undertaken in 1974. DGT was given a five-year period of budgetary freedom and a wider autonomy from the other PTT offices. At the same time, a thorough restructuring of the service provider was decided upon, to give to DGT a "more commercially minded organisation" (Cawson et al., 1990). Starting in 1975, the DGT, although acting as a public service agency, was given a wider degree of autonomy than its legal position would normally permit.[3]

The role of the state in the telecommunications framework is also clearly shown by the attention given to Research and Development in this industry.[4] In 1945, the CNET, Centre Nationale d'Etudes des Telecommunications was created. The CNET was the essential organisation involved during the first phase of the development of the digital switching system and its role was fundamental for the process of building technological competencies in the French system.[5] At the beginning, the CNET was given a double priority: research on telecommunications and technical control over the equipment to be purchased by the administration. Starting in 1964, the CNET mission also included control over prices. With the 1974 reform of DGT, the technical and price control was given to a new division of DGT, the DAII (Direction des Affaires Industrielles et Internationales).

The role of the state in the R&D within the telecommunications industry in France is shown by the high contribution of the service provider to the total R&D expenses of the industry. In 1987, R&D expenses for the telecommunication industry in France were strongly based on the contribution of the service provider. As shown by Figure 2, the French service provider was then financing more than half of the total R&D budget for civil telecommunication. In 1987, the R&D budget of France Telecom was 4.1% of revenues (610 million US$). The greatest part of the budget was absorbed by CNET. In 1987, CNET had a staff of 4,300 personnel devoted to basic research (30% of the activity) and to applied research for feature specification and the development and installation of telecommunications equipment.[6]

---

[3] The 1990 reform transformed France Telecom into a "public utility" acting as a private firm in regard to procurement, personnel and budget rules. The strategic and economic programs of France Telecom were still decided in close conjunction with the Minister of PT through a planning contract. In 1996, a project of reform was presented. Starting from 1997, France Telecom would be transformed into a private society. Part of the capital (49%) would be privatised. Thus, a deep process of restructuring is currently taking place. The new priorities of the management of France Telecom are the development of the mobile phone network and a wider diffusion of Internet facilities.

[4] As early as 1916, an R&D group on telecommunication was established inside the "Ecole superieur des postes et des telegraph", one of the organisations for higher level education traditionally characterising the French NSI. The group, whose aim was to study new technological development in the field, was formed both by administrative personnel and by external scientists and technicians known for their expertise in the field.

[5] The research lab was completed with the take-over of other existing laboratories. The new facilities included the military radioelectricity laboratory. Both the Ministry of PT and the Ministry of Defence initially provided the financing for CNET, given the strategic importance of telecommunication for the military sector (Chesnais, 1993).

[6] After the June 1996 approval of the privatisation and reform of France Telecom, the role and the position of CNET have been changing. France Telecom announced the restructuring of its research laboratory, aiming at a specialisation of CNET on "core business" research. As a result, the CNET will probably be more oriented towards applied research and standardisation dedicated to the needs of the service provider, renouncing its strength in basic research. One signal of this change has been an announcement of the abandonment of CNET's research on electronic components.

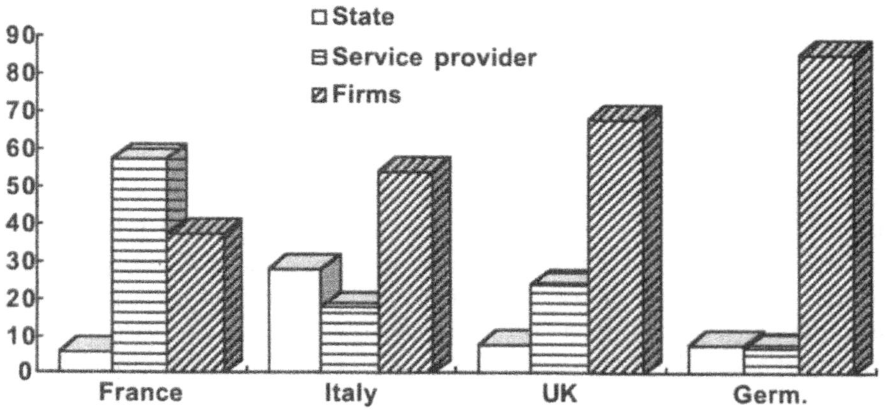

Figure 2. R&D expenses in civil telecommunication by contributors (% shares - 1987)

Source: France Telecom (1991)

1.2.2 The French telecommunication industry

We will now describe the situation on the industrial side in the telecommunications sector. At the beginning of the development of the digital switch, the market was dominated by the leading presence of foreign manufacturers (multinational subsidiaries). After the Second World War, the design and production of telecommunication products and systems was the work of three manufacturers, which were subsidiaries of foreign international groups. Two of them, "Le Materiel Telephonique" (LMT) and the "Compagnie Generale de Constructions Telephoniques" (CGCT), belonged to the ITT group. LMT concentrated on producing exchanges for large towns and very large cities while CGCT equipped smaller ones. The third, the "Societe Francaise des Telephones Ericsson", SFTE, was a subsidiary of the Swedish group, LM Ericsson. The three manufacturers shared the French public market with a fourth, CIT (Compagnie Industrielle des Telephones), part of the CGE group, later CIT Alcatel. CIT Alcatel produced switching equipment under license from its foreign partners, lacking an autonomous technological capability. Finally, a fifth French manufacturer, AOIP, was present in the market. The market shares were 60% for the three foreign subsidiaries and 40% for the French manufacturers. The main manufacturers were grouped in two supply agreements: the SOTELEC for transmission equipment and the SOCOTEL for switching equipment. The latter would play an important role in the technological development of the switching system.

During the 70s and the first half of the 80s, the industrial policy undertaken by the French government deeply influenced the industrial structure.[7] We will describe this process more fully in section 2, which studies the telecommunication policies.

1.2.3 The rules for public procurement in the telecommunication sector

The rules governing the procurement of telecommunications equipment evolved during the 30-year period considered, following the development of the institutional and technological framework. They are an important part of the national system of telecommunication and, as we will see, they were used as an instrument to complete the government's policy intervention in this industry. The main procurer was DGT, which, being part of the PTT, was subject to the general law on public procurement, "*Code de Marchés Publics*".

Until the early 60s, the DGT used to buy telecommunications equipment through a system of bidding, based on a well established negotiation with a restricted number of manufacturers, forming the oligopolistic core of the French telecommunications industry. The procurement was distributed by "historical shares", (tranche garantie, guaranteed share), perpetuating year after year the same market structure. As early as 1947, it was clear that a cartel, dominated by the US firm, ITT, had been formed to share the French market.[8] Manufacturers imposed the prices on the Administration, which was not able to negotiate, lacking an autonomous and specific competence for costs assessment (A. Costa, 1993). In 1964, the control over prices was delegated to CNET, which had the technical expertise needed to assess the costs of the equipment.[9] Despite some difficulties, the DGT, through the competence of CNET, was able to obtain great savings on the prices of the equipment.[10] Market shares were still distributed by "historical shares", despite the introduction of a new mechanism of procurement, the *tranche libre* (free share), which was too small to upset the cartel.

Starting in 1974, the reform allowed DGT to derogate partially from the strict regulatory measures connected with its legal public service role. In the case of the procurement connected with the modernisation of the network, DGT was allowed to re-

[7] The process started in 1976 - 1977, with the acquisition of SFTE (Ericsson) and of CGCT by the French Thomson, whose telecommunication activity was later acquired by CGE in 1983. As a result of this process, the market for switching equipment in France is now largely dominated by one firm (CGE, now Alcatel NV) with the presence, starting in 1986, of a second supplier (CGCT - Ericsson).

[8] In 1947, during one meeting of the National Commission for Telecommunication, for example, a CNET member denounced the existence of an agreement among the manufacturers to share the market through a partition of the French territory (P. Carré, 1996).

[9] A new instrument of control, *the cahiers de clauses comptables* (accounting rules books), including the main rules to fix prices and profit margins, was created. The controls for the costs assessment were long and difficult, and they were performed every three years (A. Costa, 1993).

[10] From 1966 - 1976, the prices of telecommunications equipment grew at an average annual rate of 1.8%, compared to the 8 % average rate of growth in industrial manufacturing prices. The controls seem to have led to a 10 - 20% "real" price cut (A. Costa, 1993).

vert to the "*Code des Marchés publics*". Instead of the procedure of bidding, DGT had recourse in that case to a price agreement with the supplier lasting more than one year.[11] After the 1976 - 1977 rationalisation of the industry, the guaranteed share mechanism was gradually substituted by the guaranteed price mechanism (tranche a prix guarantis), a rule fixing a negative correlation between prices and volume of the procurement. In the same period, the "free share" covered almost 50 % of the market for telecommunications equipment (A. Costa, 1993). After the restructuring of the industry, DGT was facing two national champions: CGE - Alcatel and Thomson. To prevent the protectionism in favour of the national firms from leading to X-inefficiency, DGT introduced new rules to select orders. The procurement was organised with a bid based on various criteria: the share of exports of the bidding firm, the quality and the performance of equipment, the effort put into the bid, and the respect given to the time and terms of the contract.

Concerning the switching system, until the mid 80s, DGT seemed not to concentrate on reducing the prices of the equipment. The problem of reducing the selling price of equipment was considered by CIT-Alcatel only for export and not for the national market (Chapuis and Joel, 1990). This situation changed in the mid 80s, with the planned expansion of investments for the digitalisation of the network. DGT promulgated new operating and maintenance standards, together with new cost requisites. The new standards had to be respected at the same time that prices had to be reduced in view of the large expansions of the size of the market offered by the administration.[12]

2. THE NSI AND THE TELECOMMUNICATION POLICY

The French NSI and the different legislative and industrial framework described in the previous section were reflected in the policy implemented in France to sustain and develop the telecommunication industry. The state intervention in this sector was mainly based on the "struggle for independence" characterising the French recovery after the Second World War. Two aims can be identified: the strengthening of the French industry and the modernisation of the network.

We can identify three main periods in which different instruments were activated:

1. 1950 - 1975: building the national technological endowment
2. 1975 - 1984: modernising the network and restructuring the industry

[11] Another important example of the enhanced autonomy of DGT in this period concerns R&D. DGT used with wide discretion the instrument of "marché d'etude". This was an external research contract normally used only in the case of applied research - also for basic research (Bonnetblanc, 1985).

[12] The new procurement policy included the strengthening and organisation of the function of price control, included after 1975 in DGT, within a new department, called "Expertise et Controle Economique et Financier" (Economic and Financing Control and Assessment). Starting in 1984, the new group, formed by technical experts, carried out an intensive analysis of the situation of the manufacturing industry. It produced a great number of studies and created a technical and economics database to monitor prices.

3. 1984 - ....: the international dimension of the "national champion" and the restructuring of the PTT

2.1 1950-1975: Building the National Technological Endowment

During this first period, the policy was directed towards the development of an autonomous technological capability in the field of telecommunications, which was also considered strategic from the defence point of view. The main aim was "the transfer of technological competence to the private industry" (Quelin, 1992). The national market was dominated by foreign technology. Thus, "in order to create an independent industry, France first needed to end its reliance on foreign technology" (P. Griset, 1993). In fact, the technological level of the French telecommunication network was very far below the foreign standard.[13] The "technocratic" features of the French national system and the role of the public laboratory (CNET) were well reflected by the policy instruments implemented in this period. Starting from 1954, the CNET gained a wide degree of autonomy and greatly expanded its activities through a substantial increase in research budget and personnel.[14] A new laboratory was opened in Lannion (Britanny) as part of a process of decentralisation and specialisation within CNET.

The main characteristic of this process of technological development was the choice to focus the research on a major innovation: a new design for a digital switching system. Instead of building incrementally on the previous research done in the US, the CNET chose to bet on a new trajectory (the time-division technology) and to try a major technological "leap-frogging" (Libois, 1983). At the same time, the French manufacturer (CIT-Alcatel) opened, under the auspices of the PTT, a new subsidiary (SLE) in Lannion, right near the new laboratory of CNET. Efforts were concentrated on the technological field and the service provider was "not very present, even if it had to face an important challenge: to build the French telecommunication network to fill the existing gap and to compete with the European neighbours" (Quelin, 1992). Since the financial, as well as human, resources were concentrated on the technological leap-frogging, the situation of the network was still lagging behind the other main industrialised countries.[15]

---

[13] In 1945, the CNET estimated a lag of at least 8 years of the French technology (P. Griset, 1995).

[14] The new personnel involved were doing research on contract and were drawn from different scientific and technical backgrounds (Chapuis and Joel, 1990).

[15] As reported by P. Griset (1993), "during the 1960s and 1970s the developmental level of the French telephone network progressed from being a problem to being a scandal". At the end of this first phase, in 1974, the telephone density (number of telephone for 100 inhabitants) of the French network, as a standard measure of the diffusion of telecommunications, was 11.8 against the 22.8 of UK and the 19.6 of Germany.

2.2 1975-1984: Modernising the Network and Restructuring the Industry

Around the mid 70s, this policy, characterised by one major instrument (CNET) and one aim (the transfer of technology), was revolutionised because of the need to modernise the network. As Quelin (1992) emphasises, "the political will to fill the French gap in the development of the telecommunication network (VII Plan - 1975 - 1980)[16] [made] unstable the technological dynamics created by the co-operation between the CNET and the manufacturers". The result was a change of leadership in the sector from the CNET to the DGT, from both the technological and the industrial points of view. National independence was no longer seen as a priority, while the main concern was now to equip the country quickly and to obtain lower prices from the manufacturers (P. Griset, 1993).[17] However, the industrial policy was used to reduce the risk of too strong a dependence on foreign firms. The French firm, Thomson, bought (for more than 1 billion French Francs) two foreign subsidiaries (LMT of ITT and SFTE of Ericsson) which were producing under licence the space-division technology. The second French manufacturer (AOIP) was incorporated in CIT-Alcatel, part of the CGE group, which was producing the time-division technology chosen to equip only the rural and peripheral network.[18] After two years (1978), the choice of the space-division technology was criticised and the time-division technology was chosen. Notwithstanding major changes in the leadership and the goals, the process of restructuring the industry confirmed the role of the state. The French belief in the role of the state was mirrored in the evolution of the French policy.[19] In 1981, the government decided for the nationalisation of the "national champion" CGE and of CGCT, in order to inject new public capital in the industry. A further step towards the restructuring was the decision by the French government, together with DGT, to impose on CGCT the manufacturing of semi-electronic switch under the license of Thomson. Finally, the process was completed at the end of 1983, with the merger into CGE of the telecommunication activities of Thomson, which was forced to give up for financial reasons and did not have the capacity to build a new time-division system. CGE became the leading supplier of DGT with 84% of public telecommunication market share.

---

[16] The VII Plan decided to allocate to the telecommunication industry 50% of the 200 billion French Francs investments.

[17] It is in this context that the adoption of the time-division trajectory (chosen by the CNET and transferred to the national manufacturer) was postponed in favour of the space-division switching system, supplied by the foreign subsidiaries. See next section.

[18] The market shares resulting after the restructuring were 40% for Thomson, 40% for CGE and 20% for CGCT.

[19] A clear example was the importance and the consequences of the Nora-Minc Rapport (1978), commissioned by the President G. d'Esteing. The report drew attention to the synergism between telecommunications and computers. The government planned the wide diffusion of telematics services (Plan Telematique) with the promotion of, and experimentation with, for example, the Minitel. The election of Mitterand and the victory of the Socialists did not change this orientation, as shown by the delivery of the "Plan Cable" for the introduction of fiber optics (50 billion FF) and the "Plan de la Filiere Electronique" for the restructuring of the equipment industry.

2.3 1984 - .....: The International Dimension of the "National Champion" and the Restructuring of the Service Provider

The third phase was characterised by a modification, driven by technological evolution, of both the industrial landscape and the institutional setting. The new competitive environment inside the telecommunication industry needed the implementation of new instruments in order to sustain the "national champion". The national market was no longer large enough for the "national champion" to recover the high R&D costs connected with the development of the digital switching system, nor was it sufficiently large to achieve economies of scale. The French government decided on a new strategy to sustain CGE. The first step was the merger among CIT Alcatel (CGE), Cables de Lyon (CGE) and ITT International in Alcatel NV.[20] With this merger, Alcatel NV became a leading world company manufacturing telecommunications systems and products, acquiring a fundamental foreign projection through the acquisition of well-established ITT subsidiaries that were present in 11 countries.[21] In 1987, the privatisation of CGE was decided upon. The strengthening of Alcatel was a source of trepidation for the service provider, which was worried by the monopoly power of the national champion. Only CGCT, still producing under the Thomson license, now survived as a second supplier. Again, the vehicle of a state-owned firm was used as a policy instrument. After a long and difficult decision process, the French government sold CGCT to the Ericsson/Matra consortium, with the aim of technologically reinforcing the second manufacturer, which was then supplying 16% of the national switching market.

3. THE ROLE OF PUBLIC PROCUREMENT AND THE DIGITAL SWITCHING SYSTEM

In this section, we will analyse in more detail the role of government intervention for the development of the digital system, focusing particularly on public procurement policies. The evolution of the digital switching system can be followed by considering three main phases, roughly corresponding to those pointed out in the previous paragraph:

1958 - 1974 : the technological leap-frogging and the role of CNET;

1975 - 1978 :the change in the leadership;

1978 - ... : the diffusion of the E10.

---

[20] The new society, Alcatel NV, officially operating from January 1987 onwards, was controlled by CGE (55.6%), ITT (37%), Societe' Generale de Belgique (5.7%) and Credit Lyonnais (1.7%). The ITT share would be acquired by Alcatel NV in 1991.

[21] With the acquisition of ITT, Alcatel took control of another switching system (the S12), doubling its installed lines but also its R&D efforts.

3.1 1958-1974: The Technological Leap-Frogging and the Role of CNET

The first phase was characterised by a complex organisational design. As we have already noted, during the first phase the essential actor of the process was CNET, the research laboratory of the French PTT. In 1958, after the opening of the switching department in Lannion, the CNET formed an alliance with SOCOTEL, the pool of the French manufacturers for switching equipment (the two French subsidiaries of ITT, CGCT and LMT, the French subsidiary of Ericsson, SFTE and AIOP and CIT-Alcatel), to do research on the new technological paradigm of electronics.

Two projects started: the first one, Socrates, involved the firms of SOCOTEL (the cartel of manufacturers for the switching market) and the CNET. The second one, Aristote, involved only the CNET and the LSE, the new society opened in Lannion by CIT-Alcatel. The Socrates project aimed at the development of an electronic switch following the space-division trajectory, while the Aristote chose to start directly on the time-division trajectory. This division of labour may be explained by two reasons.

1. On the one hand, the consideration of their technological advance justified the choice of manufacturing firms for space-division. Given their greater experience in the electromechanical paradigm, they were thought to be in a better position to introduce electronic control and managing systems gradually inside a cross-bar switch. Conversely, in the previous ten years, the CNET had accumulated a more profound knowledge of basic research in electronics, especially in transmission, and it was thus better suited to develop a new totally electronic switch.
2. On the other hand, the bet on the time-division technology was part of the project of building French independence in this strategic field. It was clear that if the time-division project proved to be successful, the French manufacturing industry would be in a monopolistic position, well ahead of foreign subsidiaries. In the case of success, the exclusivity of French efforts on the time-division technology would have pushed ITT and Ericsson outside the market.[22]

Socrates was followed by a new project, Pericles, while Aristote was followed by Platon, the project that finally led to the installation of the world's first time-division switch. The Platon prototype, later known as E10, was installed in Lannion in 1970 and followed six months later by a new bigger prototype. Between 1970 and 1972, research continued in Lannion, concentrating on how to pass from the prototype phase to the industrial one. In 1972, the E10A was ready to be produced and sold. In the same year, the Ministry of PTT confirmed the importance of the new technology for the modernisation of the French network, announcing that the E10 would cover 2% of the French switching market for 1973 and 10% in 1975 (Libois, 1983).[23]

[22] In fact, the group of researchers working on Platon aimed explicitly to "rapidly realise the prototype of [a] totally electronic switch and then to pass immediately to the industrial phase" (Libois, 1983).

[23] Moreover, the support for the time-division technique was confirmed in 1973 by the new Minister of PTT, who was hoping for a further development of the system in order to serve bigger towns as well (Libois, 1983).

The first phase of the development of the digital switching system was marked by the creation of a "specific organisational device" (Quelin, 1991), featuring the central role of the CNET and the importance of the division of labour among the manufacturers. CNET had, during this period, two main tasks: the R&D and the control over the equipment. These double functions gave the possibility to interact with both the service supplier and the manufacturers and provided the advantage of a relative autonomy with respect to both. Personal contacts and trust were the main instruments by which CNET developed a dense network of relationships with the industrial side (facilitating the technological transfer), with the political side (accelerating the financing of projects) and with the academic side (strengthening the flows of knowledge and personnel). A crucial aspect of the organisation of the technological development was the involvement of all the suppliers acting on the French market and the division of labour between the different members of SOCOTEL. The division of labour facilitated the interaction and cross-fertilisation between the two projects, given the presence of CNET researchers in both groups.[24]

3.2 1975-1978: The Change in the Leadership and the Role of PTP in the Early Adoption of an Innovation

In this phase, the role of CNET in the French industrial policy was brought into question. The criticisms of the CNET strategy concentrated on an excessive homogeneity of the environment and on too strong a concentration of the financial means and decision functions, both of which were lowering the flexibility of the telecommunication industry. Moreover, the importance of the CNET from the technological point of view was considered as a major cause of the already high dependence and scarce entrepreneurship of the French manufacturers (P. Griset, 1995).

As a result, in 1975, DGT took over the "leadership" of the telecommunication sector. The main aim was to modernise the network and to obtain a lowering of the prices for equipment. This was to be done by introducing more competition and control in a market dominated by SOCOTEL, considered by DGT to be a real cartel. This change of policy had importance consequences for the development of the digital switch.

In 1975, the DGT decided on an international bid to equip the Paris network.[25] The DGT chose, after a long and difficult decision process, to share the market between two space-division systems (the Metaconta of ITT and the AXE of Ericsson) and the E10. Space-division technology was given the priority in the planned modernisation of the French network, particularly concerning high capacity and transit switches, while the

[24] For example, the experience of the manufacturers with the reliability of the switch drove towards the adoption in the Socrates prototype of a particular principle: the "load sharing" (partage de charge), i.e., the simultaneous use of two (or more) parallel computers to enhance reliability. This principle was later used in the design of the architecture of the Platon.

[25] DGT was given a wide degree of autonomy especially with the possibility to derogate from the French administrative law for the public procurement (see section 1.3).

time-division technology was relegated to the ancillary role of equipping the rural and low density part of the network. At the same time, a "second French pole" for switching equipment was created with the acquisition of LMT (ITT) and SFTE (Ericsson) by Thomson. Three years later, in 1978, the choice of the space-division technology was brought into question. The time-division technology was seen by then as the only way to assure the planned evolution towards telematics. Thomson, which was supplying the space-division system, was no longer able to sustain the competition of CGE, which had already developed the technology with the fundamental help of the CNET. In 1983, the telecommunication department of Thomson was sold to CGE, newly nationalised.

The announced strategy of pushing the foreign subsidiaries out of the French market with the technological instrument was suddenly stopped by the 1975 decision to opt for space-division technology. Two interpretations of this change of policy can be found, each of which gives a different assessment of the role of PTP in the development of the national technology.

On the one hand, this strategy can be seen as "a major failure that delayed the international development of the French industry" (P. Griset, 1995). The decision to give Thomson the leadership, and to implement the space-division technology widely, represented the beginning of a "Franco-French competition", driving towards a loss of years of development and billions of Francs (P. Griset, 1993). The positive climate created by the CNET and the success of the strategy of technological transfer was disrupted by the intervention of the public procurement, which aimed mainly at creating a competitive environment.[26] If we accept this interpretation, we are confronted with an example of public procurement which slowed the innovation, choosing the correct option to enhance short-term static efficiency but not technological development.

On the other hand, the choice in favour of the space-division technology can be seen as an option allowing a gradual modernisation of the network. In this acceptation, the time-division technology was not ready to equip bigger towns (Libois, 1983). Moreover, it was too costly and too risky. While waiting for a mature phase of the national technology, there was a choice to be made between protecting the "national champion" or gaining from its competition with other suppliers. This "dilemma" was solved with the creation of a second "national champion" for the space-division technology. In this sense, we might say that public procurement was used as a way to enhance variety, allowing competition between two different technological trajectories in an early phase of development (Cohendet and Llerena, 1997). The three years delay in the full adoption of the E10 in the French network, coupled with the installation in rural networks, might have assisted CNET, which was still responsible for technological improvement,[27] to improve gradually the performances and the reliability of the system. In this interpretation, the choice to sustain the space-division trajectory as a

[26] According to observers of the period (*Le Monde,* July 30, 1976) quoted by P. Griset (1993.), the actions of the government were to be stopped and reconsidered because "competition is a good thing if it does not turn into anarchy".

[27] For the whole of 70s, CIT-Alcatel was responsible only for the production and the marketing of the new system while the technological improvements were still conducted under "umbrella" of CNET.

"bridge technology" was not totally incorrect. The knowledge about the time-division technology, which was totally new on the market, was not sufficient to permit the DGT to bet on this trajectory from the beginning. We recall that no other country had yet tested a massive introduction of time-division switches in its network. Even if within CNET there was great optimism about the development of microelectronics and semiconductors, the DGT did not have any guarantee that the E10 would have the capacity to handle high traffic nodes.

3.3 1978-19..: The Diffusion of the E10 and the Role of PTP in the Diffusion of an Innovation

However, starting in 1978, the strategy of the CNET was finally realised with the full choice of the time-division technology. The public procurement started having the positive effect of diffusing the new technology, which was, by then, ready to be sold and installed. According to Libois (GRINSEC, 1980), "after 1978, the situation has been clarified and the validity of CNET's advanced solutions has been proven by the prodigious development of semi-conductor technology and by the introduction of microprocessors" (our translation). The French General Director of Telecommunications confirmed, in a conference held in Paris in 1979, that "the development of time-division switching became by now a reality in the already taken or imminent decisions in most of the countries"(our translation) (Libois, 1983).

In 1980, the time-division switches covered 70% of the procurement of DGT. In 1982, the orders for E10 amounted to 8 million lines, of which 2.5 million were already installed. In the early 80s, DGT promulgated new operating and maintenance standards, together with new cost requisites. The new standards had to be respected at the same time that prices had to be reduced in view of the large expansions of the size of the market offered by the PTT. Until then, it seems that the problem of reducing the selling price had been considered by CIT-Alcatel only for export and not for the national market (Chapuis and Joel, 1990). This situation changed with the planned expansion of investments for the digitalisation of the network. Thus, the DGT confirmed, after the choice of the time-division technology, that costs and prices were a main concern of the procurement.

Starting in 1978, the time-division technology proved to be successful, thanks to the massive installation on the French market. During the 80s, the system was the basis on which the French network was digitalised. In 1989, the share of digital lines on total lines was 65.8 %, while it was 41.6 in the US, 45 in Canada, 24.9 in Japan and 31.3 in UK. In 1989, the E10 family equipped 17,000,000 lines in France. The role of the national market was fundamental for CIT-Alcatel to enhance the performance of the system. New versions of the E10 were developed in 1980 and in 1985. The latest version, Alcatel E10, was at the basis of the first ISDN trial in 1987. The success of the E10 on the national market gave CIT-Alcatel an important competitive advantage in the field of electronic switching. The process of digitalisation of the French network was a key way to prove the reliability of the system to foreign clients. At the end of the 80s, the E10 was installed in 40 international markets.

4. APPRAISAL OF SOCIETAL AND INDUSTRIAL RESULTS AND CONCLUSION

The aims of the French policy for the telecommunication sector were the development of the network infrastructure and the strengthening of the national industry. We will now briefly analyse the specific condition of the French telecommunications network, in comparison with other European countries.

A standard measure of the state of a telecommunication network is the density of telephone lines. In the late 60s and early 70s, the French telecommunication network was characterised by a significant lag in terms of density (number of lines per 100 inhabitants) (Table 1). The main part of the gap was filled during the 70s. In the first half of the 80s, France was the leading European country in terms of telephone lines per 100 inhabitants.

Table 1: Telephone lines for 100 inhabitants

	Italy	France	UK	Germany
1974	16.49	11.81	22.87	19.61
1980	23.07	29.95	31.44	33.35
1985	30.45	41.75	36.95	41.94
1989	34.90	45.23	42.10	45.90
1990	38.80	49.60	46.00	48.50

Source: OECD (1988), STET (1990 - 1991), Reseau (1991)

The evolution of the digital switching system in France resulted in the fast development of the telecommunication network of the country. The policy of procurement decided on by the DGT drove the process of numerisation of the network, giving it the possibility to introduce new features and services earlier than in other countries. Table 2 shows the period of introduction of different types of specialised network in Italy, France, UK and Germany from the end of the 60s to the 90s. The introduction of specialised networks indicates the level of technological advance in the telecommunication system of a country.

Table 2: The introduction of specialised networks

Years	Italy	France	UK	Germany
1967				DATEX
1972		CADUCEE		
1976				DATEX-L
1978		TRANSPAC		
		Transmic		
1980				DATEX-P
1981			PSS	
1982			Kilostream	
1983				
1984	RFD		OTS	
1985			ISDN Trial	
1986	ITAPAC	Telecom 1		Telecom 1
		ISDN Trial		
1987		ISDN		ISDN Trial
1988				
1989	Italsat 1		ISDN	ISDN
1990	ISDN Trial			

	Circuic switch data transmission		Packet switch data transmission		Direct digital circuit
	Satellites		ISDN Trial		ISDN

Source: Adaptation from Reseau (1991)

Table 2 confirms the path shown in Table 1, with France's earlier start and fast recovery during the 70s. Table 3 shows the data on the digitalisation of the network in the main industrialised countries from 1988 to 1993. France, thanks to the massive installation of the national switching system starting at the end of the 70s, appears as the leading country.

Table 3: Digitalisation of the network 1988-1993 (% of digital lines on total lines)

	1988	1989	1990	1991	1992	1993
France	**60.5**	**65.8**	**69.1**	**73.5**	**78.1**	**83.0**
Italy	20.1	24.7	32.6	39.9	47.4	54.5
USA	35.4	41.6	48.1	54.4	60.9	68.1
Canada	40.0	45.0	50.0	55.0	62.0	68.0
Japan	17.4	24.9	31.9	38.9	46.4	53.3
Germany	5.0	7.3	10.5	15.0	19.4	23.7
UK	23.5	31.3	39.7	41.1	53.6	59.4

Source: Zanfei (1990) from NBI and SIP Development Plan

Table 4: The top 10 firms in the telecommunication industry (market shares in %)

	1980		1985		1986		1987	
1	AT&T	26.80	AT&T	20.14	AT&T	16.66	AT&T	13.81
2	ITT	12.40	ITT	7.57	Alcatel NV	11.97	Alcatel NV	8.68
3	Siemens	6.90	Northern T	7.05	Northern T	6.74	Siemens	7.88
4	Ericsson	5.50	NEC	7.76	NEC	6.74	NEC	6.85
5	GTE	5.10	Siemens	5.00	Siemens	4.76	Northern T	6.07
6	Northern T	4.20	Motorola	4.53	Motorola	4.70	Motorola	3.80
7	NEC	3.00	Ericsson	4.42	Ericsson	4.31	Ericsson	3.57
8	Philips	2.80	GTE	3.37	GTE	3.09	IBM	3.13
9	CIT-Alcatel	2.40	CIT-Alcatel	2.72	IBM	3.08	Fujitsu	2.94
10	Thomson	2.30	IBM	2.65	Philips	2.31	GTE	1.97
	Other	28.60	Other	34.79	Other	35.65	Other	41.33

Source: Zanfei (1990)

The strengthening of the national champion was the second explicit aim of the government policy. The growth in Alcatel's market share and its accession to the top of the oligopolistic core show the success of this policy. (See Table 4 above.) It is important to note, however, that public procurement was not the only instrument used to sustain the French manufacturer.

Table 5: The evolution of Alcatel 1985 - 1991 (million US$ 1987)

	1985	1986	1987	1988	1989	1990	1991
Telecom. sales	5242.6	7783.6	8262.0	8922.5	11375.0	11889.4	16096.1
Switching sales	24.9 %	n.a.	35.4 %	34.4 %	37.3 %	39.9 %	35.2 %
Total R&D	n.a.	n.a.	9.8 %	11.6 %	10.5 %	11.4 %	11.9 %
Switching R&D	n.a.	5.5 %	6.0 %	6.0 %	5.0 %	5.7 %	n.a.
E10	n.a.	3.0 %	2.8 %	2.4 %	1.9 %	1.9 %	n.a.
S12	n.a.	2.5 %	3.2 %	3.6 %	3.1 %	3.8 %	n.a.

N.B.: Percentage share on telecommunication sales

Source: D'Orazio Zanfei (1994)

In particular, as shown in table 4, the early success of the French digital system and the procurement process of the 70s succeeded in giving CIT-Alcatel the 9th position in the market for telecommunication equipment in 1980. It was only after the acquisition of Thomson on the internal market and, above all, the joint-venture with ITT, that Alcatel became the second supplier of the market (see Table 4). Table 5 presents data about the evolution of Alcatel (total sales, sales in the telecommunication sector, sales in the switching market, and R&D expenses).

The costs involved in the process are hard to assess. Data about R&D expenses during the 70s were not reported consistently. Moreover, the strong research endowment of CNET in electronics was later used for the development of other strategic technologies (e.g., semiconductors), creating positive synergy for the entire national system of innovation.

The evolution of investments of the service provider (Table 6) provides a rough estimate of the total costs involved in the process of procurement. Unfortunately, they include investments for the total procurement and not only for the switching market and they are available only from 1975 to 1985. However, the data show clearly the support to the French national champion, starting in 1978.

Table 6: Telecommunication investments (three year average - million US$ - constant price 1980)

	1975 - 1977	1977 - 1979	1979 - 1981	1981 - 1983	1983 - 1985
Million US $	4521	5598	5190	4566	4535
(per capita)	*85.42*	*104.87*	*96.36*	*83.86*	*82.54*

Source: OECD (1988)

In France, the role of the government was primarily and mainly directed towards the creation of a technological competence in strict co-operation with the national manufacturer. An interesting feature of this process of capability-building was the choice to sustain different technological trajectories through a specific management organisation (SOCOTEL). During a "pre-paradigmatic phase", as was the case for the introduction of electronics in switching equipment, the value of diversity is particularly high (Cohendet Llerena, 1997). Moreover, in the case of SOCOTEL the public laboratory successfully played the role of "bridge" between the different technological trajectories, permitting the sharing of information and knowledge between the two groups. This was fundamental for exploiting the diversity that had been created. (See Table 7.)

The process of procurement took place following this process of developing technological capability inside the service provider and the national manufacturer. As the tender for the Paris switch shows, the procurement policy of DGT was, at first, focused on a less risky trajectory (space-division) until the time-division trajectory was in a more mature stage. Starting from 1978, public procurement had been a fundamental instrument for sustaining the diffusion of the national technology, both in the internal market and abroad. The French procurement policy showed a strong orientation towards the "national champion", following the policy orientation of the French national system. The coherence between the needs of the service providers and the orientation towards protectionism was obtained through other policy instruments, such as the nationalisation of foreign subsidiaries and the restructuring of the sector.

The French telecommunications case shows a strong interdependence between public procurement and the national system of innovation. In France, the procurement was part of a process of capability-building involving the whole high-tech sector (energy, space, defence, computers and semiconductors). The instrument of procurement was given a technology content because of this comprehensive project. The role of the state appears to have been fundamental in creating the technological endowment of the country. In telecommunications, this was accomplished through a very specific organisation (CNET) which had the technological competencies and was situated at the real "core" of the National System of Telecommunications.

Table 7: The technological evolution of the E10

Year	Major Steps
1958	Creation of the Switching Department of CNET - Creation of SOCOTEL - Project Socrate and Aristote
1967	Platon and Pericles
1970	First E10 installed
1972	E10-A (improvement in program capacity and components)
1980	E10-B (new circuits, modular architecture, improvements in program capacity)
1985	ALCATEL E10 (new access unit for network compatibility - ISDN level)
1987	First ISDN trial

Source: Our elaboration on Chapuis Joel (1990)

REFERENCES

W.S. Atkins Consultants. (1987). *The Cost of non-Europe in Public Sector Procurement.* In the European Commission (1988), *The Cost of non-Europe, Vol. V.* Brussels.

Bonnetblanc, G. (1985). Les telecommunications francaises: Quel statut pour quelle entreprise ?. *La Documentation Francaise.*

Carré, P. (1996). Le téléphone en France en 1947: Une modernisation manquée. *Les Cahiers - Télécommunications, Histoire et Société, n.3.*

Cawson, A. et al. (1990). *Competition and Closure in the European Electronics Industry.* Oxford: Claredon Press.

Chapuis, R. and A.E. Joel. (1990). *Electronics, computers and telephone switching: A book of technological history.* In North Holland studies in telecommunication, *100 years of telephone switching, Volume 2: 1960 - 1985.* Amsterdam, New York, Oxford: North Holland Publishing Company.

Chesnais, F. (1993). France. In R.Nelson (Ed.), *National Systems of Innovation: A comparative study.* Oxford: Oxford University Press.

Cohen, E. (1992). *Le colbertisme "High-Tech".* Paris: Hachette.

Cohendet, P. and P. Llerena (Eds.). (1991). *Modes of use and Diffusion of New Technologies and New Knowledge: The Case of France.* BETA, University of Strasbourg. Report for the EC MONITOR-FAST Programme. (FOP N.228). Bruxelles.

Cohendet, P. and P. Llerena. (1997). Learning, technical change and public policy: How to create and exploit diversity. In C. Edquist (Ed.), *Systems of Innovation - Technologies, Institutions and Organizations.* London and New York: Pinter Publishers.

Costa, A. (1993). Le politiche di acquisto dei gestori CEE. In E. Pontarollo (Ed.), *Politica degli acquisti e struttura dei mercati nelle telecomunicazioni.* Milano: Vita e Pensiero.

D'Orazio, A. (1993). L'evoluzione dell'industria nazionale di sistemi ed apparati di telecomunicazione: Fatti e misfatti della politica industriale italiana. *Economia e Politica Industriale, n.78.*

D'Orazio, A. and A. Zanfei. (1994). *L'industria europea delle telecomunicazioni. Relazioni cooperative e strategie competitive delle grandi imprese manifatturiere.* Milano: Franco Angeli.

Dang Nguyen, G. (1983). *Les telecommunications en Europe durant la decenie 1970 - 1980.* Unpublished Ph.D. Thesis, IUE, Firenze.

European Commission. (1996). *Green Paper on Public Procurement in the European Union: Exploring the Way Forward.* Bruxelles

France Telecom.(1991) Chercher plus, chercher autrement. *France Telecom n. 78* (July).

GRINSEC. (1984). *La commutation electronique*. Paris: Eyrolles.

Griset, P. (1993). The Centre National D'Etude des Telecommunications and the Competitiveness of French Telephone Industry, 1945 - 1980. In W. Asprey (Ed.), *Technological Competitiveness: Contemporary and Historical Perspectives on the Electrical, Electronics, and Computer Industries.* New York: IEEE Press.

Griset, P. (1995). Innover comment ? Innover pourqouoi ? Cinquante ans de recherche au Centre National d'Etudes des Telecommunications. *Les Cahiers - Telecommunications, Histoire et Société, n.2.* France Telecom.

Libois, L.J. (1983). *Genese et croissance des telecommunications.* Paris: Masson.

Llerena, P., M. Matt and S.Trenti. (1997). *Government Technology Procurement: the Case of Digital Switching System in Italy.* (Report prepared for the TSER-ISE Project, Subproject 3.2.2, 'Government Technology Procurement as a Policy Instrument '. November.)

Northern Business Information - NBI. (1989 and 1990). *World Public Switching Market.* New York: Northern Business Information.

OCDE. (1983). *Les telecommunications: Perspectives d'evolution et strategies des pouvoirs publics.* Paris.

OCDE. (1988). *Telecommunications Industry: The challenges of structural change.* Paris.

OCDE. (1989). *Indicateurs de performance pour les exploitants de telecommunications publiques.* Paris.

OCDE. (1990). *Communications Outlook.* Paris.

Quelin, B. (1992). Trajectoires technologiques et diffusion de l'innovation: L'example des equipments de telecommunication. *Revue d'Economie Industrielle, n.59.*

Reseau. (1991). Archivio delle telecomunicazioni: Tecnologie, sistemi, imprese, paesi. *Prima Comunicazioni, n. 202*, (November).

Schnoring, T. (1990). Research and Development in Telecommunications: An International Comparison. Paper presented to the VIII ITS Conference, *Telecommunications and the Challenges of Innovation and Global Competition*. Venice: March.

Trenti, S. (1992). *Acquisti pubblici e innovazione sul mercato internazionale delle telecomunicazioni e del termoelettromeccanico. Il caso italiano.* Tesi di Laurea, Universita' Bocconi, Milano, March.

Zanfei, A. (1990). *Complessita' e crescita esterna nell'industria delle telecomunicazioni.* Milano: Franco Angeli.

Zanfei, A. (1997). Domanda pubblica ed innovazione tecnologica: presente e futuro nel settore delle telecomunicazioni in Italia. (mimeo). Un. Bocconi: Milano.

8. PUBLIC TECHNOLOGY PROCUREMENT: THE CASE OF DIGITAL SWITCHING SYSTEMS IN ITALY

P. Llerena, M. Matt and S. Trenti

INTRODUCTION

The aim of this chapter is to study the role of Public Technology Procurement (PTP) in the telecommunication industry in Italy. The analysis focuses on the evolution of the digital switching system. The case study outlined is the development of the Proteo/Linea UT family. The period covered is from the 60s to the end of the 80s. In section 1, the Italian national system of innovation will be briefly described and the institutional and legislative framework for procurement and the industrial structure of the telecommunication sectors will be presented. Section 2 will introduce the evolution of the government policy interventions in this industry and section 3 will focus more precisely on the evolution of the digital switching system case, studying the specific role of PTP and other policy instruments. In the last section, we will try to assess some of the societal and industrial results.

1. THE ITALIAN TELECOMMUNICATION SECTOR

1.1 The Italian National System of Innovation

The Italian national system of innovation, developed after the Second World War, is mainly characterised by the presence of two different institutional settings (Malerba, 1993). On the one hand, there is a successful and unique network of small firms with an historical local vocation.[1] On the other hand, there is the "core R&D system" formed by larger firms, universities, small high tech firms and large public research institutes. The performance of the core R&D system is less satisfactory in terms of innovation and international competitiveness than the small firms network system. The main rea-

[1] Their capabilities are built through productive experience: Innovation stems not from formal R&D but from informal learning by doing, using and interacting

sons for this poor performance can be identified in the absence – or lack of development – of some qualitative elements (Malerba, 1993):

1. several industrial sectors do not have advanced research and technological capabilities;
2. public policy for R&D support still exhibits major flaws;
3. an advanced national infrastructure of services for R&D and an overall co-ordination of public policies are still lacking;
4. advanced basic research performed in universities and public research centres is very unevenly distributed across organisations;
5. shortages of skilled scientists and engineers are present;
6. there is no tradition of successful industry-university co-operation in research.

An important feature of the core R&D system is the presence of a small oligopolistic core, characterised by a few, limited-sized, firms.[2] This small oligopolistic core seems rather isolated from the research activities taking place in universities. The links between firms and public research laboratories are rather weak, due to the limited number of centres of excellence in Italian universities, the limited mobility in and out of the university, and the bureaucratic and institutional structure of universities (Malerba, 1993).[3] Moreover, large public research laboratories were not given the priority to build a national technological endowment and to transfer technological capabilities to firms, as happened, for example, in France after the Second World War (Chesnais, 1993, Llerena, Matt, Trenti, 1997). The lack of interfaces between industry and universities is coupled with the lack of a tradition of state support for R&D. The Italian public policy has traditionally concentrated on supporting large-scale economic sectors (steel and basic chemicals) with standard technologies. The support for high-tech industries began only in the late 60s and the 70s but only with limited resources and with an imprecise policy model. The choice of the sectors and the timing of the support was mainly influenced by the emergence of industrial and scientific lobbies in advanced sectors (Malerba, 1993). Despite these drawbacks, during the 80s the Italian R&D core has shown a new technological dynamism in R&D growth, characterised by a concentration in specific technological areas and by vertical and horizontal co-operative agreements with national and international firms.

---

[2] Part of this core is formed by public enterprises (ENI, STET, SGS) that played an important innovative role, especially during the 60s and 70s, even in the presence of political obligations to save industries in crisis and to be active in the South of the country. The other part of the oligopolistic core (FIAT, Olivetti, Pirelli) is private and controlled by family-owned holdings.

[3] The impermeability of the Italian academic world to novelties was particularly evident with respect to the development of electronics. In a 1957 Report on "50 years of Scientific Development 1907-1956", written by well-known scientists, thirty sectors were included, from physics to biology and medicine, while electronics, computer science and telecommunications were totally absent (as quoted by CSELT – Bonavoglia ed., 1994). Only by a 1961 a reform did the university come to include an electronics engineering faculty and courses in computer science. These, however, were not uniformly introduced in all technical faculties. Therefore, it was only at the end of the 60s and during the 70s that younger graduate technicians became available for R&D in telecommunications and electronics.

1.2 The Italian National System of Telecommunication

1.2.1 The regulatory framework in Italy: A hybrid situation

In Italy, the telecommunication services have been characterised by a hybrid situation with the presence of multiple service providers with different status and subjected to different rules. Figure 1 shows the institutional framework of the Italian national system of telecommunications. The government interventions were based on three main axes. The Ministry of PT directly exploited the long distance services through ASST, and had the main regulation and control functions. The Ministry of Industry intervened in the planning of industrial restructuring. The Ministry of State-Owned Firms (Ministry of PPSS) had responsibility over the financial and industrial planning of the state-owned firms grouped in STET, the Italian holding company for telecommunications. The STET group included both services providers (SIP for local calls, Italcable for international calls and Telespazio for satellite), the national research centre (CSELT) and the main national manufacturing firm (SIT-Siemens).

Because of historical contingencies,[4] the Italian telecommunication sector has been characterised by two peculiar features: the distribution of roles and functions among different societies for the provision of services, and the vertical integration of service and manufacturing facilities inside the STET group. Concerning the institutional setting, criticisms have pointed to the "anomaly" of the survival, until the early 90s, of a situation of unclear separation between the regulatory and the operational functions, and of overlapping roles.[5] The presence of different companies, together with the administration of PTT, all of which were subject to different types of regulation and behaviour, might have constituted an obstacle to an effective and prompt policy interven-

---

[4] In 1925, an international bid for the provision of telecommunications services was decided on. The tender included five local service areas and one grant for long distance service. The state aim was "not to face a big organisation with a monopolistic power"

[5] (Adunanza Generale del Consiglio di Stato, December 11 1923, quoted in CSELT, 1994). The five territories were assigned to five companies (STIPEL, TELVE, TIMO, TET and SET). Since the long distance service was still underdeveloped, no companies participated in the sixth tender. The Ministry of PT decided, then, to create an operational branch, ASST, (Azienda di Stato per i Servizi Telefonici, State Company for Telephone Services) to provide this type of services. In 1933, three of the companies providing local services (STIPEL, TELVE and TIMO) were put under the control of the state and they were integrated into the STET group, creating a holding company for telecommunications. In 1957, the process of financial consolidation was completed with the acquisition of TET and SET by STET. The STET group also included Italcable and Telespazio. Italcable was created in 1921 to test and install sub-marine cable for international links. Telespazio was created in 1961 to develop, test and install satellite services. In 1964, the five companies were unified in SIP, a hydroelectric company freed of the energy function after the 1963 nationalisation of the energy sector. SIP entered the telecommunications sector with a large financial dowry, created by the reimbursement of the nationalised energy assets. In 1964, SIP also took the control of CSELT, the research laboratory of STIPEL. Starting in 1945, the STET group also included the main Italian manufacturing firm, SIT-Siemens (from the beginning of the 80s Italtel). SIT-Siemens was the Italian subsidiary of the German Siemens, producing telecommunications equipment under licenses for the Italian market. In 1945, the company's Milan plants were occupied by the CLN (National Liberation Committee) and were later passed under state control as a German war damage recovery.

tion.[6] Concerning the presence of SIT-Siemens inside the STET group, this vertical integration, copied from the successful model of AT&T, was considered to be one of the instruments for fostering and protecting the Italian "national champion". In this sense, it might have not been seen *per se* as an obstacle to the development of the sector. However, in the particular Italian situation, this vertical integration has been the subject of many criticisms and debates.[7] The main criticisms have underlined the need for more transparency in the decision process (procurement bids, tariff levels, strategic choices) and the partial rigidity of the integrated group with respect to the different challenges faced both by the service provider and the manufacturer.

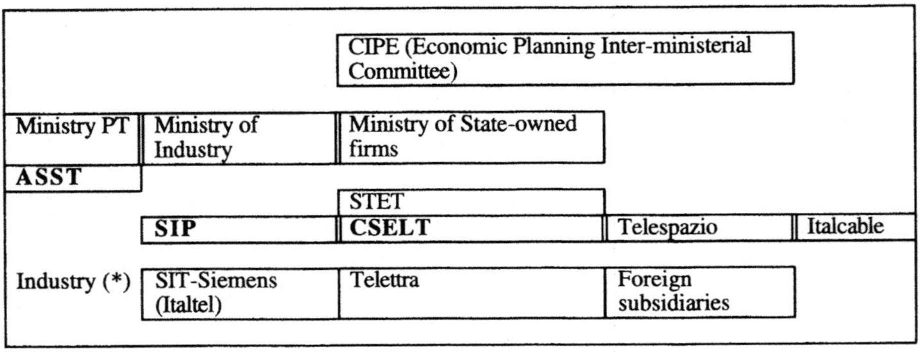

Figure 1. The Italian National System of Telecommunications

(*) The structure of the industry will be described in the next section

Concentrating only on the switching market, we can identify three main players of the government telecommunication policy: ASST, SIP and CSELT, these last two being grouped inside STET.[8] The division of labour and the distribution of tasks between ASST and SIP have evolved in the course of time.[9] The role of ASST with regard to switching equipment was marginal, with only 24 switches in the network. ASST procurement was mainly concentrated on transmission equipment, cables and installation works. SIP, now Telecom Italia, was the main procurer of the Italian market for telecommunications equipment. Its procurement may not formally be classified as public

[6] For a point of view of the 70s on this theme, see Lizzeri and de Brabant (1979)

[7] The companies grouped in the STET holding are subject to the private company laws, while ASST is a public administration. Moreover, the double control of SIP both by the Minister of PT, through the exploitation concession and the tariff control, and by the Ministry of PPSS, through the financial control, might have created problems of coordination in the government intervention. In 1992, the problem of distribution of roles was solved with the integration of ASST inside the STET group. Further restructuring is currently being decided during the process of privatisation of the STET group.

[8] See again Lizzeri and de Brabant (1979).

[9] STET is one of the main Italian patenting firms (Malerba, 1993) and it has played a major role in many technological choices and developments.

procurement, since SIP had the status of a private company. However, it was under the direct control of the government through both the Ministry of PT, for technical and tariff control, and the Ministry of PPSS for economic and financial control.[10] Concerning technological competencies, SIP had its own R&D laboratory in Rome, controlled by the General Director. The Rome group had traditionally worked in strict co-operation with CSELT, the R&D laboratory of the STET group. The vertical integration inside the STET group makes it difficult to compare the data about R&D budget and about the distribution of expenses between firms and service providers for the Italian case. However, the distribution shown in Figure 2 and the historical development of the link between SIT-Siemens and SIP may suggest a greater weight of the manufacturer in terms of technological choices and competencies inside the group. The R&D expenses of SIP were, in 1987, 90 million US$, i.e., 0.9 % of global revenues. The research personnel amounted in the same year to 160 units. The R&D data on SIP do not include the research effort of the common research laboratory (CSELT) of the STET group.

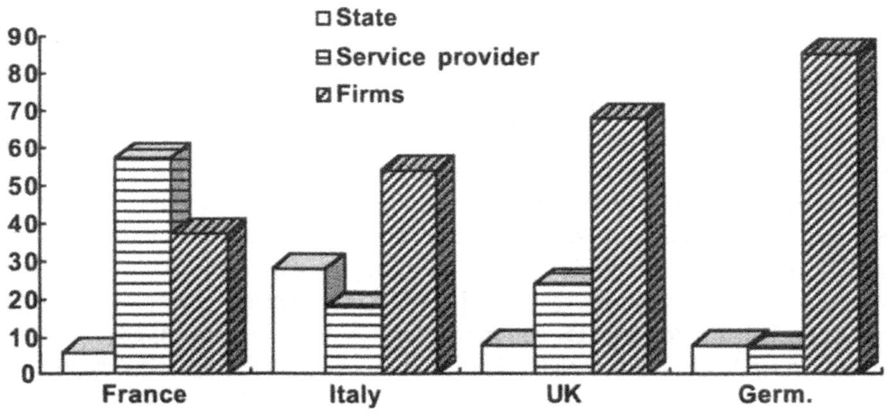

Figure 2. R&D expenses in civil telecommunication by contributors (% shares – 1987)

Source: France Telecom (1991)

[10] Until 1968, ASST was in charge of the long distance traffic, including both calls between big towns, at the head of the network districts (Milan, Rome, Turin, Naples etc.), and local calls connected to long distance calls. In 1968, a new distribution of roles was decided on. ASST was in charge of only direct links among big towns, while SIP covered all calls coming from local districts, paying 18% of tariffs back to ASST for the use of the long-distance network. Tariffs for long-distance services were very much higher than those for local calls, while investments for the network were higher for the local service. For this reason, in the early 80s, a Compensation Chamber (Camera Conguagli) was created in order to equilibrate the revenues of the two service providers according to the distribution of costs.

The CSELT (Centro Studi e Laboratori sulle Telecomunicazioni) was created in 1964 in Turin together with the entry of SIP into the telecommunication industry.[11] We can discern a double nature in CSELT. On the one hand, CSELT used to work mainly for the service provider, hosting researchers and technicians, on specific needs. On the other hand, the CSELT identified itself as one of the major Italian centres for fundamental research in telecommunications and microelectronics. However, CSELT had never been charged with the transfer of technology to national manufacturing firms. The co-operation with manufacturing firms, especially with SIT-Siemens, integrated in the same group, had been organised mainly around bilateral contracts for research, or on specific needs.[12] Until the end of the 80s, the research was almost equally split between fundamental research and applied research on contracts. Starting in 1987, contract research gained importance, responding to new needs of the service provider. The new type of activity was especially concerned with the testing and qualification of equipment for the service provider.

1.2.2 The Italian telecommunication industry

In Italy, the telecommunication industry has been historically characterised by the presence of multinational firms, since the opening, in 1909, of FACE, a Western Electric subsidiary. FACE passed, during the 30s, under the control of ITT, as part of the agreement between ITT and AT&T for the partition of the European and US market. The Swedish LM Ericsson has been present in Italy since 1927 with FATME, while the US GTE arrived in Italy in 1939 with the acquisition of a small Italian firm. The German Siemens opened its Italian subsidiary (Societa' Italiana Telecomunicazioni, SIT-Siemens) in 1921. After the Second World War, SIT-Siemens passed under the control of the Italian state as part of the war damage recovery. In 1950, SIT-Siemens became part of the STET group, the Italian holding for telecommunications. Finally, Telettra,[13] the second Italian manufacturer, historically specialised in the transmission technology, succeeded during the 70s in obtaining a small market share in the switching sector with the development of an electronic switching system, as we will see more clearly in later sections.

---

[11] The Ministry of PT control was mainly established through the concession for the provision of services in which the level and the quality of the service were fixed. Moreover, starting in 1968, SIP had to prepare an annual plan to be controlled and approved by the Ministry of PT. The annual plan was the main instrument for SIP to ask for tariff increases, since tariffs were fixed by the Ministry. Through the tariff increases and the annual planning, the Ministry had thus the indirect possibility to control the level of SIP investments. The Ministry of PT was also in charge of regulating standards and technical aspects. The Ministry of PPSS, for its part, was given the control of the financial health of state-owned firms and it thus had economic control over SIP.

[12] It was based on an already existent laboratory (CSEL – Centro Studi e Laboratori) created by STET in 1961 to fulfil the request of STIPEL managers. The main tasks of CSEL, and later – in the early years – of CSELT, were not only research and development but also the testing of the equipment and the planning of the network, concentrating in a single laboratory the technical expertise and competencies of the five companies grouped in STET. Since the presence of the five service providers had led to a fragmentation of the network into five areas with different equipment and network architectures, one of the first tasks of the laboratory was to enhance the coherence of the national network.

Table 1: Market shares of the Italian switching equipment: % of sales and of lines installed

Year	FACE	FATME	GTE	Telettra	SIT-Siemens
	sales lines	sales lines	sales lines	sales lines	sales lines
1970	20.2 19.3	21.4 24.3	14.5 15.8		43.9 40.6
1976	22.8 18.0	14.3 18.8	7.6 10.6	0.1 n.a.	55.2 52.6

Source: Lizzeri and de Brabant (1979)

Table 1 shows the leadership in the 70s of the Italian "national champion", SIT-Siemens, which was then supplying more than half of the Italian market for switching equipment. The Italian telecommunication industry had undergone a process of change during the period analysed here, because of both the paradigm shift and specific industrial policies. In the following sections we will analyse this modification.

1.2.3 The rules governing the public procurement

During the period considered here, there were two main sources of procurement for telecommunications equipment: ASST and SIP. ASST covered 13% - 15% of total investments in the telecommunications industry. The Italian market had been historically dominated by SIP, with an 80 - 82 % share of total telecommunications investments. Moreover, as already seen, ASST did not have a high level of investments in the switching market, since only 3.2 % of ASST investments were for switching equipment. ASST was a government administration. It was thus subject to the general laws on public procurement (Laws on State Accounting) established in 1923 - 24 and based on a design introduced in 1869 (L. Solimene, 1995). The aim of the regulation was primarily to assure the impartiality of the public administrations and to guarantee the state against deviations and embezzlements.[14] Unlike ASST, SIP had the status of a private company. SIP, the main purchaser of public switching, covering 30% of its investments, was thus exempted from the normal bidding procedure for public procurement.

The procurement had been historically organised with an annual agreement (Memorandum). The Memorandum included the quantities of equipment to be purchased, the level of the prices and the penalties that suppliers would be liable to pay in the case of

[13] At the very beginning, one of the major problem of CSELT was to train and educate a group of technical experts in the telecommunications field, since the Italian academic structure did not at that time provide the needed competencies. In 1964, the personnel at CSELT amounted to around 50 people. During the 70s, with the availability of the first graduate students in electronics engineering, the personnel grew steadily. In 1979, CSELT employed 600 people, one third of whom were graduates. During the 80s, CSELT grew continually. In 1989, CSELT had a staff of 800 researchers coming from all the societies of the STET group.

[14] Telettra was an independent company until 1976, when it was acquired by FIAT, the main private group of the Italian oligopolistic core.

defects and malfunctions in the equipment exceeding a fixed threshold.[15] The procurement fixed in the "Memorandum" followed an historical distribution of market shares.[16] According to SIP sources,[17] three main criteria for the choice of suppliers can be identified. SIP looked primarily at the quality and quantity needs of the service provision. Support for the vertical integrated manufacturer, SIT-Siemens, came as a second level criterion. Thirdly, maintenance of employment for the entire manufacturing industry, including foreign subsidiaries, was considered. Besides the Memorandum agreement, a second procedure of procurement was present, the so-called "extra Memorandum". The use of the extra-Memorandum was historically confined to those needs that were urgent and not foreseen in the annual planning. It was only in 1987 that the problem of reducing the price of the equipment needed to modernise the network had forced SIP to begin the practice of "extra-Memorandum" bids for the switching systems. At the beginning (1987), only a small share of the switching market was included in this open tender (5%). The results of the bid (won by FATME) in terms of price decrease were so satisfying that in 1988 SIP decided on a second call for 100,000 lines. In the 1988 bid, the discounts offered amounted to 20.5 % for Italtel, 22.5% for FACE and 27.5% for FATME. (Pontarollo, 1989).

2. THE NSI AND THE TELECOMMUNICATION POLICY

The telecommunication industry clearly illustrates some of the characteristics of the national institutional and socio-political framework, especially the absence of a coordinated and coherent public support sustained by a strategic vision for the sector (Nomisma, 1988). We can identify four different periods:

---

[15] Three main types of procedure were included:

- *licitazione privata:* an open tender in which the price was the only criterion;
- *appalto-concorso*: the supplier had to provide a project and prove its technical and organisational competence: the prices were considered only in a second phase, after which the quality of the project and of the supplier were judged;
- *trattativa privata*, a bilateral negotiation without public bid, foreseen only for exceptional reasons of urgency, complexity and social need.

ASST had used all the three types of procedure, including the last one. The suppliers agreed were those that had applied to a specific list, reviewed by ASST every three years. The ASST lacked a specific department for cost assessment. Accordingly, ASST's control over prices was limited to a rough evaluation of investments based primarily on previous contracts and bidding (L. Solimene, 1995).

[16] SIP had a Network Division (Divisione Reti) and a Procurement Division (Divisione Approvvigionamento). The procurement decision had been usually organised mainly by the Network Division, which took account of the local needs of the Regional Departments and planned the development of the network. The Network Division concentrated the main resources in terms of competence and of R&D. In a second phase, under the technical supervision of the Network Division, Regional Departments were allowed to ask for specific procurements that were later consolidated at the national level. The final "Memorandum" fixed the costs and the suppliers. (Interview at SIP , Rome, 19/5/97)

[17] When in 1964 the five companies were unified in SIP, the market was fragmented into five areas, dominated by different suppliers. At that time, because of the technical characteristics of an electromechanical switch, the costs of changing the supplier and converting a switch were so high that the traditional distribution of market shares remained almost unchanged.

1. 1964 - 1970: the entry of SIP into the telecommunications sector
2. 1970 - 1980: the "non-strategic" role of telecommunication
3. 1980 - 1987: the failure of the restructuring of the national industry
4. 1988 - .....: the modernisation of the network and the restructuring of the service provider.

2.1 1964-1970: The Entry of SIP into the Telecommunications Sector

As already seen in presenting the institutional framework, SIP entered the telecommunication sector in 1964, after the nationalisation of the energy industry in which it was originally specialised.[18] The reimbursement of the energy assets formed a big financial dowry, through which SIP started the harmonisation and the co-ordination of the national and local networks, both finally under the umbrella of a unique provider. At that time, the development of the Italian network was lagging behind the other European countries. The density of the telephone network was very low (eight telephones for every 100 inhabitants). In 1964, the complete automatisation of the local network was planned. The investments (90 - 100 billion lira) were completely in charge of STET. In 1968, a new concession was signed between SIP and the Administration of PT that fixed the plan for the complete automatisation of the network (Teleselezione Nazionale), and which also included the modernisation of part of the ASST network.[19] The plan forecast huge investments, three times greater than in the previous three years, which were to be covered by borrowing. The plan was entirely based on the electromechanical techniques. In 1970, the national automatisation of the network was completed. The Teleselezione was considered to be a great success, since Italy was the first European country to complete such a process. During this period, until 1971, telephone tariffs were revised annually, allowing SIP to cover the amortisation of investments. Despite the success of the Teleselezione Nazionale, some problems started to emerge at the end of the 70s, with a growing difficulty to fulfil the increasing demand for new telephones.[20]

2.2 1970-1980 The "Non-Strategic" Role of Telecommunication

This period is a crucial one for the case considered here. The situation can be summarised with a quotation of that period about "an atmosphere of exaggerated reservedness mixed ... with a strong uncertainty about the future of Italian telecommunication ... and with scarce ... information and a, consequently inevitable, demagogy" (Lizzeri – de

[18] Interview at SIP , Rome, 19/5/97

[19] SIP was the acronym for "Societa' Idroelettrica Piemontese" (Hydroelectric Company of Piemonte) and it later changed to "Societa' Italiana Per l'esercizio telefonico" (Italian Company for telephone service). For the history of SIP see B. Bottiglieri (1990)

[20] This was one of the reasons for the new distribution of roles described above.

Brabant, 1979). During the 70s, the telecommunication industry was mainly considered as a way to create and sustain employment and the industrial development of the South of Italy, given the labour intensive character of the production technology during the electromechanical phase. The main policy instruments activated were public procurement and the orientation of the state-owned manufacturer, SIT-Siemens, to open new plants in the South.[21] The first oil shock in 1973 arrived right in the middle of this expansion. The public purchaser was left alone to face the difficult situation created by the oil crisis. The telecommunication tariffs were blocked as a way of controlling the social effects of the high inflation rate of the period. The process of adjusting tariff rates was long, complex and characterised by bureaucratic controls and constraints. After a small decrease in 1974, SIP continued the policy of increasing investments, especially in electromechanical switching. The result was a financial crisis culminating in 1980. On the side of the manufacturing industry, SIT-Siemens was engaged in the contradictory process of developing the new digital technology while expanding the production facilities in the old one. During the 70s, attempts to rationalise the Italian manufacturing industry, involving co-operation between SIT-Siemens and Telettra, failed.

2.3 1980-1987: The Failure of the Restructuring of the National Industry

At the beginning of the 80s the situation was thus characterised by a huge amount of unsatisfied demand for telephone lines, a lack of modernisation of the national network, the financial crisis of the service provider, and the need for restructuring of the "national champion", which was burdened by the expansion of its work-force and its R&D efforts in the new digital technique.[22] Between 1979 and 1981, the investments of SIP slowed and in 1981, for the first time, SIP could not fulfil the planned program of procurement. Many instruments were activated to solve the "telecommunication problem" now at the centre of greater political attention. The first problem tackled was the restructuring of the manufacturing industry. In 1982, under the auspices of the government, GTE, Telettra and Italtel signed an agreement to create an "Italian pole" for the development of the national digital switching system.[23] The agreement was an important step towards the rationalisation of supply. However, "the decision had been taken with at least four years too late, augmenting the uncertainty about the success of the initiative and forcing the three firms to a hard labour of "'technological run' and of productive structure restructuring" (Pontarollo 1983). The planned "second pole" to be formed by FACE (ITT) and FATME (Ericsson) would never see the light.

[21] The strike-wave of 1969/1970 particularly affected the telecommunications manufacturing industry, especially SIT-Siemens. The telecommunications firms were not able to fulfil the planned supply and, therefore, SIP was not able to answer to the growing demand for telephone lines.

[22] Between 1970 and 1974, SIT-Siemens doubled its personnel, growing from 15,000 to 30,000 units with the opening of two plants in Catania and Terni; meanwhile, the volume of public procurement by SIP increased from 868 billion lira (1970) to 2020 (1973) (Lizzeri and de Brabant, 1979 and Pontarollo, 1983).

[23] This situation was well diagnosed by the Report on the "Partecipazioni Statali" (PPSS), the Italian system of State owned companies of 1980 (Morganti, 1980).

In 1985, a further step was attempted with the creation of Telit, a company formed by the state-owned Italtel and by Telettra, controlled by Fiat. The idea behind the joint venture was to create a positive synergy between the specialisation of Italtel in switching systems and of Telettra in transmission. For the Italian "national champion" the joint venture with Telettra was also an opportunity to improve the international dimension, since Telettra was already selling and producing abroad. The process of defining Telit was long and difficult. Begun in 1985, with the first talks between the two share holders, STET and Fiat, it was finally approved by the Minister of PPSS in 1986, right before the definition of the capital structure.[24] Only the management was to decide. And it was on precisely these final details that the joint venture failed, for strictly "political" reasons. The government wanted to appoint the president of the company, while Fiat refused.[25] After the failure of the joint venture, Telettra was sold to the French group, Alcatel. This operation was coherent from a strictly financial point of view, but the idea of an Italian telecommunication pole was finally wrecked.

During this period, the discovery of the "strategic role" of telecommunication drove towards a series of development plans in which were indicated "besides the amount of investments, in a not always effective overlapping of planning and forecast, a multiplicity of aims both of economic and financial nature (industrial reorganisation and tariff adjustments) and of institutional nature (the restructuring of the institutional setting)" (Nomisma, 1988).[26] After seven years of debates, in 1987 the CIPE (the Italian interministerial committee for economic policy) approved the "Piano Europa", an extraordinary program of investment, totally in the charge of STET.

2.4 1988-.....: The Modernisation of the Network

The failure of Telit and the "Piano Europa" pointed to the beginning of a new phase. Subsequently, the Italian telecommunication policy was characterised by the need to intervene rapidly to solve the "emergency" posed by the liberalisation decided on by the EU and the weakness of the Italian telecommunication system. The approval of the Piano Europa marked the beginning of the modernisation of the network and of a new policy for procurement in the sector. The main aim was to reduce the price of the equipment. Despite the technological change in the electronic switching system and the change in procurement policy towards consistent lowering of prices (Morganti, 1990) the market for switching equipment in 1990 was still mainly distributed with the same "historical market shares" as it was during the 70s (Table 2).

---

[24] Italtel has been the new name of SIT-Siemens since 1980, which marked the end of the license agreement with the German Siemens.

[25] The capital structure was: 48% to Italtel, 48% to Telettra and 4% to Mediobanca, the Italian private financial "vault".

[26] In this episode, some characteristics of the Italian institutional framework can be discerned. For example, the conflicts inside the Italian financial system, characterised by a small group of families on the private side and by "political" families on the public side, are well illustrated. The failure of Telit shows the presence of a conflict between an "industrial" point of view and a "financial" point of view. (D'Orazio, 1993).

Table 2. Distribution of market shares for switching equipment in 1990

Italtel	FATME (L.M. Ericsson)	FACE (Alcatel)	Siemens Telecom. (ex GTE)	Telettra (Alcatel)
50 %	24 %	12 %	10 %	4%

Source: Our estimates on Annual Reports

The new needs expressed by the service provider in order to face the new competitive challenge accentuated the difficulties of the national manufacturer. Its market share was maintained, thanks to vertical integration inside the STET group, but the manufacturer was losing its price-maker role. With the failure of Telit, the possibility of finding a national partner to strengthen its technological and market position was no longer available. The STET group then decided to follow the strategy of the main leading companies in the sector and to look for a foreign partner. The choice was the US firm, AT&T.[27]

3. THE ROLE OF PUBLIC PROCUREMENT AND THE DIGITAL SWITCHING SYSTEM

In this section, we focus attention on the innovation of digital switching and on the role played by procurement in sustaining and diffusing this new technology. We can identify three main phases, roughly corresponding to those of the previous section:

1. 1968 - 1980 : SIT-Siemens, Telettra and CSELT: The lack of co-ordination of the national research activities
2. 1980 - 1986: The search for external national and international competencies
3. 1987 - ....: The Linea UT and the role of PTP in sustaining the national technology

3.1 1960-1980: SIT-Siemens, Telettra and CSELT: the Lack of Co-ordination of the National Research Activities

This long period is a crucial one for the development of the digital technology. During these twenty years, three main actors (SIT-Siemens, Telettra and CSELT) researched digital switching systems in Italy. Despite some attempts at agreement (between SIT-Siemens and Telettra and between SIT-Siemens and CSELT inside the STET group), and despite the strong complementarities among the different actors, the research was

[27] In 1984, the CIPE approved the Ten-Year Plan of the Ministry of PT (Plan 1985/1995) and, in the same year, it commissioned, from the Ministry of Industry, a study on the restructuring of the supply. In 1986, the plan of the Ministry of Industry was presented. Together with the plan for restructuring supply, it included a "new investment plan" for the modernisation of the network. The 1984 Ten-Year Plan of the Minister of PT was judged as insufficient and new investments were proposed – without, however, clearly indicating the way to finance them (Nomisma, 1988).

mainly conducted in an isolated way. We will now briefly present the three axes of research and then present the procurement policy of SIP and ASST in this period, in order to assess the role of procurement in the technological development.

In the early 60s, a small group of researchers at SIT-Siemens started working on a prototype of electronic PBX (Private Branch Exchange). We recall that at that time SIT-Siemens was producing the majority of its equipment under the license of the German Siemens. Starting from this first prototype, a project for the development of an electronic public switching system was conceived. The research project took place completely inside the manufacturer.[28] The STET top management took part in the decision process, but no participation from the vertically integrated service provider was planned. In 1969, a first draft of the project was presented to SIP, in order to have the approval of the service provider concerning the architecture and the main characteristics of the switch. The SIP R&D department approved the main characteristics of the switch.[29] Its vertical integration inside the STET group and its historically dominant position in the Italian market for switching equipment gave SIT-Siemens considerable certainty about the market for the new product, especially after SIP's technical approval. The decision was taken to "jump" directly into the time division trajectory, following the successful example of France (Llerena, Matt, Trenti, 1997). The project was extremely ambitious for a firm not having an autonomous technological capability and without experience in the new electronic paradigm. SIT-Siemens was in fact concentrated in switching systems and not in transmission, where electronics was first introduced into telecommunication equipment. The new digital system presented a peculiar architecture, formed by a central transit switch linked with small peripheral switches. The whole design project would never be realised. Instead, one small prototype of the peripheral switch was installed in 1972 in Milan. The service provider was involved only in the installation phase. The prototype became operative only two years later. During the 70s, other prototypes of the Proteo were installed (1975 in Rome, Pordenone, Florence and Messina) and activated only at the end of the decade.

Beginning in the early 60s, the second Italian supplier of telecommunications equipment, Telettra, historically specialised in transmission technology, started research on electronic switching (Sintel project). Telettra was convinced of a convergence on the microelectronics paradigm of the whole span of the telecommunications equipment. Therefore, Telettra researchers "felt the hazard of being excluded from the world of future communications" (A. Bellman, 1976). Unlike SIT-Siemens, Telettra had already accumulated competencies in electronics, thanks to its historical specialisation in transmission technology. However, Telettra had no practical experience in electromechanical switching and no historical market share in the switching sector. During the first

[28] Currently, Italtel is 50% owned by Siemens. On the side of the service provider, the process of adaptation to the EU rules has been long and impeded by the difficulties posed by the confused institutional setting. In 1992, a first step was undertaken with the reform of the Ministry of PT. Further reforms are currently being undertaken as part of the process of privatisation of STET.

[29] An early participant of the research affirmed that the project was kept almost "secret" for fear of the reaction of the German licensor. Interview with a SIT-Siemens engineer – January 1992 (Trenti, 1992).

decade of research, efforts were focused on the acquisition of knowledge about switching problems and on the training of researchers in the new paradigm, working on small time-division switches and on hardware and software equipment.[30] Starting in the early 70s, a second phase of the research started, aiming at the realisation of marketable switching products (A. Bellman, 1976). The initial Sintel project, aiming at the development of a complete time-division switch, such as the SIT-Siemens Proteo, was judged as being too ambitious. The research was then focused on two main products: the DST1, a space-division switch with an electronic cross-point network and a distributed control; and the DTN1, a transit switch with a PCM switching network and stored-program control, fruit of the early research on the Sintel system. The DST1 was installed for the first time in 1975 in the Italian telephone network, while the DTN1 won an ASST bid for the international switch of Verona. The DTN1 was to be installed in 1977 but the installation was postponed and the switch was finally installed only in 1979.[31] The Telettra digital switching equipment was later modified (AFDT1) to equip the voice and data transmission network of SIP.

Meanwhile, CSELT was also researching the electronic switch technology. The aim was primarily to acquire a competence in the new paradigm and to respond to the need of the service provider. Different products were developed. In 1967, a small semi-electronic switch was successfully put into service. During the 60s, answering to a specific need of SIP concerning the documentation of traffic, a prototype for a small time-division switch (TECA) was developed. Finally, the prototype of a small time-division switch (GS – Gruppi Speciali) with advanced features (PCM technique and stored-program control) was installed in 1971 in Mestre (Venice), only one year after the French success with the Platon. Further improvements of the GS switch were developed and installed during the 70s (CSELT, 1994).

The brief description of the research in course during the 60s and 70s shows clearly the lack of co-ordination among the three actors. Given the shortage of highly skilled personnel in the Italian NSI, the novelty of the technology, and the complementarities between the three actors, this lack of co-ordination seems particularly important. During the 70s (in 1972, 1976 and 1978) Telettra and SIT-Siemens tried without success to co-ordinate their R&D activities and their development of digital systems.[32] Moreover, even inside the STET group, the co-ordination between SIT-Siemens and CSELT was scarce. A clear involvement of the CSELT can be proven only for an improved version of a Proteo switch (CT-2) installed in 1982, wherein the Turin laboratory provided an advanced component (Chapuis and Joel, 1990). Taking the French case as a benchmark (Llerena, Matt, Trenti, 1997), where the R&D laboratory of the service

---

[30] Interview with the General Director of CSELT, who at that time worked at the SIP R&D Department (Torino, 4/7/97).

[31] One of the first versions of the system (Sintel III) was presented in an international symposium in Boston in 1972. The Sintel III featured two main characteristics (a time-division PCM signal switching and a stored program control) in order to be economically attractive and technologically valid (A. Bellman, 1972).

[32] Interview to the Vice-President of Italtel, who was at that time responsible for the R&D at Telettra (Milano, 25/6/97)

provider was given the task of co-ordinating the national research, we might suggest that CSELT had the technical expertise to become the catalyst of national efforts in the telecommunications field, as shown by the technical success in digital switching. However, CSELT was never given a clear mandate to develop a national technology and to transfer it to the national firms.

During the 70s, the two service providers continued to buy and install electromechanical switches. Only ASST choose to buy semi-electronic switches, while SIP decided to leap-frog the semi-electronics phase and wait for fully electronic switches. This decision was based on an evaluation done by SIP. Semi-electronic switches were judged better suited to equip bigger transit nodes, such as the one in the ASST network. Moreover, the passage to the electronic switching system was considered by SIP to be a means of reducing the number of suppliers. Given the rigidity of the supply of electromechanical switches, it was very difficult for SIP to change the historical shares in the switching market.[33] The advent of an Italian switching system was seen at the beginning, as a way to simplify the market and sustain the vertically integrated "national champion". The process of development was long and difficult, demonstrating the problems of the Proteo system. The presence of a second Italian supplier, Telettra, was not considered as a possible alternative (only ASST tried to install a Telettra switch, but the switch was installed only after a delay of two years). The result was that SIP was "waiting for the Proteo" (Morganti, 1980). The Telettra system was instead chosen only to equip the national voice and data-transmission network, pushing Telettra to develop a specific switch for this purpose. The first trial for the voice and data-transmission network (Rete Fonia Dati) started in 1976. By the end of the 70s, the network was ready for the service.[34] At the end of the 70s, given the late and the partial failure of the Proteo system, SIP was aware that the way towards the "unique system" was still long and difficult. For this reason, SIP decided to fix, in specific documents, the characteristics of the switching systems to be purchased in the future. The feature specifications were done to permit the installation of different systems in the Italian network.[35]

The role of the procurement policy in this period should be interpreted carefully. Two choices had been the object of criticism and debate: the choice to pass directly to the fully electronic switch, and the choice to wait for the Proteo. Concerning the first point, the choice to leap-frog the semi-electronic phase was criticised because the trial of the semi-electronic systems could have enhanced the capacity of SIP in handling the new microelectronics paradigm (Pontarollo, 1983). However, the technological leap-frogging was also successfully chosen in other countries (for example, Canada). Concerning the "waiting for the Proteo" strategy, it might be considered to have been a "providential" help for the vertically integrated manufacturer (Cozzi and Zanfei, 1996).

---

[33] In 1976, Telettra passed under the control of FIAT.

[34] Interview with the General Director of CSELT, who was working at that time at the SIP R&D Department (Torino, 4/7/97)

[35] However, the RFD was activated only in 1984, because of a long "querelle" between the Ministry, SIP and ASST over tariffs and charges.

During the 70s, the Proteo showed many technical problems and it was thus too risky for SIP to install it. The delay in the modernisation of the network gave SIT-Siemens the possibility of being ready with a new product, as we will see below. However, the choice to push Telettra into the development of the specialised switch for the voice and data-transmission network, together with the difficulties encountered by Telettra in installing its digital switches, shows that public procurement was not used, in this period, as an instrument to co-ordinate the already scarce national technological resources.

3.2 1980-1986: The Search for External National and International competencies

At the end of the 70s, while Telettra was successfully equipping the voice and data-transmission network, the Proteo finally entered the industrial phase. A group of engineers of SIT-Siemens was sent to Dallas to start a co-operation with a US company, Advanced Business Communication (ABC), in order to develop a new generation of switch. The decision to find a foreign partner was motivated by the difficulties encountered in finding adequate personnel, especially for software, in the Italian context (Chapuis and Joel, 1990). Great progress was made in two years: the first prototype of the new system (UT10/3) was installed in Milan in 1981. One year later, Telettra, GTE and Italtel signed the agreement to build a complete switching system, finally pooling the national resources.[36] The industrial production began in 1983 while the production of the old system was cut in 1984. The UT10/3 evolved very rapidly in this period from an isolated switch to a family based on a modular architecture (Linea UT). The initial architecture, proposed by ABC, was completely reviewed to reach a truly modular solution, which allowed a low first cost and an almost flat curve for the cost per subscriber (A. Bellman, 1987). Started as an isolated project, the Proteo/Linea UT system development was strongly marked, in this period, by one major organisational device: the search for external technological competencies at both a national level (the Telettra R&D) and the international level (ABC).

After the difficulties encountered during the 70s, the UT was developed in a short time. The system was ready to be sold as early as 1984, but a full development of the orders on the Italian market had to wait until 1987, when the "Piano Europa" was finally approved.[37] There was thus a three-year delay in the beginning of the national plan for procurement. This delay could be interpreted as a "lack of [support by] the national public procurement" for the new system, the Linea UT, which was by far "more reliable and performant" than the Proteo. (Cozzi and Zanfei, 1996).

---

[36] Interview with the General Director of CSELT, who was working at that time at the SIP R&D Department (Torino, 4/7/97).

[37] The distribution of roles inside the system saw Telettra supplying a peripheral switch, GTE the GTD5 as a transit switch, while Italtel (the new name of SIT-Siemens from 1981 onwards) supplied the new switches that became known as UT10/3. In the same period, the responsibility of the Telettra R&D department for the switching system passed to Italtel.

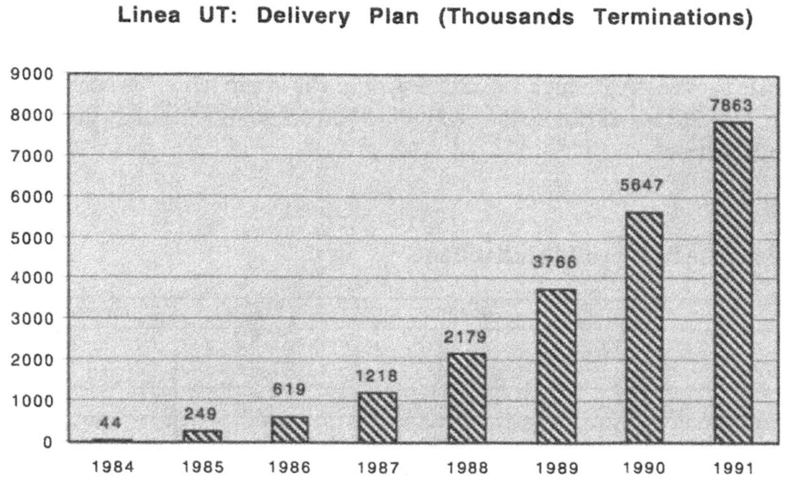

Figure 3. The diffusion of the Linea UT – 1984-1991

Source: A. Bellman (1991)

3.3 1987-..: Approval of the "Piano Europa" and Diffusion of the Linea UT

During this period, the digital switching system evolved rapidly. In 1987, a new and bigger switch (UT100), the upper end of the modular architecture, was presented. The industrial phase of UT100 began in 1989, the year of the Joint Venture between Italtel and AT&T, adding new features and improvements to the Linea UT. At the same time, the investment plan for the modernisation of the network was finally approved. Starting in 1988, the level of investments steadily grew. Around 30% of investment was devoted to digital switches. The number of digital lines installed in the Italian network passed from 1.5 million in 1986 to 8.5 million in 1990. In relative terms, the proportion of digital lines on total lines passed from 11.8 % in 1986 to 33 % in 1990. The diffusion of the Linea UT followed the same path. Figure 3 shows the diffusion of the Italian system in terms of cumulative lines installed, pushed by the high volume of investment on the national market.

4. APPRAISAL OF SOCIETAL AND INDUSTRIAL RESULTS AND CONCLUSION

The aims of the policy towards the telecommunication sector were the development of the network infrastructure and the strengthening of the national industry. A standard measure of the state of a telecommunication network is the density of telephone lines. In Italy in the late 60s and early 70s, the telecommunication network was characterised by an important lag in terms of density (number of lines per 100 inhabitants). As shown in Table 3, the telephone density was in the early 70s lower in Italy than in Germany and UK. In 1990, the situation is substantially reversed for France, while Italy continues to show a gap, though a decreasing one. The improvement is concentrated at the end of the 80s, as shown by the evolution of investments and policy described in the previous sections.

Table 3: Telephone lines per 100 inhabitants

	Italy	France	UK	Germany
1974	16.49	11.81	22.87	19.61
1980	23.07	29.95	31.44	33.35
1985	30.45	41.75	36.95	41.94
1989	34.90	45.23	42.10	45.90
1990	38.80	49.60	46.00	48.50

Source: OECD (1988); STET (1990 - 1991); Reseau (1991)

The more traditional use of public procurement and the decision to await the building of the national system, together with the lack of a clear and coherent policy and institutional framework, had led to a gap in the process of modernising the network. Table 6 shows the period of introduction of different types of specialised network in Italy, France, UK and Germany, from the end of the 60s to the 90s. The introduction of specialised networks gives an indication of the technological advance of the telecommunication system of a given country.

Table 4 confirms the path shown by Table 3, with France having an earlier start and fast recovery during the 70s and with Italy exhibiting a slow process of modernisation during the 70s and early 80s and an acceleration at the end of the 80s. Table 5 shows the data on the digitalisation of the network in the main industrialised countries from 1988 to 1993. The Italian data confirm the importance of the most recent investments and efforts of SIP to modernise the network.

Table 4: The introduction of specialised networks

Years	Italy	France	UK	Germany
1967				DATEX
1972		CADUCEE		
1976				DATEX-L
1978		TRANSPAC		
		Transmic		
1980				DATEX-P
1981			PSS	
1982			Kilostream	
1983				
1984	RFD		OTS	
1985			ISDN Trial	
1986	ITAPAC	Telecom 1		Telecom 1
		ISDN Trial		
1987		ISDN		ISDN Trial
1988				
1989	Italsat 1		ISDN	ISDN
1990	ISDN Trial			

	Circuic switch data transmission		Packet switch data transmission		Direct digital circuit
	Satellites		ISDN Trial		ISDN

Source: Adaptation from Reseau (1991)

Table 5: Digitalisation of the network 1988-1993 (% of digital lines on total lines)

	1988	1989	1990	1991	1992	1993
France	60.5	65.8	69.1	73.5	78.1	83.0
Italy	**20.1**	**24.7**	**32.6**	**39.9**	**47.4**	**54.5**
USA	35.4	41.6	48.1	54.4	60.9	68.1
Canada	40.0	45.0	50.0	55.0	62.0	68.0
Japan	17.4	24.9	31.9	38.9	46.4	53.3
Germany	5.0	7.3	10.5	15.0	19.4	23.7
UK	23.5	31.3	39.7	41.1	53.6	59.4

Source: Zanfei (1990) from NBI and SIP Development Plan

Table 6: The evolution of Italtel 1984-1991 (million US$ 1987)

	1984	1985	1986	1987	1988	1989	1990	1991
Telecom. sales	717.1	803.1	936.2	1103.4	1342.6	1759.9	2085.3	2444.5
Switching	63.1 %	55.5 %	55.0 %	64.1 %	62.7 %	61.2 %	59.0 %	56.5 %
Total R&D	8.9 %	10.8 %	11.3 %	12.2 %	11.5 %	10.5 %	11.7 %	11.9 %
Switching R&D	n.a.	n.a.	5.9 %	7.0 %	6.2 %	5.28 %	5.7 %	n.a.

N.B.: Percentage share on telecommunication sales

Source: D'Orazio Zanfei (1994)

The strengthening of the national champion was the second explicit aim of the government policy. In Italy, the position of the national manufacturer has been reinforced by the development of the digital switching system. Even if the development encountered early problems during the 70s, Italtel has succeeded in obtaining a market share in the digital switching market. Table 6 presents data about Italtel (total sales, sales in the telecommunication sector, sales in the switching market and R&D expenses).

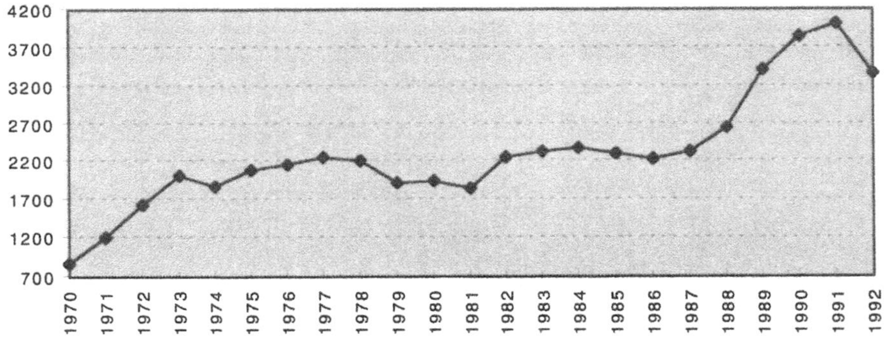

Figure 4. SIP investments 1970 - 1992 (Billion lira – constant price 1980)

Source: Our elaboration on Pontarollo (1983) and Solimene (1993)

The costs involved in the process are hard to assess. However, an evaluation of the costs of the Proteo I and the Linea UT systems is available. The costs of the development of the Proteo family of switch amounted in 1983 to around 3800 employee-years of research, including the entire family of switches (CT2, CIMA, TN16) and combining the costs for research, development and feature proliferation. The costs of UT were lower (amounting to around 1,000 employee-years).[38] The evolution of investments of SIP (Figure 4) can give an estimate of the total cost involved in the process of procurement

[38] Despite the low level of investments and procurement of SIP, in 1986, 335 switches were delivered and 294 were already in service.

and in modernisation of the network. Unfortunately, they include investments for the total procurement and not only for the switching market, which was on average around 30% of total investments. Figure 4 shows clearly the constant level of procurement, concentrated on the old electromechanical technique, during the 70s. Moreover, the low level of procurement in the early 80s, (and until 1988) when the Italian digital system was ready to be installed, is shown.

Table 7, below, describes the technological evolution of the Italian switching system. This trajectory can be contrasted with that of Italian public procurement of telecommunications technology.

During the 70s the procurement policy was used as an instrument to create and sustain employment while the technological content of the procurement was evident only in the second half of the 80s, when the national digital system was already available and the need to modernise the network was urgent, given the competitive challenge facing the service provider. The path followed by the procurement in the Italian case shows a strong orientation towards the "national champion", identified with the vertically integrated manufacturer, during the 70s. However, when the national digital system was ready and needed to be sustained by the internal demand, the public procurement was not able to start a massive introduction of the new technology because of the obstacles and impediments to a prompt and effective decision making. In the absence of a national pole of competence (public or private), together with the lack of a clear commitment of the service provider and/or of the government to sustain the process of capability building, the Italian manufacturer chose, starting in the early 80s, a strategy of research dependent on external and international resources. The case study shows a strong interdependence of the role of public procurement with the national system of innovation. The lack of a coherent political framework for the sector mirrors the absence, until the 80s, of a comprehensive framework for R&D policy. Thus, the procurement instrument has rarely been used as a stimulus for innovation in other sectors as well (for example energy and trains) (Pontarollo, 1986).

Table 7: The technological evolution of the Italian digital switching system

Year	Major Steps
1970	Beginning of Proteo
1972	Proteo CTA installed
1979	Proteo CT-2 (improvements in memory) – Beginning of UT
1980	TN-16 and TN-5
1981	UT10/3 prototype
1983	UT10/3 industrialised
1989	UT100/60
1992	First ISDN trial

Source: Our elaboration on Chapuis and Joel (1990)

[39] Interview with the Vice-President of Italtel (Milano, 25/6/97)

REFERENCES

W.S. Atkins Consultants. (1987). *The Cost of non-Europe in Public Sector Procurement*. In the European Commission (1988), *The Cost of non-Europe, Vol. V*. Brussels.

Bellman, A. (1972). An Integrated Time Division PCM Exchange. Paper presented to the *International Switching Symposium*. Boston.

Bellman, A. (1976). *Syntel System: Evolution, Development, Feature*. ISS: Kyoto.

Bellman, A. (1987). The Linea UT Concept: Architecture, Technology, Evolution. Paper presented at the conference, *Italtel Telecommunications for Today and Tomorrow*. Geneva: October.

Bellman, A. (1991). The Evolution of Communication Technologies. Paper presented at *TELECOM '91*. Geneva: October.

Bottiglieri, B. (1990). *SIP – Impresa, tecnologia e Stato nelle telecomunicazioni italiane*. Milano: Franco Angeli.

Bragho', A. (1988). La concorrenza globale nel settore della commutazione pubblica. *Economia e Politica Industriale, n. 58*.

Campo Dall'Orto, S. e S. Mariotti. (1986). Il settore delle telecomunicazioni. In R&P Ospri (Eds.), *L'Italia multinazionale, Vol. II* (Chap. 3). Milano: Il Sole 24 Ore ed.

Cawson, A. et al. (1990). *Competition and Closure in the European Electronics Industry*. Oxford: Claredon Press.

Chapuis, R. and A.E. Joel. (1990). *Electronics, computers and telephone switching: A book of technological history*. In North Holland studies in telecommunication, *100 years of telephone switching, Volume 2: 1960 - 1985*. Amsterdam, New York, Oxford: North Holland Publishing Company.

Cohendet, P. and P. Llerena (Eds.). (1991). *Modes of use and Diffusion of New Technologies and New Knowledge: The Case of France*. BETA, University of Strasbourg. Report for the EC MONITOR-FAST Programme. (FOP N.228). Bruxelles.

Cohendet, P. and P. Llerena. (1997). Learning, technical change and public policy: How to create and exploit diversity. In C. Edquist (Ed.), *Systems of Innovation – Technologies, Institutions and Organizations*. London and New York: Pinter Publishers.

Commissione Morganti. (1984). *Rapporto al Presidente del Consiglio sulle Telecomunicazioni*. Milano: Franco Angeli.

Confindustria – Centro Studi. (1990). *Le Telecomunicazioni*. In F. Morganti (Ed.), *Progetto Infrastrutture, Vol. II*. Roma: SIPI.

Cozzi, G., A. D'Orazio, A. Zanfei. (1991). *Evoluzioni istituzionali e tecnologiche nell'industria delle telecomunicazioni*. (mimeo) IEFE – Divisione Economia e Tecnologia.

Cozzi, G. & A. Gambardella. (1985). La commutazione pubblica e le trasmissioni in Italia: Effetti del progresso tecnologico, dimensioni del mercato e struttura dell'offerta. Atti del Convegno, *Le strategie imprenditoriali nel campo delle telecomunicazioni e dei servizi telematici*. Universita' L.Bocconi Milano: November.

Cozzi, G., A. Ninni & A. Zanfei. (1988). Le licenze in due comparti produttivi dominati dalle commesse pubbliche: Elettromeccanica per la generazione di energia e telecomunicazioni. In II Rapporto CESPRI – Camera di Commercio di Milano, *L'Europa come opportunita' e come vincolo alla mondializzazione dell'economia italiana* (Chap. VI). Milano: CESPRI – Camera di Commercio di Milano.

Cozzi, G. and A. Zanfei. (1990). Il settore dei sistemi e apparati per telecomunicazioni: Struttura industriale e percorsi strategici. *Quaderni di Economia e Politica Industriale*. Milano: IEFE – Universita' L.Bocconi

Cozzi, G. and A. Zanfei. (1996). La commutazione elettronica in Italia: Il mancato decollo di un'innovazione di frontiera. In C. Bussolati, F. Malerba and S. Torrisi (Eds.), *L'evoluzione delle industrie ad alta tecnologia in Italia: Entrata tempestiva, declino ed opportunità di recupero*. Bologna: Il Mulino.

CSELT. (1994). *CSELT – Trent'anni*. L. Bentivoglia (Ed.). Torino: CSELT.

D'Orazio, A. (1993). L'evoluzione dell'industria nazionale di sistemi ed apparati di telecomunicazione: Fatti e misfatti della politica industriale italiana. *Economia e Politica Industriale, n.78.*

D'Orazio, A. and A. Zanfei. (1994). *L'industria europea delle telecomunicazioni. Relazioni cooperative e strategie competitive delle grandi imprese manifatturiere.* Milano: Franco Angeli.

European Commission. (1996). *Green Paper on Public Procurement in the European Union: Exploring the Way Forward.* Bruxelles

France Telecom.(1991) Chercher plus, chercher autrement. *France Telecom, n. 78* (July).

ITALTEL. (1982). *Piano Quinquennale 1983/1987.*

ITALTEL. (1990). *Atlante delle Telecomunicazioni.*

Lizzeri, G. and F. de Brabant. (1979). *L'industria delle telecomunicazioni in Italia.* Milano: Franco Angeli.

Llerena, P., M. Matt and S. Trenti. (1997). *Government Technology Procurement: the Case of Digital Switching System in Italy.* (Report prepared for the TSER-ISE Project, Subproject 3.2.2, 'Government Technology Procurement as a Policy Instrument '. November.)

Malerba, F. (1993). Italy. In R. Nelson (Ed.), *National Systems of Innovation: A comparative study.* Oxford: Oxford University Press.

Malerba, F. (1997). *Public Policy in Industrial Dynamics: An Evolutionary Perspective.* Paper prepared for the ISE Workshop. Athens: March.

Morganti, F. (1980). Le prospettive nel settore elettronico. In Ministero PP.SS., *Rapporto sulle Partecipazioni Statali* (Annex 6). Milano: Franco Angeli.

NOMISMA. (1988). Politica industriale e grandi progetti: Il caso delle telecomunicazioni. *Laboratorio di Politica Industriale, n. 3* (December).

Northern Business Information – NBI. (1989 and 1990). *World Public Switching Market.* New York: Northern Business Information.

OCDE. (1983). *Les telecommunications: Perspectives d'evolution et strategies des pouvoirs publics.* Paris.

OCDE. (1988). *Telecommunications Industry: The challenges of structural change.* Paris.

OCDE. (1989). *Indicateurs de performance pour les exploitants de telecommunications publiques.* Paris.

OCDE. (1990). *Communications Outlook.* Paris.

Pontarollo, E. (1983). Le politiche di acquisto nel settore delle telecomunicazioni in Italia. *L'Industria, n. 2.*

Pontarollo, E. (1988a). Il caso delle telecomunicazioni. In P. Bianchi (Ed.), *Antitrust e gruppi industriali.* Bologna: Il Mulino.

Pontarollo, E. (1988b). Gli standard tecnologici nelle telecomunicazioni: Aspetti generali ed implicazioni pratiche. *L'Industria n.1.*

Pontarollo, E. (1989). *Domanda pubblica e politica industriale: Fs, Sip, Enel.* Venezia: Marsilio.

Reseau. (1991). Archivio delle telecomunicazioni: Tecnologie, sistemi, imprese, paesi. *Prima Comunicazioni, n. 202,* (November).

Schnoring, T. (1990). Research and Development in Telecommunications: An International Comparison. Paper presented to the VIII ITS Conference, *Telecommunications and the Challenges of Innovation and Global Competition.* Venice: March.

Solimene, L. (1993). Le politiche di acquisto in Italia. In E. Pontarollo (Ed.), *Politica degli acquisti e struttura dei mercati nelle telecomunicazioni.* Milano: Vita e Pensiero.

Trenti, S. (1992). *Acquisti pubblici e innovazione sul mercato internazionale delle telecomunicazioni e del termoelettromeccanico. Il caso italiano.* Tesi di Laurea, Universita' Bocconi, Milano, March.

Zanfei, A. (1990). *Complessita' e crescita esterna nell'industria delle telecomunicazioni.* Milano: Franco Angeli.

Zanfei, A. (1997). Domanda pubblica ed innovazione tecnologica: Presente e futuro nel settore delle telecomunicazioni in Italia. (mimeo). Un. Bocconi: Milano.

9. PUBLIC TECHNOLOGY PROCUREMENT: THE CASE OF DIGITAL SWITCHING SYSTEMS IN GREECE

L. Tsipouri

1. INTRODUCTION

Greece is one of the less favoured member states of the European Union. After a period of very high industrial growth in the '60s and early '70s and two decades of economic stagnation accompanied in many cases by de-industrialisation, Greece saw its competitiveness eroding and trust relations among major actors in economic life deteriorating. Economic recovery did not start until 1996, and was based on strict macro-economic policies to cope with the Maastricht criteria.

In this deteriorating and uncertain environment the behaviour of the key players in the economy and the interaction among them evolved differently than in the other EU member states:

- Industrial companies based their production and export performance on cheap labour, imported productive equipment and a protectionist inertia. The role of foreign clients and suppliers was very important and, conversely, interaction among Greek companies themselves was more limited than in the "learning economies" of the core European Union countries.
- Universities and research centres were concentrated on teaching and basic research, financed almost exclusively by the state budget and remaining isolated from the local productive activities until the late '80s. University-industry interaction was virtually absent. The market failure justification was used to introduce a technology policy only a decade ago – that is, 10 - 20 years later than recommended by theory and international organisations.
- Finally, the state concentrated its industrial intervention in investment subsidies and did not (with fragmented exceptions) apply any demand instruments. Technology considerations were limited to isolated cases, always combined with infant industry protection argumentation.

The accession of Greece to the European Economic Community in 1981 triggered new ideas and policies, and a gradual modernisation process started in the mid '80s, strongly reinforced by the adoption of the first (1989 - 1993) and the second (1994 - 1999) Community Support Frameworks. The creation of industrial research associations provided bridging relationships between industry and academia, increased funding for national and EU research projects, and inserted the evaluation mentality into the culture of the research system. The massive (compared to earlier days) opportunities to fund industrial research paved the way for interaction among the main economic and technological actors. While a lot remains to be accomplished and a lot of criticism can be applied, Greece now has an innovation system with increasing local, national and transnational linkages.

1.1 The Procurement Case Studied

In the above-described context public procurement was always characterised by limited interaction.[1] This chapter presents a case study of the procurement of public switching centres by the Greek telecommunications operator. It is mainly based on interviews and press articles, since the case is not documented by scholarly research and publication.

The Greek telecommunications sector has lagged behind the rest of Europe in technological terms. For several reasons related to political and economic problems and the structure of external trade (i.e., the importance of barter trade with the former Eastern European countries, before Greece joined the EEC) the telecommunications infrastructure was for a long time characterised by an analogue network, including so many different technologies from so many different providers that it was ironically called the "mosaic". In the late '70s the government decided to proceed with the digitalisation of the network, implementing what was at that time an emerging technology. In order to achieve that end, a new company, ELVIL, was created in 1978. ELVIL was founded as a subsidiary of the Hellenic Industrial Development Bank (H.I.D.B.) and the Greek Telecommunications Organisation (OTE). Its statutory objective was the project management of production of public electronic switching centres and telephones, and its main activity focused on invitation and evaluation of tenders in order to choose systems that could connect at least 1 million subscribers. For ELVIL, the procurement aimed at the creation of a turn-key plant and the selection would include the bid for both the plant and the product (electronic switching centres) it would produce. The original idea was that a public plant should be built in a joint venture with a foreign technology provider, since the Greek private sector did not seem willing to take the risk of investing in this new, still unstandardised technology – especially given an environment that was fairly uncertain at the time. Three major areas for competition were suggested for the bidders:

1. The local value-added content in the production of the proposed plant should include a package of raw materials, components, education and services.

[1] Formally, at least, because the accusations of corruption and state-corporatism have been very frequent, in both academic papers and the press.

2. The productive capacity of the production plant should be higher than the needs of the local market; exports would be promoted under the responsibility of the foreign supplier.
3. Know-how transfer should be assured with the development of local RTD activities, which ideally would be part of the corporate strategy of the supplier.

The procurement itself thus aimed at the digitalisation of the telecommunications network, preferably through the creation of a plant producing standardised systems, with no major element of international novelty in it. Nevertheless, the integrated system (including software, educational needs, legal background etc.) would need adaptation and many tailor made specifications because of the specificity of the Greek "mosaic" system. The local plant was thus expected both to improve the overall performance of the system and to contribute to the mastering of a new technology in the country. The Greek side had one major advantage: *At that time digital technology was in its infancy and the bidders needed the market testing that would allow them to improve their own knowledge and capabilities. Mass production at that stage would provide references and allow the winner not only to acquire the Greek contract but to become more competitive in the world market, which had very good prospects.* From the winner's perspective installing a plant in Greece would be used as a demonstration case. Thus, despite the high international industrial concentration, competition worked more efficiently than in other oligopoly cases and gave additional negotiating power to the board of ELVIL in 1980 - 82.

As originally conceived, the procurement studied contained an element of novelty and adaptation processes that would allow the potential new local plant to learn, accommodate technological change and create new skills. For the international partner it contained a basic demonstration effect and a possibility to further test and adapt a very new system, helping to impose standards on the market.

At this stage, the procurement was a case of "adaptive public procurement" (APP), as:

- new software was needed for adapting to local circumstance, and
- it was a test opportunity for the later stages of development for the technology provider.

As later explained, the switches were finally produced in a double joint-venture between transnational and local companies keeping the same technological parameters as suggested by the original ELVIL proposal. However this occurred at a much later stage, when the technology was already mature. But even at that stage the technology was not totally standardised and new features were continually introduced to improve the performance of the switches.

The problem to be solved was the improvement of the performance of the telecommunications network, a need perceived both for social and economic reasons.[2] The

[2] The digitalisation of the network was expected to improve access to the basic service for all remote rural areas. At the time, companies stated that a key disadvantage for locational choice was the very low quality of the telecommunication services.

political authorities at the highest level,[3] in co-operation with the telecommunications monopoly and the industrial development bank, formulated the problem. In a sense, the vision was political and the administration asked to implement it did at first sight not share it, or at least was reluctant to proceed in a procurement governed by strong uncertainty conditions.

1.2 Some Theoretical Elements of the Specific Case

Technology procurement is a type of user-producer interaction, and in that sense it is important to try to identify whether regional characteristics of users and producers matter in this process. The key question asked here is whether the regional dimension matters and, if yes, then how can that be explained and how does it affect policy implications? In an effort to answer this question, we first try to analyse the characteristics of producers and users in less favoured regions,[4] and then apply a product life cycle methodology to explain the differences stemming from regional characteristics.

By definition, from *the point of view of the producer,* technology procurement is of interest to big, technologically leading firms, which:

- are able to research for breakthroughs, integrate complex technologies into new products and at least partly finance such undertakings; and, at the same time,
- are willing to take additional risks, as they envisage the global market and expect high profitability resulting from an ultimately very large scale of production.

We may thus argue that from a producer's point of view regional characteristics do matter, because only in countries where a critical mass of such companies exists is it worthwhile to create the institutional environment, culture and skills that would facilitate technology procurement. Exceptions exist, since some semi-industrialised countries demonstrate a pattern of sectoral specialisation that permits technology procurement in niche markets, but the typical cases of infrastructure products, like telecommunications, information technology equipment, and mass transportation (trains, planes) are not locally produced. Totally new products in these sectors, those with future market growth potential and high profit margins, are imported.

In many, though not all, cases there is a differentiation *from the point of view of the user* as well. The 'lead user' concept, implicit in technology procurement, suggests that the future user is able to co-design product specifications and to share the project monitoring. In other words lead user organisations are expected to be well informed, flexible and efficient. But while in less favoured countries the educational capabilities of engineers can increase rapidly, the national administration is usually neither flexi-

---

[3] The Prime Minister himself at the time.

[4] We use the term in the sense of the European Development Fund: EU regions with GDP per head lower than 75% of EU average. In an international context, these regions belong to the most developed in this world and are industrial countries with established basic infrastructures and a welfare state but rather specialised in mature industries.

ble nor efficient. The public sector faces austerity programmes and is unable to compete with the private companies in offering remuneration packages that would attract high quality engineers

For these reasons, i.e., insofar as both producers and users do not demonstrate key characteristics that constitute the parameters of an ideal procurement case,[5] it is unlikely that this can be a generalised practice in less favoured regions.

Nevertheless, if we drop one assumption – namely, novelty of the product at world level – we are confronted with a new, broad spectrum of products liable to become objects for technology procurement:

1. Less favoured countries are interested in the early procurement of major new equipment, which can improve their infrastructure and amortise rapidly, either through direct costing or through its influence on the improvement of economic and social conditions. These countries are in a position to afford expensive infrastructure,[6] without having to wait for prices to decline to the levels of the maturity or standardisation phase (Vernon 1966).
2. Most new products of this type need local adaptation, particularly for their software, either in order to link to broader infrastructures (like the new digital exchanges having to take connectivity to the previous network into consideration) or to adapt to local culture (like computers for schools, which have to take language and educational practices into consideration). In that sense, the user has an interest in co-operating with the producers to achieve the best possible adaptation, while the producer has every interest in promoting co-operation between the headquarters (innovators for the initial product) and the local subsidiary (responsibility for adaptation). Every additional order gives the producer the flexibility to increase its potential for adapting its mass production to customised markets.
3. From the point of view of international producers, this early-stage procurement presents a particular attraction: Usually, this type of equipment goes through a long pre-paradigmatic phase, where success is a matter of survival; only companies that integrate technologies appropriately stay in the market (Utterback and Suarez 1993). In that sense, competition for these companies takes on a different perspective, since they need to sell, even at relatively low margins, even before they are sure of staying in the market. To some extent, this can be seen as an exception to the conventional inverse correlation between profit margins and technological maturity. In Vernon's terminology, this early procurement better covers their "need for swift and effective communication" (Vernon 1966: 195), since their customers are bound to be spread throughout the global market.

[5] As defined in the first chapter of this volume.

[6] This is in general the case in semi-industrialised countries, but even more so in the EU, where less favoured countries benefit from massive transfers of resources for their modernisation. It is likely that in the next round the Central and Eastern European countries will benefit from the same conditions.

Taking these remarks into consideration, two types of technology procurement seem to be of interest to less favoured countries: One concerns products related to niche markets, where both local industry and public authorities demonstrate above average expertise; the other concerns products early in their life cycle with a high local value-added for adaptation purposes. The rules for the former do not differ from those of the development technology procurement encountered in more advanced countries, whereas the latter case constitutes a specific case for policy intervention with some differentiation in characteristics and policy implications. We consider the latter to be the framework for the project discussed here.

Our main argument is that products eligible for adaptive public procurement

- are still in the early stage of their life cycle, and as such
- have a high market potential, while
- a number of incremental innovations are added, which
- may be of interest to a new group of client-countries.

If the user (public agency) in a country that is a pioneer in a new product (or technological trajectory) can identify this opportunity early enough, it will be in a much better position to negotiate an exceptionally (for the stage of development) low price with the potential suppliers. (In the present case, the 'suppliers' constitute pairs composed of a pioneering multinational and its local subsidiary or subcontractor.) This is because the successful supplier is expected to gain two potential benefits:

- use of this locally adapted product as a demonstration and reference for other countries with similar needs which are expected to follow in the market growth phase
- if the selection of timing is adequate, the supplier will benefit from each additional market by increasing the likelihood of adoption of its standard and thus increasing its chances of survival when concentration due to high economies of scale will occur in the maturity phase.

These benefits constitute an increased power of negotiation from the side of the user. It is argued that, if well-prepared, the user may be able to obtain a price which would correspond to growth or even maturity conditions, while benefiting from the use of better infrastructure at an earlier stage than potential competitors.[7] The components that are very important are thus

- the selection of the right timing to start the process,
- the ability to react rapidly, in order not to allow the novel technology, for reasons of time elapsed, to proceed to the growth phase – where the supplier is not interested in the fringe benefits of demonstration and diffusion of its standards, and
- the ability of the administration to negotiate a fair price.

---

[7] Countries that have the same level of development and manufacturing specialisation.

This type of user does not have the same characteristics as 'lead users', conventionally conceived, but has abilities to identify early enough which of the existing technologies is likely to become a winner, as well as the ability to select the right timing in both design and execution. We will therefore call the ideal type of this user a *pioneering user*. The producer is no longer one single supplier but a *producer pair*.

2. THE INSTITUTIONAL FRAMEWORK

Public procurement in Greece needs to be analysed in the perspective of a less favoured region with an inefficient administration. During the rapid industrialisation of the post war years, public procurement was dominated by strong, though not selective protection (Mitsos 1989). Because of the nature of Greek industry the highest share of production equipment was imported and the press frequently reported attempts of bribery, indicating that the level of training and the organisation of the state did not allow for the creation of an efficient public procurement system. In the early '80s, an attempt to create a new team and modify the rules to use public procurement as a means of leverage for industrialisation and promotion of technological change was thwarted in a change of government.[8]

Ever since that time, the system has been dominated by public tenders, and selection has followed price considerations and local value-added criteria, as long as the transition period for adaptation to the EU legislation has allowed for it. Partly because in the second half of the '80s and in the early '90s many cases went to court, the public service is dominated by a reluctance to take on any responsibility that would, fairly or not, risk creating allegations of corruption. Thus, decisions are referred from one commission to the other, and this has often delayed the absorption of the very generous funds for big construction and procurement projects that have been approved for co-financing by the Community Support Frameworks.[9]

At the time of the case studied the procurement framework for the telecommunications sector was governed by the regulations of OTE, still a state monopoly with a minority of shares traded in the Stock Exchange. The internal regulation of procurement was adopted by the board of OTE in 1970. Although it is not formally valid anymore, the processes it entailed are still the basis of the decision process. This regulation distinguished among three kinds of processes: procurement through an official public call for tenders, through a short list, and through a direct contract. The regula-

[8] This involved 'promotion of technological change' in the sense of increasing the technological capabilities of local companies by including in the eligibility criteria the local value-added content, not in the sense of public technology procurement as a demand instrument for technological change in the way it is used in the present study. It goes without saying that this policy was only possible due to the transition period that allowed Greece to use protection measures in public procurement.

[9] As an indicative example: IBM has calculated that the Greek market for large computer equipment, which was the largest among less favoured European countries, was surpassed by both Portugal and Turkey, which had a faster decision process in the late '80s and early '90s.

tion foresaw the creation of three different committees and stipulated the various processes to be followed in an open bid.

In general terms, the last 15 years have been a hybrid situation of transition from the totally autonomous system to that of the European Union regime of procurement. Many negotiations took place in an effort to extend the protectionist elements of the Greek system, but the discretion to use protection was not effectively exploited. In the case of new technologies, the rule has been that the national industry is usually unable to offer the required products, at least not alone. OTE has thus had to open its bids for new equipment to the European or the international market. Many non-European providers have been selected in the past.[10] In recent procurement of high technology equipment (including the selected case study), US and Japanese producers were also considered as potential bidders.

3. THE KEY ORGANISATIONS INVOLVED

3.1 Public Organisations

The public purchasing organisation (procurer, buyer and final user) was – and remains – OTE. An open call for tenders was initially formulated by ELVIL, the company mentioned above. ELVIL was created ad hoc for the selection of the technologies to be used, and was later dissolved when its responsibilities were taken over by OTE. The OTE directorate for electronic switches now appraises the need and the funds available and, in co-operation with the directorate for procurement, proceeds to a tender. The technological standards required are based mainly on international bibliography and European norms adapted to the needs of OTE for compatibility, training, installation and familiarisation. OTE's staff has many experienced engineers with excellent skills, but they are scattered among the various services of the organisation. There is no single research or development department and. Thus, issues of technological change are not well represented, nor is there a unit able to advise the board on these specific matters. OTE has to rely largely on the expertise of university professors sitting on its board. The organisation is repeatedly accused of inefficient management and has managed to uphold protectionism longer than its counterparts in other EU member states. The pressure for OTE to modernise is constantly increasing, but this has only very recently led to the provision of new services and joint ventures.

OTE is formally supervised by the Ministry of Transport and Telecommunication, which does not have the necessary skills to control important processes of procurement effectively. Thus, although in some cases its intervention is strong, in the particular procurement studied here OTE did not have the skills to judge properly and intervene accordingly. The serious trouble that accompanied this (and other procurements) led, in early 1996, to the Minister's announcement of the establishment of an "interparty"

---

[10] In particular, a lot of Eastern European equipment was bought during the '70s, partly because of its low price, partly because of the possibility of barter trade.

public procurement committee, composed of 5 members of scientific and moral authority, selected by a Parliamentary Committee. It was the Minister's hope that this committee would reverse the current mistrust and eliminate unnecessary friction and accusations. No moves to implement this decision have been made as yet.

More government organisations (including the Department of Justice, the Prime Minister's Office and the opposition) were involved in the particular procurement studied, as it attracted tremendous attention from the media. A whole literature in the press has denounced inter-linked interests between the media, local companies and representations of multinational companies. In summer, 1997, the hearings in court were terminated (decision 1920/1998) and the press announced that the case would close (Eleftherotypia, 8.7.97 p.53). In December, 1997, the district attorney officially proposed dismissing all charges accumulated against 68 people working for OTE, Intracom and Siemens. An independent committee, composed of experts and examining magistrates, concluded that there has been no damage for OTE and the Greek State and that, on the contrary, the procurement has been highly beneficial.

The number of the potential supplying firms was limited by definition, because of the high market concentration. In the first bid, five main producers reacted, all of them major transnational corporations. Of these, four had a local subsidiary marketing their products, while Siemens also disposed of producing facilities in Greece.

ELVIL evaluated the proposals practically three times, under very strong political pressure, which reached successively higher levels of intervention, including eventually the level of the prime ministers of the interested countries. Five years elapsed (including one major change of government, whereby the ruling party changed) during the process of the initial selection. Every time, it produced the same results: a combination of Siemens and Ericsson, GTE being third. In 1985, after a new change of government (the same party remained in power, though) the ELVIL evaluation was adopted with one major amendment: no new plant would be built, since the private sector demonstrated substantial interest. The selected technologies would be retained but production would be undertaken by Siemens-Teleindustries, a Greek subsidiary of Siemens, and Intracom, a Greek telecommunications equipment producer, in co-operation with Ericsson. Following this arrangement, Ericsson bought shares in Intracom. ELVIL thus had no reason for existence anymore and was dissolved; subsequently, the procurement process was handed over directly to OTE, where it is still conducted. It is clear that in the time elapsed the buyer lost every advantage in terms of timing. After the selection of the AXE-10 and EWSD models of Ericsson and Siemens, respectively, and the bankruptcy of GTE, only Alcatel has remained a major contestant of the decision to continue the contracting process.

3.2 Private Organisations

The profile of the two partnerships is different in some respects: for Siemens and Ericsson, the Greek market is gradually losing its interest. In the initial call for tenders by ELVIL there was an important element of uncertainty sharing, although the systems were already operational, because the Greek market would be a good demonstration case. For the international technology providers, the long delays in the implementation turned the switches produced for the Greek market into a standard product procurement. But the case was different for the local companies. Both companies were fairly small compared to the international giants and could not compete on standard products by benefiting from high economies of scale. For them, the provision was vital for their growth and technology acquisition. Siemens-Teleindustries, established in 1964, is a subsidiary of Siemens with the major Greek commercial bank participating in its share capital. The company has a plant in Thessaloniki which is not viable without the OTE procurement. The entire Siemens group employs 1700 people in Greece. Intracom is the biggest Greek telecommunications equipment producer, created in 1977. It has had a considerable growth momentum, which has helped it multiply its production and exports by orders of magnitude: Sales grew from 2.9 billion drachmas in 1987 to 41.5 billion in 1995, while from the first year of exports of 611 million drachmas in 1990 the corresponding amount is now 11.5 billion. In 1995, the company outperformed every other Greek company in personnel statistics, since out of 1484 employees 523 held a university degree, 165 held graduate degrees, and 33 held Ph.D.s. Of these people, 427 worked in the R&D department and 220 engineers were employed in the development centre. The company is ISO 9000 certified. The balance sheet demonstrates that the company succeeded in executing the OTE procurement without incurring any major financial burden.

Both selected local manufacturers had to create special development units to respond to the software needs of the electronic switching procurement. This process was supplemented by the original clause of export obligations. It was only through this adaptation procurement that they were able to reduce uncertainty and finance their respective software development centres. For Siemens, a subsidiary of one of the biggest transnational companies in the area, the effect was a reinforcement of the local branch vis-à-vis the headquarters. The headquarters would not have authorised the cost of the creation of a Greek software centre, which was high, since it included not only the investment on equipment and current wage cost but a significant investment in training. Nevertheless, once established, the centre became competitive in intra-firm costing and now sells to the headquarters its services for other European and non-European countries.

For Intracom the investment in a software centre has had the same story of origin and seems to have paid off even more. While the initial investment was financed because of the OTE contract, the skills acquired have enabled Intracom not only to become an accredited supplier for Ericsson (its partner in the OTE bid) and IBM, but also to diversify the acquired skills and apply them in bids in Eastern and Central European countries. The experience from the 1,300,000 switches has given the company the opportunity to gain: a bid for new software in the Kremlin for 10,000 lines, an installation in Moldavia which is currently under way, negotiations for the telecom ring of Sophia, and other bids still under negotiation or lost. The procurement has helped Intracom build up a competitive advantage of skills because it has had to master all the necessary technologies to assure compatibility in the mosaic of OTE, in addition to acquiring expert knowledge in Network Management Systems (NMS), which is the basis for technology transfer in other areas (e.g., power control at the Public Power Corporation). The direct benefit from the export clause has been 18 million US dollars worth of software export to Ericsson. Now, although there are no more contractual obligations, Intracom products are more and more incorporated in the Ericsson list of sales (including both software and hardware production).

For Siemens Tele-industries and Intracom, OTE had, and continues to have, the tremendous market power of a monopsonist. This power was used initially (in the old regime) to negotiate the highest possible local value-added and export obligations. Only in recent bids has there been an effort to use this power in order to upgrade the product.

3.3 Competence and Other Resources

Despite the inadequacies of the human capital supply in Greece, OTE has, as mentioned above, a good but not concentrated team of well-trained engineers.[11] In the case of the electronic switches, procurement standards were drafted in respect of European norms. Some training was necessary, and expensive studies from specialised international think-tanks provided the necessary supporting knowledge. An issue of interest, which will also arise later, is that there is no autonomous technology department in OTE. Engineers are organised into functional divisions, each one following the interests of its mandate. Thus, there is little interaction and no capability for taking decisions based on strict technological strategy considerations. This role is entrusted ad hoc to the board of OTE, which in the latest part of the period studied has included academics with records of having excelled abroad.

For their part, the local suppliers have acquired the necessary competence during the period studied. The two international technology providers are both at the leading edge of competence and their knowledge is incontestable. The two Greek suppliers succeeded in very rapidly acquiring the necessary knowledge from a variety of sources. The core technology was supplied by their respective partners. While initially there were gaps in their local skills, because it was the first time they had produced electronic switches, they very rapidly built up high-performance and, compared to international standards, low-cost teams of engineers. In the case of Siemens, their training took place in Germany; in the case of Intracom, they were trained in both Sweden and Greece. As time went by and the procurement was upgraded, the engineers also started to increase their co-operation with local universities. Both companies have created dense contacts with the telecommunications research teams in the universities of Athens, Patras, Crete and Thessaloniki. These links were partly used to improve the firms' insight into the elements needed for the electronic switches and partly went beyond this particular project. Overall, research institutions were of importance for capability development, rather than very concrete projects directly related to the AXE-10 and EWSD produc-

[11] Some public utilities like OTE, the Public Power Corporation, and Olympic Airways benefit from a salary regime different from that of the public service in its narrow sense.

tion. The adaptation part of the procurement (which is partly original) has concerned the signalling software and the compatibility (co-operation) with existing equipment for switching and transmission.

The financial means necessary for the procurement provided have not been any problem: the procurement has been, and remains, a priority both for Greece and the EU, as it constitutes major infrastructure in one of the growing sectors. It thus constitutes a perfect match for selection of funding under the Community Support Framework.

4. ORGANISATION AND MANAGEMENT OF THE PROCUREMENT PROCESS

The most characteristic feature of the procurement process was the long delay in the decisions described earlier, reflecting the Greek informal rules for public procurement. Delays turned out to be extreme in this case, for two reasons:

1. The higher the value of the procurement, the higher will be the technological lock-in element it brings with it.
2. The higher the profits of the participating companies, the higher will be the fears of allegations.

The procurement studied was not a once-off procurement but a series of procurements, partly changing their characteristics though remaining within the same basic technology. The idea was that the first procurement would select among the technologies offered with the view of digitising the whole network of OTE. Once the two technologies were adopted, subsequent bids would be organised with specified numbers of circuits and ports each time. The cheaper bidder in each individual procurement would gain 60%, while the second would get the remaining 40%. That would protect OTE from future monopolist threats from the producers' side, and would allow the producers to count on a minimum purchasing platform. In every new procurement OTE would set the conditions for technological upgrading.

The specific content of the formal contracts (following one template, with slight amendments each time) includes the description of the product to be purchased, a set of clauses for quality assurance and a set of clauses for the benefit of the Greek economy. The pricing system includes clauses for delays and the drachma parity. The innovativeness of the product is not explicitly described. Nevertheless, the contract requires that the technology providers guarantee that the product offered incorporates the latest technology available in the international market. The innovative element is thus only moderate for the technology providers, but is automatically generated by the local contractors in their effort to meet the services required in every new contract.

The contracts include provisions according to which the hardware supplied is the latest version offered in the international market (e.g., art. 26 of the contract 7440/28.6.94 with Intracom). This provision was possible because Greece still had some degrees of protectionist freedom in transitional legislation preceding its full accession to European Union procurement regulation requiring open competition. The idea behind the provision was that all newly acquired competencies should be given the opportunity to be exported, in order to improve the competitive image of the Greek producers.

A complication arose from the fact that in one case OTE opened up the bid to the international market and new offers were invited from technologies alternative to those provided by Siemens and Ericsson. Alcatel, which took part in this bid with much lower prices and was then eliminated on technological grounds, accused the whole process of lack of adequate planning and transparency. This protest, nurtured by the general climate in public procurement, further deteriorated the public image of the case.

Table 1 presents the original data taken from the subsequent contracts between the Hellenic Telecommunications Organisation and the bidders, as documented by the responsible office of OTE.

Table 1: Procurement of electronic switches by OTE

	Reference contract	Total capacity of switches in all centres	Total capacity of circuits	Ports (sum of 3,4)
Intracom/ Ericsson for AXE-10	7160/30.3.88	10,000	53,000	63,000
Siemens Teleindustries/ Siemens for EWSD	7170/30.3.88	10,000	47,000	57,000
Intracom/ Ericsson for AXE-10	7230/1.2.90	285,000	82,000	367,000
Siemens Teleindustries/ Siemens for EWSD	7220/1.2.90	280,000	76,000	356,000
Intracom/ Ericsson for AXE-10	7310/5.12.91	460,000	60,000	520,000
Siemens Teleindustries/ Siemens for EWSD	7300/5.12.91	430,000	70,000	500,000
Intracom/ Ericsson for AXE-10	968/1.2.93	106,000	245,000	351,000
Siemens Teleindustries/ Siemens for EWSD	1003/8.7.93	50,000	15,000	65,000
Intracom/ Ericsson for AXE-10	7440/28.4.94	472,000	113,000	585,000
Siemens Teleindustries/ Siemens for EWSD	7430/28.6.94	479,000	129,000	608,000
Siemens Teleindustries/ Siemens for EWSD	968 as a subcontractor of Intracom			

In 1996, a further order placement was discussed, with the expectation that it would lead to 590,000 new switches (Kathimerini 29.9.96). The needs until the year 2,000 are estimated at 2,220,000 switches and 100 billion drachmas. In February, 1997, a new order was placed for 335,000 new switches and upgrading of older equipment to include ISDN features.

Nepotism was not identified in this case, but as mentioned earlier the accusations about corruption involved in the procurement became a central element of the whole process. Investigations have been open for a long time, but as described elsewhere the current state of affairs is that they will be rejected by the district attorney.

What is most interesting about the way the development and design work have been organised is a lack of close interaction between OTE and the suppliers after a bid has been concluded. Two especially important causes for that are:

- the absence of an organised R&D department in OTE, which would have the capabilities and the interest to follow the development closely; and
- the fear of the OTE personnel of being accused of having too close contacts with one or both of the suppliers.

It is characteristic that in all interviews it was reported that OTE people visited the companies only twice a year and, whenever they did, they had to organise control visits following the terms of the contract.

Wider interaction was limited in the Greek case, because:

A) Surveillance of the market was assured by OTE, mainly through its interaction with corresponding organisations outside Greece and the work undertaken in international fora, plus the acquisition of the necessary literature.

B) Formulation of the product requirements followed the standard literature suggested by European standards.

C) Different specifications by the suppliers referred to the high number of different technologies in the Greek system, which had to be encompassed in the digital switches.

D) Negotiations concentrated on price issues both ex ante and ex post, when price reductions and cash refunds were required.

5. APPRAISAL OF SOCIETAL AND INDUSTRIAL RESULTS AND POLICY ASPECTS

5.1 Societal and Industrial Results of the Case Studied

The procurement was a success with regard to technical development and diffusion. It was important for the diffusion of a generic technology in the country, one which greatly facilitated communication. It was an important innovation policy instrument for the local producers, since it enabled them to improve their skills and capabilities, develop related products successfully, and open new markets. Technical diversity considerations followed the same pattern of all European markets of the size of Greece, where two technologies were considered necessary to avoid losing negotiating power, but three would not permit amortisation as this number would multiply compatibility costs and not allow interested parties to benefit from economies of scale.

As far as the local companies were concerned, the procurement also worked as an industrial policy instrument, because it allowed the suppliers to increase their turnover and benefit from economies of both scale and scope. Cost reductions and increased competitiveness were also possible because of the improved reputation of the suppliers and their potential to negotiate better input prices for themselves in the international market.

The cost has changed over time and it is difficult to estimate it. What is available is the price of every subsequent bid but it is almost impossible to establish a good "value for money" argument for each stage. Invoicing by each company is so different that it is not easy to make comparisons: The price of the contract itself is always low, in order to get the contract; then, companies invoice differently their maintenance, spare parts, training and software. So, at the end of the day, only a very transparent and thorough study ex post can identify the real price paid. However, it would be difficult to compare with competitors, since one doesn't know how they would invoice post contract needs. As a rule of the thumb, one may say that 50% of the cost is the initial contract and 50% subsequent services. An additional problem is related to the exchange rate of the drachma: Bidding prices have to be in drachmas but devaluation expectations affect the price a great deal. At any rate, the report of the investigators to the case, submitted at the beginning of 1998,[12] states that the contract has been beneficial for the Greek State and OTE in particular

A few points arising from interviews and publications in the press indicate a certain direction, which is not the same as the report of the investigators. One of the supervising secretaries of state (Ministry of Transport and Communications) has published an article in the press, suggesting that the process of locking-in the technology created a major problem in terms of pricing. Since bids were accepted only from the short list of the two originally selected companies/technologies, the secretary of state argues that a clause should have been incorporated to payback to OTE the difference between offers in a lower bid and the actual prices paid by the procurers. This resulted in a refunding of 1.5 billion drachmas and arbitration is pending for more, up to 10 billion. The article concludes that although technological or even protectionist arguments are valid, it is important to restructure and modernise OTE in a way that will guarantee the safeguarding of its interests. At the same time, the ALCATEL complaints suggest that their bid was lower, though in different technical features. But the most thorough investigation in court suggested that OTE and the Greek market have ultimately benefited from the contract.

In quantitative terms Intracom calculated the opportunity cost of the delay to amount to 3 billion drachmas (Intracom 1999).

The benefits of the suppliers were discussed earlier in terms of new products produced and new markets opened. The major beneficiaries have been Greek telecommunications users. The reliability and speed of Greek telecommunications have increased tremendously and queues for getting a telephone line are in constant decrease. This was

---

[12] This is not a publicly available source and quotations come from the press.

a major advantage for industry (the bad telecommunications infrastructure was stated as one of the key barriers to the attraction of foreign direct investment in a study conducted in 1991) but also for the average citizen, who formerly suffered from long waiting time and a very bad quality of services. The procurement has not only improved the services offered by the digitised part of the network, but has also helped compatibility throughout the system.

5.2 Policy Aspects

The case studied above deals with procurement under technological and institutional uncertainty.

In our view, the case study indicates that it is worthwhile to establish the rules for technology procurement even in a less favoured country: Pure price considerations based on auction mechanisms are unlikely to be successful in an uncertain environment. But the case also demonstrates that it is very difficult to exercise technology procurement effectively, even if initially the vision exists and subsequently the conditions are fulfilled. This is because the skills required on the part of the buyer are very demanding. These skills are unlikely to be found in technologically less favoured regions. Nevertheless the procurement can be beneficial, even if not totally successful.

The main lesson learned from the case study is that it is vital to find ways to avoid delays, in particular when dealing with complex, state-of-the art systems. It also suggests that it is vital to identify development pairs, rather than individual technology providers, and to make sure that the necessary skills are available to the buyer organisation, when such an undertaking starts. Competence-building, trust relationships, and transparency are the elements that will forge the quickest and most beneficial technology diffusion within a country. Particularly with systems that are early in their trajectories, any standard processes are suspect. They may not only fail to guarantee lowest price bids; they may even risk inhibiting the country from making an early entry into highly beneficial investments (thus reducing its competitive position in the international arena).

From the Greek experience, then, it is suggested that it is unlikely that isolated initiatives will succeed and that the creation of a whole new environment is necessary. Important aspects allowing a new approach in policy are:

- *In terms of skills*, it is necessary to create adequate skills for identifying the best timing for APP, as well as adequate negotiation skills. While the former are of a specific nature, particular to each procurement, and can be assured by the creation of permanent research and technological observatory committees in the major utilities, the latter are horizontal and can be acquired through hiring specialised consultants or trained business staff.
- In terms of the establishment of *formal rules,* it is important to create the necessary framework that will allow for transparency, accountability and accurate cost identification, in order to re-establish the image of public procurement and allow public officials to control suppliers without being accused of corruption. Until then, it

is unthinkable to conduct procurement with considerations other than price minimisation.

- This will also involve changing the *informal rules and management of the whole process* through the institutionalisation of continuous interaction among the key organisations involved in each procurement.

6. SUMMARY AND CONCLUSIONS

The Greek case concerned the procurement of electronic exchanges by OTE. Two suppliers were selected, offering alternative, comparable technologies, each one in a joint venture with foreign technology providers and notably the key transnational companies in switching technologies. The whole process took place in an institutional framework where the EU directives on telecommunications procurement did not yet apply, but the bid was open to an international tender, because local companies did not have the skills and the size necessary to produce the required product. Local value-added was one of the evaluation criteria.

The problem to be solved in this case study was the digitalisation of the Greek telecommunications network, necessary both for social reasons and for the competitiveness of the Greek economy. The *need* was identified fairly early by the Greek political authorities and the telecommunications administration. The latter was inspired partly by the need to leapfrog in order to reverse the, at the time, miserable telecommunications infrastructure in the country and partly by the personal *vision* of the then prime minister (who saw himself as the founder of modern Greece after his success against the dictatorship). The *translation into functional requirements* was originally entrusted to a company founded for this particular procurement and in particular to its board composed mainly of academics in the fields of telecommunications and economics. At a later stage, when this company was dissolved, the functional requirements and the adoption of specifications became a task of the national telecommunications organisation. The *mode of procurement* at all stages was a formal public tender, which included specific standards for the necessary incremental improvements and provisions for strengthening the technological position of local companies. The *product procured* was electronic switching systems, originally conceived to be state of the art, but ultimately becoming a rather standard product, as delays in the procurement were high. The open bid process and the importance of the project, because of the situation of technological lock-in it would create, ended up in a substantial delay. Although the evaluation committee finalised its selection, re-evaluations related to political sensitivity delayed the project by almost a decade at the first stage. Eventually, the decision was made to transfer the results to two companies in the private sector.

An outdated institutional framework, an unstable environment trying to extend the period allowing protectionism and a delayed deregulation of the telecommunications sector were additional factors inhibiting rapid decisions.

The buyer's competence was satisfactory in terms of engineering, but apparently less so in terms of business decision-making. The local suppliers' competence, which was not fully available from the beginning, built up very rapidly in both companies. Through a mix of foreign acquisition, training and co-operation with local research institutions, the two suppliers benefited strongly from their contracts in terms of being able to improve their skills and to promote interaction in a key sector, previously extremely underdeveloped in Greece.

The process was organised in subsequent stages with very similar characteristics. Management was not tight and interaction was very limited. Moreover, corruption allegations have been a most disturbing element in the process.

The key elements of the procurement relate to:

- a bad organisation, which for various reasons resulted in very long delays, turning the procurement from a leading-edge-technology product into a standardised system, and thus degrading constantly the benefits the buyer could obtain;
- the strong benefit of local suppliers from the procurement, in terms of improved R&D capabilities, which they could ultimately also exploit in other products and markets;
- a significant reduction of the uncertainty of investing in people for software production, which allowed skill upgrading; and
- an astonishingly low interaction occurring among the relevant elements of the national innovation system, i.e., the actors that were (or should have been) involved in the process

Interdependence was a crucial element in terms of:

- interaction between the procurement organisation and the Greek suppliers, which was limited to price negotiations;
- interaction among the Greek suppliers themselves, which was practically non existent; and
- interaction of Greek suppliers and their technology providers (this is not a national innovation system anymore, but one open to the rest of the EU), which has been the most dense and mutually beneficial.

REFERENCES

Best, Michael and Robert Forrant. (1996). *Creating Industrial Capacity: Pentagon-Led versus Production-Led Industrial Policies.*

Intracom (1999). Economic Assessment of the Digitalisation of Greek Telecommunications. Athens: (Internal paper).

Mitsos, Achilleas. (1989). *The Greek Industry in the International Market*. Athens: Themelio.

Utterback, J. and R. Suarez. (1993). Innovation, competition and industry structure, *Research Policy, 22.*

Vernon, Raymond. (1966). International investment and international trade in the product cycle. *Quarterly Journal of Economics, Vol. 80*,(2).

10. ON IMPLEMENTING THE AUSTRIAN COMPUTERISED DIGITAL SWITCHING SYSTEM (OES)

M. Husz

1. INTRODUCTION

The present study describes the introduction of the Austrian digital switching system, OES, and evaluates the technological, economical, and societal effects of this step. The system has been procured by the public Austrian telecom provider, *Österreichische Post-und Telegraphenverwaltung* (ÖPTV), and is thus an example of public technology procurement.

In 1976, ÖPTV faced serious problems: The demand for telecommunication in Austria was still quickly increasing. The number of lines was to increase from 1.390.456 in 1974 to 2.033.351 in 1979. Forecasts rightly predicted an increase of at least further 2 million by 2000. ÖPTV already employed about 19,000 people in its telecommunication department and could not afford any further increase in cost. Most of the equipment in densely populated areas still stemmed from the efforts of the post-war reconstruction phase and had become obsolete. Switching was expensive because its maintenance tied up a lot of room and personnel. On average, ÖPTV needed 3,6 employees per link then, a number reduced to 0.9 in the nineties.

ÖPTV reacted to this situation by digitalising its switching system. Having little technological know-how in the matter, it founded a company to develop criteria for the choice of a digital system and to elaborate specifications for an adaptation of the system to Austrian requirements. ÖPTV held 50.4 per cent of the shares of this company named *Österreichische Fernmeldetechnische Entwicklungs- und Förderungsgesellschaft* (ÖFEG). The remaining 49.6 per cent were divided equally among the four Austrian suppliers of telecom equipment, ITT Austria, Kapsch, Schrack and Siemens Österreich. In their joint company ÖFEG, R&D-staff of all five partners co-operated in finding a solution to the above problems.

Following recommendations of ÖFEG, ÖPTV decided that ÖFEG should take a licence from Northern Telecom and adapt their system, DMS 100, to Austrian needs.

By political influence this decision was reversed. According to our interviews and the literature available, political influence had produced a variety of implicit goals, most of which were of an economic nature. The basic idea was to import technology for Austrian firms co-operating in increasing exports and employment in the telecom sector. From the standpoint of national security, a major goal was preservation of independence in an infrastructure considered vital for the Austrian economy.

In 1981, the Minister of Transport finally decided to procure two systems, DMS 100 from Northern Telecom, and EWSD from Siemens. OES now stood for *Österreichisches Elektronisches System*, the word 'uniform' having been changed to 'electronic', and the two subsystems were called OES-D, based on the DMS-solution, and OES-E, based on EWSD. Both systems were to be licensed abroad and partially produced in Austria. The Austrian digital switching system would operate on the Pulse Code Modulation (PCM) principle, introduced for increased reliability and efficiency in use. It would also use the Stored Program Control (SPC) principle, which would allow for greater flexibility and value-added services. Digitalisation of the analogue signal would not happen at the client level but in the exchange. PCM-30 would allow for 30 simultaneous calls and signals in time multiplex and for the common channel signalling system no. 7.

Having co-operated closely before, the Austrian firms Kapsch and Schrack founded a joint development company named *Austrian Telecommunication* (AT). To produce OES-D on the basis of the Canadian DMS 100, AT bought the respective licence from Northern Telecom and adapted the system according to the specifications elaborated by ÖFEG. The Austrian subsidiaries of Siemens and ITT were less natural partners. They co-operated on the basis of an development association named *Arbeitsgemeinschaft OES-E, Siemens, Alcatel* (AOSA). They took a licence from Siemens in Germany and adapted EWSD to Austrian specifications, again conforming with the specifications formulated by ÖFEG.

By 1985 the first two switching exchanges went into operation. The plan was to complete digitalisation within thirty years. Actually, the digitalisation of the Austrian telecom switching system will be completed by 1999.

2. THE AUSTRIAN TELECOM SECTOR

2.1 The Procurer

After the second world war, rebuilding the Austrian economy seemed a natural task for the public hand. Reconstruction was planned centrally. Most of the Austrian heavy industry and infrastructure remained in public hands. Steel, mining, chemicals, energy, automotive industry and public transport were state-owned. Moreover, the public hand held the majority of all major banks. Most important for our study, the sole Austrian post and telecom provider (ÖPTV) was organised as a section of the Ministry of Transport and, thus, subject to public administration. ÖPTV itself acted as a public institution within a framework of goals far exceeding profit optimisation. It perceived itself as an instrument of social policy rather than a business company. Thus, the decision had to be made in favour of a domestic supplier.

In the post-war period the independence of Austria hinged on close collaboration of all Austrian interest groups. The coalition of the two large parties, the socialist and the people's party (SPÖ and ÖVP), in the years between 1945 and 1966, laid the foundation for the *Social Partnership*.[1] Its most important political instrument came to be the *Paritätische Kommission (PK)*, where the representatives of the Economic Chamber, the Agricultural Chambers, the Workers' Chamber and the Federation of Trade Unions have traditionally found compromises on all potential economic conflicts in Austria. Most Chamber memberships are compulsory, which implies that the degree of organisation in Austria is extremely high.

There are sub-commissions of the PK elaborating recommendations not only on prices and wages but also on most other important issues of economic and social life in Austria. Despite being an unofficial instrument without formal constitutional backing, the PK has been of tremendous influence in the country. The *Social Partnership* is an expression of an Austrian tradition of seeking to evade conflicts and finding compromises. It has meant letting left- and right-wing interest groups participate in the management of the nationalised industries as well as solving the wage struggles in private industries without strikes. The decision on procurement of a digital switching system in Austria has to be related to this economic culture of close personal contacts and joint orientation.

One typical result of close personal contacts was the partition of markets. In 1976 this situation also prevailed in the telecom sector and had its influence on procurement. There was a certain amount of competition between the suppliers trying to increase their respective quotas by intruding on a competitor's market segment. This was no vital threat to the competitors, however. The government had a strong commitment to involve all national firms in order to maintain a high employment level. The government did not face restrictive legislation demanding specific procurement procedures like public tenders, at least not in the telecom sector. Nor did consumer preferences play a role. The Workers' Chamber, considered as Austria's main consumers' representative, basically voted for employment, not for low prices or inexpensive technical solutions.

Telecommunication was not considered a particular topic for Austrian research policy. It was a responsibility of the Austrian Ministry of Transport. Following the logic of budgetary competence, telecommunication was thus a topic already taken care of. Governmental planners saw no need to devote further financial efforts of other ministries to technological lags, which, by definition, were not supposed to exist. In case of financial shortage, it was the duty of the Minister of Transport to take care of sufficient promotion of research on telecommunication. Results of this logic were the fragmentation of Austrian technology policy and a lack of incentive compatibility with respect to innovation.

---

[1] In the meantime the former Socialist Party of Austria (Sozialistische Partei Österreichs) has changed its name to Social-democratic Party of Austria (Sozialdemokratische Partei Österreichs).

2.2 The Suppliers

There existed four supplier companies in Austria. Their present key figures are shown in Table 1. Siemens Österreich is obviously the largest of the four, even if only the core company, Siemens AG Österreich, is considered. Only two of the companies, Siemens and Kapsch, have survived in their original form. ITT became Alcatel, and Schrack has been bought by Ericsson. For an estimation of the market shares before digitalisation of the Austrian telecom net, compare Table 2. The market shares do not represent the development capacities of the firms with respect to digital systems, however. In 1976, Kapsch had about 40 designers working on PCM (digital) and SPC controlled equipment. Kapsch and Schrack had no digital systems of their own, however. Kapsch therefore started an intensive search for a system licence, visiting, among others, the development departments of NSI, Stromberg & Carlson, Siemens, Philips, ITT Belgium, and Northern Telecom. It was known that the Austrian Chancellor had a preference for large firms, whose capital he wanted to attract to Austria. There was an inclination towards Philips, who had promised to found a software-house in Austria if successful in selling. Furthermore, the Austrian state owned 43.6% of Siemens Österreich, which made Siemens a strong candidate.

Table 1: Key Figures of the Suppliers (1996)

Key Figures	Siemens Group Austria	Siemens AG Österreich	Kapsch	Alcatel Austria (former ITT)	Ericsson Austria* (former Schrack)
Year of foundation	-	1879	1892	1871	19th century
Employment	20,173	9,684	2,019	1,630	1,350
Turnover (mio ATS)	37.984	21.530	3.714	3.380	3.900
Exports (% turnover)	48.9%	36.1%	9.9%	11.4%	32%
Investments (% turnover)	7.8%	3.4%	5.0%	2.8%	3.5%
R&D (% turnover)	14.6%	20.9%	12.5%	17.8%	8.4%

* Ericsson: Data from 1995

Sources: Siemens, Alcatel, Ericsson, Kapsch (some information taken from the companies' respective Internet pages.

One of the major turning points at the beginning stages of the procurement process was the licence choice by Kapsch. They did not choose Philips or another European partner, but Northern Telecom and its digital system, DMS 100. It was the most advanced system in terms of commercialisation and there were operational switching exchanges ready to visit. The Philips system PRX was considered as technologically inferior and actually was no commercial success world-wide. As a result of the decision by Kapsch, Philips retreated from the Austrian public telecom market and is not a main player in this sector today.

Table 2: Shares in the Austrian Switching Market

Estimates	1970	1975	1980
Siemens	26%	33%	30%
ITT	24%	21%	21%
Kapsch	35%	34%	33%
Schrack	15%	12%	16%

(Total Switching: public, private and subscriber equipment)

Sources: Estimates based on interviews and informal literature

Here is a good example how Austrian *Social Partnership* works: By personal contacts Kapsch gained the support of high representatives of the Workers' Chamber and the Austrian Federation of Trade Unions. An official government delegation, consisting of the Chief of the Chancellor's Cabinet, the Secretary of the Minister of Transport, and the Head of Transport Division in the Workers' Chamber, visited Northern Telecom for a hearing. The workers' representatives favoured a solution which gave Northern Telecom a chance to intrude the European market dominated by national champions. The result was a good licence contract for Kapsch and Schrack, who from then on lobbied the system in Austria.

According to interviews, the established market leaders, ITT and Siemens, did not take Kapsch and Schrack seriously. This applies particularly to the parent companies. ITT was considered a technologically advanced Austrian telecom company. Within ÖFEG, ITT insisted in lobbying for its System 12-30. This strategy was influenced by the parent company. It was not a wise decision, however, since it brought their own subsidiary, ITT Austria, into a difficult position. The system was not yet fully developed and already deficient in capacity. Just one month before the closing date, ITT withdrew System 12-30 and replaced it with System 12-40. Their entry was not as well elaborated as the others, however, and the change in system itself provoked criticism on behalf of the competitors.[2]

As to Siemens, they considered themselves as natural favourites and had good reasons to be optimistic. Siemens is one of the oldest telecom suppliers in Austria. Siemens had the largest employment of Austria's telecom industry and an active works council acting in favour of their digital system. Under a socialist government this was a definite advantage. Furthermore, their system, EWSD, was further developed than System 12-40 at the time and had less capacity problems. Most important of all, there existed trial facilities which could be visited.

EWSD was not yet fully developed, however. According to European tradition, it was a system needing highly skilled personnel. Integration of the central operation,

---

[2] It should be mentioned, moreover, that ITT Austria had a second problem shortly after the Minister´s decision on OES. In 1982, a scandal concerning bribery in the market of private automatic branch exchanges resulted in the arrest of the general manager of ITT Austria and a total change of management. Named after the journalist discovering the story, the case has become known as 'Worms-Skandal´.

administration and management system (OAM) was realised later than in the case of DMS 100. According to ÖFEG-experts, the two systems are considered as more or less equivalent today and require a portion of about 65 per cent of semi-skilled personnel. There had been no extensive testing in the case of EWSD, however, which raised doubts about the reliability of the system under workaday conditions. Visits of Austrian decision makers in Munich were not attended to by high-rank Siemens technicians but by marketing experts unable to explain technical details. Some managers of Siemens and ITT thought that Kapsch, having no know-how of its own, would not be able to adapt DMS 100 to Austrian specifications, anyway.

3. THE DECISION PROCESS

3.1 Getting Organised: Foundation of ÖFEG

The first phase of the decision process, starting in 1976, was a period of gathering information about possible steps towards rationalisation of the Austrian telecom network. It ended in 1978, with the foundation of ÖFEG. Primary goals of ÖPTV were:[3]

- to renew certain components of the net which were obsolete.
- to increase the capacity of the net.
- to introduce value added services for the public.
- to rationalise maintenance by automation and centralised management.

In this phase, ÖPTV laid down the rules of the game: The most important decision was to introduce a *purely digital system* and to use the new opportunities of computerised operation. This decision was made in 1977. It explains why certain prominent solutions in other EU-countries were not considered. Most American systems, as well as the French E10, were not candidates because the systems were not purely digital.

A second decision was to introduce *one uniform system* completely compatible with functionalities already existing in the Austrian network. Most managers of ÖPTV realised the role of their public company for Austrian economic policy. They were mostly engineers, however, and the decision for just one system seemed straightforward to them at the time.

The third decision was to *adapt an existing system* to Austrian needs instead of developing an indigenous system or of buying a turn-key system. ÖPTV thought that there were too few Austrian resources for the development of an indigenous system, which would thus have been too expensive as an alternative. Similar experiences in Switzerland show that this intuition was right. The second alternative, payment for adaptation abroad and taking a licence for the finished product, was ruled out for political reasons. This alternative would have resulted in lower domestic employment and a higher degree of technological dependence, consequences contrary to the policy goals

[3] Latzer e.a., 1994, p.27.

of the government. Furthermore, there was a commitment to retain all Austrian services which had been introduced before digitalisation. The consumer should not notice the dramatic change in the network, at least not by missing already existing functions of the net.

A fourth decision was to *include the suppliers* of ÖPTV in the decision process. As there was no further information available at Austrian universities and research institutes, ÖPTV used its traditional contacts to the Austrian telecom industry to fill its information gaps. At that time, Austria had no significant production of digital components. In the case of Kapsch and Schrack, the technological know-how concerning the development of digital systems was limited. Based on a joint development with AEG-Telefunken, Kapsch put a semi-electronic pilot-exchange (OHS/fixed wired logic) into service in 1967. It started to develop and produce PCM 30 technology based products in 1973 and SPC controlled equipment in 1976 (ZM1, ZUP, C-wire multiplexer). There was not enough competence to offer an integrated network, however, nor the ability to develop one within a reasonable time span. Through their contacts to their parent companies, the Austrian subsidiaries of Siemens and ITT disposed of some rudimentary know-how. The parent houses of Siemens, Philips, and ITT worked on their own digital Systems: Siemens worked on EWSD, Philips on PRX, and ITT on its Systems 12-30 and 12-40. None were produced in Austria, however, and only Siemens had started first capacity testing in Germany and South-Africa. The European digital systems existed mainly on drawing boards, especially with respect to the software.

To enhance competence, ÖPTV decided to build up technical know-how on a national basis before taking the system decision. In the tradition of a centrally planned economy, ÖPTV bundled all national competence by founding the *Österreichische Fernmeldetechnische Entwicklungs- und Förderungsgesellschaft*, ÖFEG, in 1978. It was a partnership among:

- The Federal Republic of Austria
- ITT Austria Gesellschaft m.b.H. (now Alcatel Austria)
- Telephon- und Telegraphen-Fabriks Aktiengesellschaft KAPSCH & SÖHNE in Wien
- E. SCHRACK Elektrizitäts-Aktiengesellschaft (now Ericsson Austria)
- SIEMENS Aktiengesellschaft Österreich

The Austrian Republic held 50.4 per cent of the shares of ÖFEG. The other four parties held the rest, share and share alike (12.4 per cent each). It was, at the same time, a good example of outsourcing, a rather modern concept then. ÖFEG was to provide the necessary preconditions for the production of electronical digital telephone systems in Austria. ÖFEG was, moreover, responsible for the gradual implementation of OES into the existing network, which implied an evolutionary process. Further long-term tasks were the introduction of a common channel signalling system, actually introduced in 1990/91, and of an Integrated Services Digital Network (ISDN) and an Operation, Administration and Maintenance System (OAM), both introduced in 1992/93.

It is also interesting to note which actors did not participate in the digitalisation of the Austrian switching system:

- Austrian universities
- other public research organisations
- research organisations abroad
- consumer organisations (apart from the Workers' Chamber)
- private consultancies

In the case of digital networks, Austrian universities and research centres did not have enough technological know-how to be considered helpful by ÖPTV. Given the strong influence by politicians trying to build up national competence, however, it is interesting that there were no attempts to involve academia. Until the nineties there were practically no professional links between the personnel employed by ÖFEG and the universities. This indicates that the digitalisation process implied little research but, rather, pure development. Apart from some private experts giving lectures at universities there was no attempt by public science organisations to complement the efforts of ÖPTV and the suppliers. The same applies to potential links between the research & development departments of the parent companies abroad and Austrian academia. The chosen strategies were pragmatic in the sense that a given technology was to be installed in Austria. The idea was to adapt an advanced international standard to the Austrian situation. Further research in order to become an international leader was not part of the strategy. As to further developments, ÖPTV waited for international standards.

Another possibility not considered was to entrust national or foreign consultancies or research organisations with feasibility studies or technological research facilitating the decision on the system choice. It seems that this idea had never been discussed. Finally, there had been no market analysis, and no consumer representatives had been involved in the decision process. Although some experts in private firms hint at the differences in viewpoint this might have caused, most decision makers in ÖPTV and ÖFEG still think today that consideration of consumer preferences would have led to qualitatively inferior and technologically conservative solutions. Leaving the decision to technicians has, in their view, led to quicker and more efficient introduction of high technological standards.

The priority of ÖPTV was to solve their capacity and cost problem. The ÖFEG contract, however, also shows how the original goals of ÖPTV were adapted. A national dimension entered via the goals formulated in §3 of the ÖFEG contract:

- Maximisation of the Austrian value added.
- Acquisition and realisation of technological know-how.
- Establishment of genuine development.

All suppliers were invited to present system solutions within ÖFEG. ÖPTV threatened to develop its own solution in case the other partners did not co-operate. A closing date for entries was fixed. After this date, by joint team work, evaluation criteria were to be developed as a basis for the choice of one system. These criteria, combined with weighting factors provided by ÖPTV, were used to make the final choice.

Collaboration within ÖFEG was sometimes hard because of the different economic interests of the respective parties. As to the staff from the four companies, there were many 'engineering artists'. They had considerable knowledge of electro-mechanical systems and were used to solving technical problems from the very basics. Adaptation of an existing digital system by such highly trained personnel led to some 'radical' solutions and basic criticism of the systems they were to adapt. As a consequence, discussions and sometimes conflicts with the parent companies abroad evolved. The ÖPTV-fraction within ÖFEG, consisting of selected ÖPTV-staff members, tried to stay neutral. One effect of the conflicts was a natural selection process. Engineers with higher ability to adapt and to be satisfied with conservative solutions, had better chances to stay.

ÖFEG itself was kept out of all price negotiations. There was some bargaining on the technological criteria formulated as a basis for the choice of a digital system but no discussion of prices. In the end, ÖPTV dominated the decision process, reflecting the majority in shares held by the Ministry of Transport. The exclusion of price considerations was necessary to create a culture of open talk, which resulted in a more or less harmonic elaboration of the criteria needed for the system choice. This in itself, as well as the rather smooth collaboration with the development firms, can be considered a success.

3.2 The Decision Criteria Elaborated by ÖFEG

The first task of ÖFEG was to describe the systems already existing in Austria and to formulate criteria for the choice and adaptation of a uniform digital system named OES. All in all, it took about 40 employee-years to describe about 100 interfaces. Of these 100 interfaces, 30 were retained. The new system thus had to be adapted to these 30 interfaces. Some experts say that this number could have been reduced further which would have saved some cost. Others say that the chosen strategy allowed for a smooth transition and preserved employment, factors more relevant than an insignificant cost reduction. It should be mentioned in this context that most employees of ÖPTV were permanent government officials. If they had lost their positions or responsibilities by rationalisation, they would have had to be employed somewhere else within ÖPTV. Both management and works council saw limited possibilities beyond the above reduction of interfaces.

Considerable effort was mobilised to achieve an objective choice based on internationally acceptable rules and decision practice. As to the choice criteria, five decision trees were elaborated, containing more than 650 single criteria. After acceptance of the decision procedure by ÖPTV, ÖFEG continued by describing the competing digital systems in terms of the criteria elaborated in advance. The suppliers participated in this process and a consensus was achieved, at least on major topics. There was a secret trial of the decision algorithm executed by the ÖPTV-group within ÖFEG. The result was very similar to the actual decision taken later by ÖPTV officials following the same algorithm. In 1981, the results were passed on to the Minister of Transport. According to this algorithm, DMS 100 was the best system, followed by EWSD. The crucial argument for the ranking was the higher degree of maturation of DMS 100, which again shows the pragmatism of the decision makers.

3.3 The Revision of the Decision by Political Influence

The decision by ÖPTV had been based mainly on technological criteria. After the decision, Siemens made its presence felt, however. On the levels of both the management and the works council, intensive lobbying was conducted. Siemens threatened to stop all joint-ventures in Austria. In addition, the Austrian State owned 43.6% of Siemens Austria. The Siemens arguments fell on fertile soil and the Chancellor intervened. The Minister of Transport decided to procure a licence both for DMS 100 and EWSD. The Austrian government and the *Social Partnership* preferred a co-operative solution, anyway. There was no inclination to harm Kapsch or Schrack which were genuinely Austrian, though standing in a tradition of family businesses as opposed to socialist economic policy. The idea was, rather, to induce Siemens to invest in Austria. Siemens was considered as an important economic factor in Austria and, actually, plays a particular role in Austria's technology policy up to the present day.[4] Co-operation represented a 'culture of friendship' as an expert said. The alternatives were considered as fatal to home production and human capital formation. The government intended:

- to import technological know-how on a broad basis.
- to foster co-operation between the involved firms and to bundle technological competence in the telecom-sector.
- to increase employment by national production.
- to enhance the export opportunities of the Austrian telecom industry.
- from the standpoint of national security, to preserve independence in an infrastructure vital for the Austrian economy.

One interesting aspect of this list is the lack of a consistent technology policy. Historically, there exists a certain tendency in Austria to use technology policy as a synonym for export policy. Even substitution between the two policies can be observed. Export-oriented production was an explicit goal of OES-procurement. Research in the field of telecommunication was not. There was a clear commitment to strengthen the Austrian telecom industry but no commitment to endow it with technological superiority. The international technological standard was accepted as a benchmark. In this sense, the decision of the government was half-hearted. Procurement was not co-ordinated with

[4] The general manager of Siemens Austria is the speaker of the private shareholders in Austria's major research centers and participates in the formulation of papers on Austrian technology policy together with leading officials of public institutions.

other policy instruments like, for example, the installation of an university institute on digital telecommunication or the harmonised subsidisation of private research activities. Considering the approach of National Innovation Systems, the decision can be considered as 'incomplete'. It was not taken on grounds of a long-term plan for the technological development of Austria but rather as a reaction to political necessities. The part of the Chancellor, especially, was a reaction rather than a step in a long-term technology policy.

4. THE IMPLEMENTATION PROCESS

Up to the decision of digitalising the switching system, the procurement procedure usually used by ÖPTV was to invite the suppliers to solve a certain technical problem specified by ÖPTV and to prove applicability by erecting a trial facility at their own cost. ÖPTV then decided on whether the system fitted its needs. If it did not, the facility had to be removed and there would be no financial compensation for the supplier. This shows that ÖPTV had enough monopsonistic power to shift all risk to the producers. In the case of digitalisation, however, ÖPTV chose a completely different procedure, which implied riskless participation for all major suppliers. The foundation of ÖFEG can partly be explained by ambitions beyond the primary goals of capacity increase and cost reduction. It is difficult to decide how important this aspect has been, however, because ÖPTV was also in urgent need of technological know-how, which led to early involvement of private firms in the decision process.

4.1 The Development Organisations

At the time when ÖPTV still looked for one system, ÖFEG was intended to take the licence and to develop the Austrian adaptations. All supplying firms were to co-operate within ÖFEG, not only to find criteria for the choice of the system, but also to elaborate Austrian specifications for the adaptation of the system and to carry out joint development according to these specifications. Due to the two-systems decision, however, two licences were needed for OES. Two development organisations were founded instead of one organisation, ÖFEG, being responsible for joint development. Two suppliers co-operated in each of these development companies, respectively. In the case of Kapsch and Schrack, this was easy as the two had co-operated closely before. They founded *Austrian Telecommunication* (AT) in order to adapt DMS 100 to Austrian standards. Only by close co-operation with Northern Telecom and through the efforts of a young, hard-working team, did they succeed in adapting the system. At present, about 250 people are employed for the development of OES-D.

The licence of EWSD, sold by Siemens, was bought by its Austrian subsidiary and by ITT Austria. Among representatives of Alcatel Austria (formerly ITT Austria), the humiliation still can be felt. ITT had the reputation of being technologically advanced, although the Austrian branch had no research field of its own. ITT had its own system, which was the only one not accepted in the end. The shock led to a complete re-orientation of the Austrian subsidiary. Development was concentrated on private automatic branch exchanges, financed by the surplus from OES-sales. Co-operation between Siemens and Alcatel is considered as having been productive by experts of both firms. Trust never reached a level to allow for the foundation of a joint firm, however. Only by regulatory influence, that is, by procurement of OES, could the two competitors be induced to form the development association named *Entwicklungszentrum OES-E*. This association was later renamed *Arbeitsgemeinschaft OES-E, Siemens, Alcatel* (AOSA). Currently the association AOSA has about 160 employees, and about 65 employees work on the development of OES-E. At the time of the foundation of the association, the number of developers had been higher (about 85) whereas the remaining staff, mostly marketing, had been smaller (about 20).

A closer co-operation between the AT-group and the AOSA group would have been possible if the *System of Operation, Administration and Maintenance* (OAM) of the digital net, procured after 1989, could have been made a joint effort. Such a plan faced strong resistance by all suppliers, however. Kapsch argues that the high degree of system specific OAM features necessary for the DMS 100 switch to run on an outside OAM computer platform would have made a joint effort futile. In the end, ÖPTV conceded to this heavy opposition and each group developed its own version of OAM. This produced higher costs because of a doubled development effort, and will also be a long-term burden due to higher maintenance costs and lower compatibility within the system.

4.2 Development and Actual Procurement

As mentioned above, development and actual procurement started with the foundation of the two development organisations, AT and AOSA. These two organisations were responsible for the development of OES along the pathway given by the specifications elaborated by ÖFEG. In 1985, ÖFEG decided to risk an ambitious start with two large and outdated exchanges in Vienna, whereof one was equipped with OES-D, and the second with OES-E. Each exchange served about 10,000 lines plus about 2,000 trunks. The trunk function went into operation immediately, the user lines were operational by January 1986. The immediate functioning of these installations was the first success and a source of relief at the same time. According to PTA, by 1992 about 1,468,000 of 3,466,500 main lines were served by digital exchanges, that is, the portion of main lines on digital exchanges was 42 per cent. The most developed countries in 1992 were New Zealand (95 %), the Netherlands (86 %) and France (83.2 %). Austria's position in the range was 19, and Germany was placed 24th (12 %).[5]

[5] This finding has to be taken with some caution. The Austrian numbers have been taken from the 1992 annual report of ÖPTV. The international numbers stem from ITU (1994), table 3.3, p.36. According to this table Austria had only about 936,000 digital main lines, which gives a digitalization per centage of 27% and puts Austria in position 21.

The planning horizon for the introduction of OES was 30 years. This careful planning was due to a worst-case presumption according to which a new central control would be needed every week. Today most experts expect finalisation of the project by 1999. As it turned out, there were fewer technical difficulties than expected. Still more important was the unexpected quick decline of prices, which fell by about 80%. The numbers of ports purchased till 1995 are depicted in Figure 1. By now ÖFEG has elaborated about 1,000 specifications and the removal of obsolete structures is as important as the specification of new ones. Austrian peculiarities concern loudness and sound quality, high reliability, and value added services.

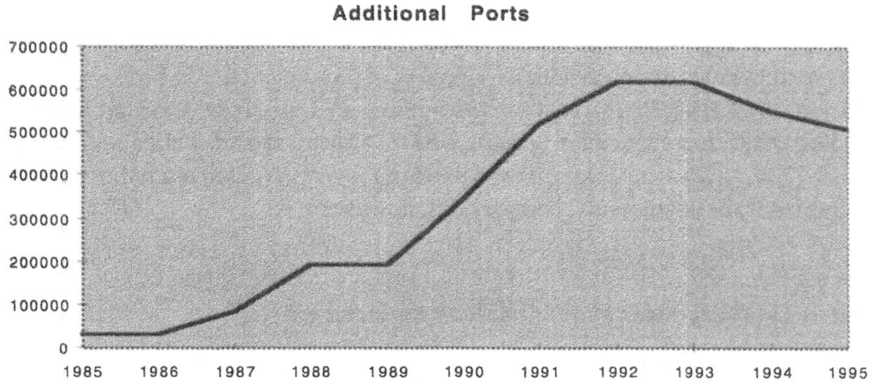

Figure 1. Ports purchased by ÖPTV

Source: Interviews with PTA-Officials

In the first years, the installation of OES seemed to follow the principle of indiscriminate distribution. The main reason was the federal constitution of Austria. All individual states demanded to be served first. There are voices saying that it would have been technologically simpler, and thus cheaper, to start with an international system in one part of Austria and to expand gradually from this spot, installing new gateways with each expansion. The decision for evolutionary growth and slow adaptation to given interfaces implied less compatibility with international standards and a late liberalisation of the market for terminal equipment. One of the problems of OES seems to have been that there were too many Austrian specialities. Traditionally, all Austrian specifications were given by the provider. Only in recent years have firms been invited to co-operate in the elaboration of new specifications. Users are still not involved in the decision process.

Procurement of OES was a compromise following from personal contacts, governmental economic policy, and *Social Partnership*. What role did nepotism play in this process? Personal contacts resulted in individual career opportunities and increased status, rather than in direct monetary compensation. This Austrian particularity is especially characteristic of higher hierarchical ranks, above all, if the decision makers must act in public. In the present case, decisions have been taken on the highest level possible. The foundation of ÖFEG, moreover, with its transparent decision algorithm was easy to control by all competitors, thus reducing the opportunities of just one supplier to influence the decision. Thus, it is not surprising that no hints of bribery could be found in the present case.

4.3 Costs, Prices, and Quotas

Compared to international sales, the Austrian switching market is very small, only about 10 per cent of German figures. Partition of the market following from the two-systems decision even aggravated this problem. The small demand implied that ÖPTV had a weak position in negotiations with its suppliers. According to our information, the installation of OES until 1995 cost 34.877 million ATS for about 3.616 million ports including OAM, supply, support, and upgrades (OAM cost about 6.4 billion ATS).[6] (See Table 3.) This gives an average price of 9.645 ATS per port. That is a rather questionable indicator for international comparisons, however.

Table 3: Total Payments by the ÖPTV (VAT included)

	1985	1986	1987	1988	1989	1990	1991	1992	1993	1994	1995
Payments (mio ATS)	321	227	631	1.684	1.934	3.175	5.461	4.766	4.896	5.762	6.020

The numbers include ports, OAM, supply, support, and upgrades
Source: PTA (Austrian telecom provider, former ÖPTV)

We already mentioned that ÖFEG never intervened in price negotiations. Price cutting could have hampered the free flow of information and the team spirit. Prices were settled directly between representatives of the ÖPTV and the supplying companies. In the years from 1985 to 1989, prices were determined by annual price contests (*Jahrespreiswettbewerbe*) between the two teams, AT and AOSA. This procedure was adopted from German examples. In principle, the market was divided equally, that is, each team had 50 per cent. The winner of the contest, however, could win up to 4 per cent of the opponent's market share. No matter what the shares looked like, all suppliers had to pay the lower price resulting from the contest. This practice followed a model similar to a first-price sealed-bid auction with variable quantities. Efficient in theory, it only works if the bidders do not collude. Unfortunately, this is exactly what happened in the case of OES. Most interviewed experts agree that the suppliers formed an informal cartel, at least in respect to their pricing strategies.

[6] Source: PTA (Austrian telecom provider, formerly ÖPTV)

From the first delivery until the present, OES-E was cheaper than OES-D from AT, owing to the higher cost of adapting the Canadian system. On the other hand, Kapsch also claims higher reliability for OES-D. On account of the better offer, AOSA always had a share of 54 per cent of the Austrian switching market. Being dependent on selling switches and becoming even more dependent by entering the market for digital switches, Kapsch avoided the risk of a conflict with Siemens, who were much larger and more powerful.

Within the two teams, AT and AOSA, there existed fixed arrangements on how to divide earnings. The two partners of AOSA, Siemens and ITT, decided on dividing their share in the following manner: 55 per cent for Siemens and the remaining 45 per cent for ITT. Delivering the lower bid, Kapsch and Schrack had to settle for 46 per cent of the switching market. From these payments, two thirds were cashed by Kapsch and the remaining third was left for Schrack. From the above statements, the following quotas of the respective firms, adding up to unity, can be derived:

Siemens	0.54 * 0,55 =	0.297
ITT	0.54 * 0,45 =	0.243
Kapsch	0.46 * 0,66 =	0.307
Schrack	0.46 * 0,33 =	0.153

Figure 2: Revenue Shares of the Suppliers

The prices of OES-switches were about 5-8 per cent higher than of Swiss switches and about 15 per cent higher than the switches which Siemens adapted for the German post. For ÖPTV, this seemed acceptable because of the small size of the market. Nobody doubted, however, that it was a good bargain for all Austrian suppliers.

In 1989, the AT-group intervened with the Minister of Transport to end the annual price contest. They argued that the contest resulted in high and unnecessary transaction costs. By then the market shares were fixed, anyway. ÖPTV agreed to open negotiations. The idea was to provoke open bidding and bargaining in order to reduce the adverse effect of collusion on the price. This accords with auction theory. Another strategic advantage for ÖPTV was the possibility to start negotiations with a low price offer dependent on foreign prices. ÖPTV-officials used personal and unofficial links to colleagues of other European telecom providers in order to get a conception of 'correct' prices. On the other hand, they never tried to minimise the price beyond macro-economic goals like, for example, fostering national production or reduction of unemployment. All in all, the Austrian digital system was rather expensive, owing to:

- the weakness of ÖPTV because of the small size of the market.
- the system licences and the adaptation of the systems to different interfaces owing to the diversity of exchanges. Only by CCITT no. 7, could 17 signalling systems be reduced to 10 (including interfaces to Romania and Bulgaria). This point accounts for about 50 per cent of adaptation costs.

- the fact that electromagnetic meters had to be 'rebuilt' in a virtual way. This accounted for about 15 per cent of the price increase in the first years; today, the effect is nil.
- the existence of very old terminal equipment with, for that time, rather advanced functionality. Adaptation made for about 35 per cent of extra costs, a percentage which increased to nearly 50 per cent, owing to the requirement of absolute compatibility.

For an estimation of development costs, we again rely on numbers given by PTA. They estimate that development cost about 1.76 billion ATS for OES-D and OES-E, respectively, plus 0.72 billion ATS owing to factors stemming from the two-systems solution. This yields 4.24 billion ATS, which is regarded as 'extra cost' resulting from the adaptation of the digital systems to Austrian requirements.

4.4 Market Power and the Number of Systems

The two-system-decision cost the Austrian consumer twice the development cost and nearly doubled the support and maintenance cost, not to speak of logistic problems remaining after completion of the digitalising process. According to estimates of the Austrian telecom provider, PTA, the development of OES cost approximately 4.24 billion ATS, whereof about 17 per cent was due to the doubled development effort only.[7] The two-system solution's extra cost was probably more than two billion ATS.

On the other hand, procuring two systems implied competition between the suppliers. Over time, the price of the hardware becomes more and more irrelevant. Updates and extensions of a digital system depend on the software. There is a potential danger for customers, buying cheap hardware, to become dependent on a supplier selling software as a monopolist free to dictate prices. It is mostly agreed that the Austrian suppliers acted as a cartel, a fact confirmed by the negotiation representatives of ÖPTV.[8] Thus, competition had probably little effect on the price. It is more commonly accepted that the decision to implement two systems has produced significant competition in terms of quality. This refers mostly to the speed of development and to the reliability of the two systems. There was a permanent 'reliability contest', which at least in one case changed the behaviour of a supplier. In 1990, ÖPTV threatened to take exchanges away from Siemens if they did not succeed in solving a software problem. This threat was obviously not taken seriously by the parent company in Germany, which actually

[7] According to other experts´ estimations, an additional labor force of about 200 technicians has been needed which amounts to approximately 140 million ATS per year if we take an average annual gross wage of approximately 0.7 million ATS per technician. This is to be seen as a lower boundary and reflects our caution on the matter. If we calculate a time span of 17 years (from 1982 to 1999), then this estimation would give extra costs of about 2.4 billion ATS, which is only slightly higher that the number estimated by PTA (2.12 billion ATS).

[8] During price bargaining, there were cases when the suppliers' representatives left the room together to establish a common position outside before returning to negotiations.

led ÖPTV to take an exchange away from Siemens and to commission AT with it. It was an order worth only 30 million ATS, which did not really hurt Siemens financially but led to a painful loss in reputation. According to ÖPTV representatives, this measure improved the performance of Siemens considerably.

5. APPRAISAL OF SOCIETAL AND INDUSTRIAL RESULTS

Considering the goals of ÖPTV, the introduction of OES can be judged a success: Renewal of the network will be finished earlier than planned. The number of ports has increased enormously and is still increasing. OES has been a basis for ISDN and value-added services. Finally, ÖPTV has saved millions by reducing their requirements for room and personnel. But what about the goals of the government?

The effects of the introduction of OES on Austrian telecom *know-how* have been strong and positive. Detailed Austrian knowledge of digital systems had been practically non-existent before. By means of the construction of ÖFEG as a firm and stable platform gathering all Austrian resources, Austrian knowledge of digital systems has been created, allowing for genuinely Austrian developments and technological independence. Furthermore, the foundations for further developments like ISDN, broadband ISDN, GSM, and Intelligent Networks have been laid. AT and AOSA co-operated with foreign companies on high technological level and also developed genuinely Austrian products. The basic structural problems in the Austrian telecom sector have remained unchanged, however:

- The private suppliers are too small to compete with the national champions of larger countries.
- Research, if existent, is strictly private. Until the beginning of the Nineties, universities and other public or semi-public research organisations did not participate in digital telecom research. As in the time before OES, students have to spend at least two years in a private company's R&D department to learn the basics allowing for their own research in the telecom sector. Co-operation in research between the Austrian suppliers is weak and diminishing.
- ÖFEG and the two development firms, AT and AOSA, have not developed into incubator organisations. There is little mobility in staff and the three organisations have not succeeded in incubating new firms: There are no technological spin-offs.
- Innovation has not allowed for significant exports. We return to this point in the next part of this chapter.

These critical points are enforced by a patent analysis. Comparisons of the absolute numbers of European patents for selected countries and the respective EU patents per population show that Austria is not a leading developer of telecom equipment. The Austrian telecom sector is less innovative than its competitors in other countries. It is obvious that the introduction of OES has been an innovation, at least for Austrian standards. Still, the primary incentive was to reach international standards rather than achieving technological superiority — that is, innovation has been diffusion-oriented. Thus, the introduction of OES has been a case of Adaptive Public Procurement (APP).

Estimation of the effects on *employment* will be rather crude here.[9] To be competitive on the world market, national champions need reference projects in their home country. Without participation of all Austrian competitors the survival of at least the smaller ones would have been at stake. Experts have expressed different opinions regarding this point, some believing that specialisation in the private telecom sector would have been a safe line of retreat, especially when there was still an excess demand and less competition. Most experts agree, however, that if, for example, Siemens had got an exclusive contract for developing OES, then ITT Austria, Kapsch, and Schrack would not have survived. According to Table 1, their closure would have implied a loss of about 5,000 jobs. Exclusion of just one firm would have exposed its respective number of jobs to danger. Due to the habit of keeping production domestic and of partitioning the market by a quota system, however, such a solution would have been rather improbable. If ÖPTV had decided to produce in Austria but to leave all development work to the licenser, then about 400 to 500 jobs at AT, AOSA, and ÖFEG would have been at stake.

The effect of the introduction of OES on aggregate Austrian telecom *exports* can be analysed on two levels: First, did the Austrian telecom industry succeed in developing a digital system of its own that was exportable to other countries? Second, did the participation in OES have a positive effect on exports of more general telecom equipment? Table 4 dsiplays some relevant data.

Table 4: Shares in the World's Market of Telecommunication (1992)

	Share in World's Exports (%)			Share in World's Imports (%)		
	Exchange Equipment SITC 764.15	Total*	Annual Change 1979-'89	Exchange Equipment SITC 764.15	Total	Annual Change 1979-'89
Austria	0,23	0,57	-4,16	1,56	1,39	-2,39
Belgium	3,41	1,35	-4,27	2,32	1,74	-2,99
Switzerland	0,26	0,98	-8,70	1,88	1,53	5,73
Germany	14,54	7,10	-5,77	1,98	7,28	3,17
France	10,14	3,43	-2,11	1,46	3,40	1,51
Great Britain	3,35	4,80	1,07	2,78	4,95	8,23
Italy	0,71	2,03	-0,46	2,16	2,69	0,12
Netherlands	6,43	2,29	-4,55	3,94	3,18	-0,79
OECD Europe	50,08	33,18	-2,51	23,84	35,95	1,97
Japan	5,72	20,05	4,94	1,62	1,94	11,72
USA	37,67	21,02	ca.-4,00	9,53	21,00	n.d.

* Total includes SITC 764.11, 764.13, 764.15, 764.17, 764.19, 764.3, 764.81, 764.9

Source: Leo et al. (1994), p.104 - 107

[9] We discount the effects of OES on employment at ÖPTV. The telecom department of ÖPTV has reduced its workforce from about 18 900 to about 17 600. This loss of jobs due to rationalisation is considered as inevitable and could have been worse.

Exports of an OES-version or even of an Austrian version of DMS or EWSD have been insignificant. In the case of AOSA, Siemens intensified its activities in Austria. There were few exports of telecom equipment produced by Siemens Austria, however. On the contrary, Siemens Germany used its superior market power to compete against Austrian products on the European market. AT has initiated several attempts to export. In 1989, just before the political change in the former Eastern Block, AT sold 300,000 lines in Hungary and prospects looked good. There existed, moreover, a contract between AT and the Polish telecom provider. The client paid in advance for the adaptation development of a Polish digital system. In both cases, Siemens took over the market. Two factors were decisive: The first was a complete change in management of the eastern telecom providers. Former suppliers were considered as politically unacceptable. The second factor was the early privatisation of the German telecom provider and its massive support abroad by German politicians. In the case of Hungary, the liberalisation of the market allowed an intrusion of the provider market. Massive political support by the German government allowed Siemens Germany, as the main supplier of the German provider, to take over the AT-share. In Poland, Siemens simply bought AT out of the market. It might be said that the small size of Kapsch and the late privatisation of ÖPTV had more effect on the Austrian export opportunities than a flawed procurement policy. It remains to be seen whether the loss of monopoly by the traditional national providers will allow for new opportunities.

Let us now turn to the second question, i.e., did the introduction of OES increase exports of telecom equipment in general? Austria is a net importer of telecom equipment. Table 4 gives the export and import shares of Austria on the world market for telecom equipment. The figures are quite disappointing. The Austrian export share is smaller than in comparable countries (e.g. Belgium). From 1979 to 1989, both exports and imports have decreased. Imports have decreased less than exports, however. This indicates that one of the most important goals has not been reached: The introduction of OES has not endowed Austrian firms with the ability to participate in the international telecom game. Sales of OES are restricted to PTA.

6. CONCLUSIONS

6.1 Dimension of Time

Austria had little research in the field of digital systems. Thus, it took a conservative strategy, trying not to lose touch with the established technical standard, rather than choosing unique or particularly innovative solutions. In case of new technological developments, decision makers often waited until a standard emerged and only then started to procure. As a consequence, procurement of OES did not enable Austrian firms to establish innovative brands and to intrude foreign markets. On the other hand, the risk aversion of the Austrian telecom provider in most cases prevented a lock-in into unsuccessful technological trajectories.

6.2 Competence and User-Producer Interaction

From the Austrian tradition of centralism and *Social Partnership*, user-producer interaction evolved naturally. With low competence on behalf of the procurer and most suppliers, a national solution like the foundation of a joint development company, ÖFEG, went well with this tradition. The telecom market had already been divided up earlier and the government made clear that all suppliers would get a share, even when these shares were a matter of permanent bargaining. It should be noted, however, that consumers, university researchers or management consultancies were not involved in the actual procurement decision. Most decisions were prepared by technicians from the Austrian telecom provider in co-operation with the suppliers. Owing to licensing of two systems, moreover, actual development took place in two separate development companies, AT and AOSA, implying that the two main competitors, Kapsch and Siemens, demonstrated no combined effort.

6.3 Political Influence and Levels of Intervention

The Austrian telecom provider, ÖPTV, was not a private company but a section of the Ministry of Transport. Rather than maximising profits, ÖPTV executed the industrial policy of the government. A purely technical decision in favour of the Canadian System was revised by governmental intervention, resulting in procurement of two different systems. In the end, policies concerning increases in employment, attraction of foreign investment, building up Austrian competence, and improvement of the Austrian trade balance were decisive. Austria probably paid about ten to twenty per cent more for its digital system than necessary, but preserved quality competition and actually attained the first three of the above goals. As an export policy, procurement of OES has not been successful, perhaps because too few innovative risks were taken. Procurement had not been accompanied by a long term innovation policy.

6.4 Interaction Environment

Up to the present day, Austria's economy has been influenced very fundamentally by the concept of *Social Partnership*, although liberalisation and participation in the EU have reduced its influence considerably in recent years. *Social Partnership* in a small country implies close personal contacts of the representatives of large firms, the government, and the unions. In the Austrian telecom sector this co-operative approach implied the partition of the market. As the procurer had little competence in respect to digital systems, the suppliers participated in the decision process. This co-operation was institutionalised by foundation of ÖFEG, which determined the technical principles of the new network.

APPENDIX: LIST OF ABBREVIATIONS

AOSA	Arbeitsgemeinschaft OES-E, Siemens, Alcatel (development association Siemens, Alcatel)
ARCS	Austrian Research Centre Seibersdorf
AT	Austrian Telecommunication (one of the two OES-development firms)
ATS	Austrian Schilling(s)
CCITT	Comite Consultatif International Telegraphique et Telephonique
ETSI	European Telecommunications Standards Institute
EU	European Union
EZ OES-E	Entwicklungszentrum OES-E (later renamed AOSA)
GDP	Gross domestic product
GSM	Global System Mobile (pan-European digital mobile telephone system)
ISDN	Integrated Services 4Digital Network
ITT	International Telefone and Telegraph
ITU	International Telecom Union
OAM	Operation, Administration and Maintenance
OECD	Organisation for Economic Co-operation and Development
OES	Österreichisches Elektronisches System (the Austrian digital switching system)
ÖFEG	Österreichische Fernmeldetechnische Entwicklungs- und Förderungsgesellschaft (Austrian Telephone Development and Promotion Company)
ÖPTV	Österreichische Post- und Telegraphenverwaltung (Austrian post and telecom)
ÖVP	Österreichische Volkspartei (Austrian People's Party)
PBX	Private Branch Exchange
PCM	Pulse Code Modulation
PK	Paritätische Kommission (Basic organisation of the Austrian *Social Partnership*)
PTA	Post- und Telekom Austria AG (Austrian telecom provider after privatisation)
R&D	Research and development
SPC	Stored Program Control
SPÖ	Sozialistische Partei Österreichs (Socialist Party of Austria, later Social Democratic Party of Austria)
TBP	Technology balance of payments (receipts minus payments)
VAS	Value Added Service
WIFO	Austrian Institute of Economic Research

REFERENCES

Bauer Johannes M. & Latzer, Michael (Eds.). (1993). *Nützliche Verbindungen; Österreichs Telekommunikationsdienste im internationalen Kontext.* Vienna, Munich: Oldenbourg.

Bundesministerium für öffentliche Wirtschaft und Verkehr. (1995). *Geschäftsbericht 1995.* Vienna: BMin f.Öff. Wirtsch.u.Verkehr.

Communications and Information Technology (CIT) Research Limited. (1993). *Yearbook of European Telecommunications 1993.* Woodbury (CIT)

Günther, Johann. (1992). *Telekommunikation in Österreich.* Vienna: Literas Universitätsverlag.

Günther, Johann. (1989). *Digitale Telefonie; Technik und Organisation.* Vienna: Literas Universitätsverlag.

Hansen, Robert. (1988). Auctions with Endogenous Quantities. *Rand Journal of Economics, vol.29* (1), 44 - 58

International Telecommunication Union. (1994). *World Telecommunication Development Report 1994; World Telecommunication Indicators.* Geneva: ITU.

Kapsch AG. (1977). *Das digitale Jahrhundert der Telephonie.* Series of lectures held at ÖPTV after a study trip through North America, Nov. 30thm 1977.

Knoll, Norbert G. (1992). *Europäische Integration und Auswirkungen auf das Fernmeldewesen in Österreich.* Mimeo.

Latzer, M., Ohler, F. & Knoll N. (1994). *Neue Rahmenbedingungen in der Telekommunikation: Perspektiven für die Post.* Seibersdorf, study for the Bundeskammer für Arbeiter und Angestellte, OEFZS-A-3052.

Leo, Hannes et al. (1994). *Telekommunikation im Umbruch; Innovation - Regulierung - Wettbewerb.* Vienna, joint study of the program „Technologie, Information und Politikberatung" (tip) of the Austrian Institute of Economic Research (WIFO) and the Austrian Research Centre Seibersdorf (ARCS).

Leo, Hannes (1997): Beschäftigungstendenzen im Telekommunikationssektor. *WIFO-Monatsberichte, vol.70*, 7/97, 431 - 439.

Lundvall, Bengt-Ake (Ed.). (1992). *National Systems of Innovation: Towards a Theory of Innovation and Interactive Learning.* London: Pinter Publishers.

Mowery, David C. (1994). *Science and Technology Policy in Interdependent Economies.* Boston, Dordrecht, London: Kluwer Academic Publishers.

Nelson, Richard R. (Ed.). (1993). *National Innovation Systems: A Comparative Analysis.* New York, Oxford: Oxford University Press.

Nelson, Richard R. & Winter Sidney G. (1982). *An Evolutionary Theory of Economic Change.* Cambridge/Mass., London: Belknap Press of Harvard Univ.Press.

Nowotny, Ewald. (1991). Wirtschafts- und Sozialpartnerschaft in Österreich. *Wirtschafts- und sozialpolitische Zeitschrift, vol.14*, WISO 2/91, 21 - 49.

OECD. (1995a). *Information Computer Communications Policy: Vol.35. Telecommunication Infrastructure: the Benefits of Competition.* Paris: OECD.

OECD. (1995b). *Information Computer Communications Policy: Vol.36. International Telecommunication Pricing Practices and Principles; a Progress Review.* Paris: OECD.

OECD. (1995c). *Industry and Technology: Scoreboard of Indicators 1995.* Paris: OECD.

Pachlatko, Michael. (1997). Empfehlungen zur Regulierung des österreichischen Telekom-Sektors. *Wirtschafts- und sozialpolitische Zeitschrift, vol.20*, WISO 2/97, 105 - 134.

Wirtschaftskammer Österreich. (1994). Sozialpartnerschaft im internationalen Wettbewerb. *Wirtschaftspolitische Blätter, 5/6*, 41.Jahrgang.

11. ANALYSIS, FINDINGS AND CONCLUSIONS

C. Edquist, L. Hommen and L. Tsipouri

INTRODUCTION

In the Overview beginning Part II of this book, we discussed our selection of cases in relation to theory, concepts, and policies. We also discussed making analytical generalisations from a theoretically based sample of comparable cases. Now, after reviewing relevant theoretical, conceptual and policy issues in Part I of this book, and presenting a series of empirical case studies in Part II, we turn to a comparative analysis of the case studies. As discussed in our Introduction, our analysis is theoretically informed, and its object is to produce analytical generalisations. These will later provide the basis for developing policy implications.

1. ANALYTIC FRAMEWORK

Our analytic strategy – comparative case study analysis – is explained at some length below, in part 2 of this chapter. There follow, in parts 3 and 4 of this chapter, accounts of the findings and conclusions. Before turning to those parts of the discussion, however, it will be useful at the outset to identify and elaborate briefly the main 'themes' or 'dimensions' of our analysis. These comprise our analytic framework, and are used to compare the cases of public technology procurement presented in Part II of this book. Our purpose here is to outline the themes and explain how the analytic framework as a whole is related to our theoretical standpoint, the systems of innovation approach.

1.1 Themes

We analyse the cases in relation to several key themes. These themes represent important aspects of public technology procurement that are touched upon in the initial theoretical discussion (Chapter 1). Our themes are:

- The Dimension of Time
- Competence and User-Producer Interaction
- Political Influence and Levels of Intervention
- Interaction Environment

The heterogeneity of our various case studies, together with our primary interest in focusing on 'dynamic' aspects of systems of innovation, has required us to use abstract categories in constructing a framework to compare our cases.[1] We have based our analysis on categories that can integrate various aspects of disparate 'systems of innovation', such as those discussed in the different case studies, without lapsing into a 'comparative statics' approach.[2] We have then integrated these categories into themes that reflect some of the most fundamental dimensions of systems of innovation. These dimensions are discussed in earlier theoretical work describing correspondences between systems of innovation approaches (as conceptual frameworks) and actual (empirical) systems of innovation (Edquist, 1997).[3] They are also addressed in the analysis of public technology procurement in the initial chapter (Chapter 1) of this book. Below, we offer a more comprehensive account of the relationship between 'themes' and 'dimensions', proceeding on a theme-by-theme basis.

1.1.1 The dimension of time

A concern with 'time', or the temporal dimension of innovation processes, is an essential aspect of systems of innovation approaches. In Edquist's (1997: section 3) enumeration of basic characteristics of such approaches, it is indicated that "a historical perspective is natural" to them (ibid., point 3). We can briefly explain this statement as follows: Since processes of innovation develop over time and involve the influence of

---

[1] See, for example, the discussion of the difficulties involved in arriving at a common standard for determining 'success' or 'failure' in part 1 of the Foreword to this book.

[2] Such an approach could have been used, for example, to describe alternative 'network' patterns with typical 'nodes' and 'links', to identify the 'weak links', and to relate outcomes to the limitations imposed by these links.

[3] Edquist (1997: section 3) identifies nine key characteristics of systems of innovation approaches. Of these, three refer primarily to the character and current state of theoretical development. These are, namely: point 2, that systems of innovation approaches are "holistic and interdisciplinary in nature" (ibid.: 17 - 18); point 8, that they are "conceptually diffuse"; and point 9, that they are "conceptual frameworks rather than formal theories" (ibid.: 26 - 29). The remaining six characteristics are related to corresponding aspects of the object of study (i.e., systems of innovation). Of those, one is treated as a 'given' in our work. We define public technology procurement as a form of product innovation – i.e., as involving the acquisition by a public agency of a product or system which does not exist at the present time, but which could (probably) be developed (Chapter 1, part 1). Thus, we incorporate as a basic assumption of our empirical research Edquist's (1997) point 6, that systems of innovation theory "encompasses product technologies and organisational innovations" (ibid.: 22 - 24). Taking this point 'for granted', we employ the remaining five characteristics of systems of innovation enumerated by Edquist (1997: section 3) as the basis for constructing analytical categories and themes. We provide further explanation below.

many factors and feedback processes, they are best studied in terms of the historical 'co-evolution' of knowledge, innovation, organisations, and institutions (ibid.: 18 - 19).

We follow this approach by considering the 'Dimension of Time' in theoretical terms and including it in our analytical framework. In the first chapter of this book, 'time' is addressed primarily by discussing the technology life cycle as it relates to processes of technology procurement (Chapter 1, parts 3 and 4). This discussion covers both private sector contexts (Chapter 1, section 3.3) and public sector ones (Chapter 1, section 4.3).

All of the case studies included in this book employ a historical approach. In different ways, the various case studies demonstrate the importance of processes of 'co-evolution' among knowledge, institutions and organisations, showing their significance for technological innovations taking the form of public technology procurement. Many of our case studies cover processes of technological development and learning spanning several decades. To do justice to the case studies, as well as to emphasise 'dynamics', we include the 'Dimension of Time' among the main themes of our comparative case study analysis

1.1.2 Competence and user-producer interaction

A focus on the relational character of knowledge, or 'competence', is another fundamental aspect of systems of innovation approaches. In support of this assertion, Edquist (1997: section 3) makes the following two related points about systems of innovation approaches: First, they focus on "innovation and learning at the centre" (ibid.: point 1). Second, they place "emphasis on interdependence and non-linearity" (ibid.: point 5).

The first of these points expresses an understanding that technological innovation is a matter of producing new knowledge or combining existing elements of knowledge in new ways. It is thus, in the broadest sense, a 'learning process' (Edquist, 1997: 16 - 17). The second point is based on the realisation that firms almost never innovate in isolation but interact more or less closely with other organisations, through complex relations that are often characterised by reciprocity and feedback mechanisms in several loops (ibid.: 20 - 22). This interaction occurs in the context of institutions – e.g., laws, rules, regulations, norms and cultural habits.[4] "Innovations are not only determined by the elements of the system, but also by the relations between these" (ibid.: 21).

In keeping with this emphasis on the often complex dynamics of knowledge-creation, the first chapter of this book pays much attention to the relational character of knowledge and its vital importance to innovation processes (Chapter 1, parts 3 and 4). In particular, sections 4.2 and 4.3 of Chapter 1 both follow systems of innovation approaches by relating knowledge, or 'competence', to structural and developmental aspects of interaction and learning in public technology procurement. These are also cen-

[4] The concepts of institutions and organisations are specified and their roles in innovation are discussed in Edquist and Johnson (1997: sections 3.1 and 3.2). For further discussion in this volume, see Chapter 1, notes 13 and 18 and corresponding parts of the main text.

tral concerns of our case studies. Accordingly, we include 'Competence and User-Producer Interaction' among the main themes of our comparative analysis.

1.1.3 Political influence and levels of intervention

In our Introduction and also in Chapter 1, both of which discuss public technology procurement as a policy instrument, we indicate a fundamental concern with public policy. In systems of innovation approaches, this concern is typically dealt with from an 'institutional' perspective, since policy is both based on institutions and often acts to change them. According to Edquist (1997: section 3) "institutions are central" in systems of innovation approaches (ibid.: point 7).

We can briefly explain the importance of institutions to systems of innovation approaches as follows: These approaches stress the social patterning of innovative behaviour – i.e., its typically 'path-dependent' character. They therefore seek to understand the role played in such 'patterning' by institutions in the sense of norms, rules, laws, etc., and by organisations (Edquist, 1997: 24 - 26). From a systems of innovation perspective, innovation processes, like almost all learning processes, are "influenced ... by the institutional set-up of the economy" (Johnson, 1992). This leads to a basic concern with the possibilities for influencing the institutional set-up itself, through public policy. That concern is at the root of the distinction made between 'formal' and 'informal' institutions: "Formal institutions are changed by design (policy or strategy) while informal institutions are more likely to develop spontaneously" (Edquist & Johnson, 1997: 56).

Our work on public technology procurement as a policy instrument reflects these essential theoretical and practical interests of systems of innovation approaches. Accordingly, the case studies we have conducted are, on one hand, concerned with how formal and informal institutions 'shape' innovation processes. On the other, they suggest how lessons drawn from these instances could be used by policy-makers to reshape the institutions affecting innovation. In particular, policy-makers can address themselves to the modification of formal institutions, which can be changed more directly by policy. The theme of "Political Influence and Levels of Intervention" reflects these considerations. It does so by developing a central focus on policy 'actors', their actions, and the institutional structures through which they act.

1.1.4 Interaction environment

As already mentioned, the concern with institutions and their effect on interactions among elements within a system leads systems of innovation approaches to include both 'formal' (or 'designed') and 'informal' (or 'spontaneously evolved') institutions as analytical categories. The relationship between these two and other types of institutions is highly complex in the economic sphere. An understanding of the institutional aspects of markets is required for "analysing allocation of resources in a static context", and to grasp the dynamics of innovation processes "an institutional specification of the market becomes even more necessary" (Edquist & Johnson, 1997: 49).

In addition, many patterns of economic relations characterising the environment within which innovations occur cut across existing taxonomies of institutions (such as the formal-informal dichotomy), and require their own nomenclature. A case in point is that of demand (or 'buyer market') structure, which is a main focus of discussion in part 3 of this book's initial chapter (Chapter 1, section 3.2). Part 4 of that chapter also demonstrates that many previous analyses of public technology procurement have paid serious attention to demand structures, their determinants and effects on innovation processes (Chapter 1, section 4.2). In our analysis, we address demand structures under the heading of 'Interaction Environment'.

Systems of innovation approaches "stress the differences between systems, rather than the optimality of systems" (Edquist, 1997: section 3, point 4). They make the differences between systems of innovation a main focus, rather than something to be abstracted away from. This means conducting comparisons between existing systems, rather than between real ones and an ideal or 'optimal' one (ibid.: 19 - 20). Accordingly, systems of innovation approaches do not employ static concepts of market structure in abstract modelling exercises. Instead, they offer a more concrete, developmental perspective, in which a structural pattern such as monopsony or oligopsony may be associated with very different functional effects, according to variation in other, related features of different systems of innovation. This use of concepts of demand structure is demonstrated in several of our case studies. It is also reflected in the central importance of the theme of 'Interaction Environment' for our analysis of the case studies.

2. ANALYTIC STRATEGY

Our analytic strategy, as mentioned earlier, is a comparative one. It is also oriented towards the identification of problems and possible solutions to them. Under each of the themes outlined above, key problems in public technology procurement are identified and explained in terms of causal relationships evident in the cases. The cases are then re-examined for solutions to these problems. Here, again, particular attention is paid to causal relationships demonstrated in the cases. The main findings are presented below, in the following part of this chapter (part 3).

It is not possible, here, to provide an extended explanation of why this approach of 'problem' and 'solution' identification was chosen. However, some general remarks by an authority on 'systems' approaches in the social sciences may provide some insight into the relationship between "systems theory" and this particular analytical approach, which is consistent with a 'functional methodology'. According to Luhmann (1995), "systems theory" and "functional methodology" are close relatives. The latter is particularly useful as a method because: "It can ... grasp precisely the relation between a problem and its solution and avoid the uncertainty resulting from comparing what exists with functionally equivalent other possibilities or block this uncertainty by fixing values" (ibid.: 56). Thus, functional methodology involves comparative analysis (Luhmann, 1995: 55 - 57).

More briefly, then, functional methodology, essential to 'systems research', is at base a comparative method involving a search for common 'problems' that take specific forms in particular systems. Further, it involves the identification of alternative solutions (however successful) that are specific to particular systems but may, in principle, be transferred to others. This is essentially the character of the analysis that we have performed in comparing our case studies of public technology procurement.

As pointed out in part 1 of our Introduction, our selection of cases of public technology procurement includes instances of both 'success' and 'failure'. The inclusion of cases belonging to the latter category facilitates an initial identification of problems. Thus, the majority of problems addressed here are defined, not in relation to more successful cases, but in relation to less successful ones. Conversely, the solutions to these problems are defined primarily in relation to the more successful cases. However, our analysis can not, and does not, assume that in their original form of appearance the 'solutions' found in more successful cases can be directly applied to the 'problems' found in less successful ones. In order to establish true functional equivalencies between the cases, it is necessary for our analysis to reach a higher level of abstraction, based on insights gained from an overview of all the cases.[5] At this level of abstraction, 'descriptive' categories have to be replaced with 'analytic' ones, as we suggest above. The main benefit of using such an approach is that "more complexity becomes visible than is accessible to the observed system itself" (Luhmann, 1995: 56).

In methodological terms, our analysis follows a 'replication logic', as indicated in our Overview of Cases (part 2). Replication logic in multiple case study analyses is similar to that used in multiple-experiment research designs. In the latter, if similar results are obtained from the repetition of critical experiments, replication is said to have occurred (Hersen & Barlow, 1976). The application of replication logic in multiple case study analysis has been described as follows:

> The logic underlying the use of multiple case studies is the same [as that in multiple experiments]. Each case ... either (a) predicts similar results (a *literal replication*) or (b) produces contrasting results, but for predictable reasons (a *theoretical replication*). ...
>
> ... An important step in all of these procedures is the development of a richer theoretical framework. ... The theoretical framework later becomes the vehicle for generalising to new cases, again similar to the role played in cross-experiment designs. (Yin, 1994: 46)

The manner in which this logic is applied to our case studies on public technology procurement has, in effect, already been described in the earlier passages of this section. It involves deriving general propositions about 'problems' and their possible 'solutions' from an initial review of the cases, then retesting these propositions against the vari-

---

[5] Such an overview was necessary precisely because of the lack of any clear 'best practice' models of complete success, or 'worst practice' models of total failure in all aspects of public technology procurement. Nearly all cases exhibited some problematic features, but none of the less successful ones represented 'complete' failures. Thus, general 'problems' had to be constructed in relation to all cases, even though they might be best illustrated with references to particular cases. This was also the case with 'solutions', since the more successful cases did not all exhibit the same characteristics. Some commonalities were found, but alternative solutions to the same problems also emerged from different cases.

ous cases. Thus, our framework finds 'theoretical' as well as 'literal' replications, and context-dependent variations of its basic propositions. These findings are presented in the following part of this chapter (part 3).

It should be borne in mind that, for reasons we outline above, the following summary of findings is not simply an account of 'best practices'. It is not, in other words, a descriptive listing of "what works" in public technology procurement, buttressed by carefully selected 'success stories'. Rather, the analysis can be said to produce a critical, policy-relevant account of actual practices. It deals, not only with "what works" but also – and of more fundamental importance – "why" it works and "how". It provides explanations of how certain problems are solved, and it identifies mechanisms for implementing such solutions by referring to exemplary features of particular cases. Accordingly, a word of caution is called for with respect to the interpretation of findings and the conclusions based on them. They should not be treated as simple 'recipes' for success, but rather as broad principles to be followed in relation to general problems in public technology procurement.

The following presentation of findings stresses their general character by using the terms 'problem' and 'solution'. The presentation also stresses the 'abstract' character of results from the case studies by keeping empirical illustration to a minimum. The discussion, therefore, does not relate its main points to any more exemplary cases than necessary.

3. SUMMARY OF FINDINGS

Before formulating conclusions regarding the various case studies, we first summarise in a succinct way the main findings from them. We begin by presenting, in a schematic way, the major problems we encounter under each of our main themes, the alternative solutions to these problems that we find contributing to the success of public technology procurement, and specific aspects of cases which exemplify these solutions. On this basis, the findings are presented in a way that lends itself to drawing particular 'lessons' about the conditions and strategies that promote effective public technology procurement. This, in turn, provides a basis for the formulation of more general conclusions, which are subsequently presented in part 4.

3. 1 The Dimension of Time

'The Dimension of Time' is the first theme addressed in the case study analysis. Under this heading, a common problem in all cases is the importance of 'early and sophisticated' demand. Two Swedish cases, in particular, demonstrate this point. In the case of Compis (Chapter 4), the demand was articulated at a relatively late point in the technology's development. Further, it was difficult for suppliers to develop the necessary sophistication, or competence, in the short period of time required for completion of the project. Stating that, generally, "the requirements seem to have been well timed", the study goes on to point out that "the most optimistic feature of the specification was the time-table requesting delivery of a prototype" (ibid., part 1). Delays due to a lack of supplier competence turned out to be one of the main factors preventing Compis from achieving complete success. In the case of the X2000 (Chapter 2), a lack of sophistication in the form of inadequate 'buyer competence' on the part of the procuring agency, SJ, acted as a brake on the early formulation of demand (ibid., part 4). In both cases, the trade-off between 'timing' and the development of competence led to disappointing results. Compis eventually failed, even on the domestic market, and the X2000 did not succeed in capturing export markets, at least not in its original form.

Regarding this trade-off between 'timing' and 'competence', two alternative kinds of solutions can be identified. In the first of these, the procuring agency monitors 'leading edge' technology through collaboration with leading firms. This was exemplified in the case of telecommunications procurement in Sweden (Chapter 5). There, the procuring agency monitored the leading edge both directly, through contacts with AT&T, and indirectly, through the formation of a 'Development Pair' with a leading domestic producer, LME (ibid., section 2.5). Thus, the public service provider, STA's, 'foreign relations' with AT&T were complemented by LME's links with foreign subsidiaries. The second kind of solution is based on 'early competition', a situation wherein the procuring agency is exposed to a competitive 'buyer's market' – that is, the presence of other competing buyers on the 'demand' side of the market – and the early presence of leading foreign firms in the home producer's market. This was the case of telecommunications procurement in Finland (Chapter 6), where there was a division of the buyers' market between the PTT and local telephone companies, plus an 'open' producers' market for telecommunications equipment (ibid., section 4). These circumstances led the PTT to create an early 'experimental' demand for innovative technology.

A second problem that emerges from the cases with respect to 'Time' is the threat of being 'locked in', too early, into a risky (i.e., uncertain) new technological trajectory. This difficulty appears, in different ways, in both the Italian and Swedish telecommunications studies (Chapters 8 and 5, respectively), to name but two examples. In the case of Italy (Chapter 8), for instance, the manufacturer, SIP's, decision to 'jump' directly into the time-division switching trajectory was "extremely ambitious", given the firm's lack of an independent technological capability and prior experience with electronic switching (Chapter 8, section 3.1). Due to technical problems with the Proteo system, this process turned out to be "long and difficult", demonstrating the risks associated with early commitment to an untried system (ibid.).

The appropriate solution that can be identified from comparison with other case studies is a strategy of 'pre-paradigmatic diversity'. In this strategy, public procurement and other instruments are used in 'early' phases of development to support multiple trajectories, in a manner that allows choice and strategic alternation between them. The clearest example is provided by the case of France (Chapter 7). There, various policy instruments served, during a period of uncertainty, to sustain the 'space division' switching system as a 'bridge technology' (ibid., section 3.2). At the same time, there was continuing support for the riskier and longer-term 'time-division' switching technology project (ibid.: section 3.3). Another example comes from the Swedish study of high voltage direct current (HVDC) transmission (Chapter 3). In that case, too, a less risky technology (alternating current) was used to respond to immediate market demands, while the less well-established technology (direct current) remained under development (ibid., part 3).

A third problem arising in connection with the theme of 'Time' is the need for buyers to develop bargaining power by selecting the right timing in design and execution. This problem is articulated especially clearly in the Greek case study (Chapter 9). There, it is argued that if a pioneering user can identify a new technological opportunity early enough, this user will be capable of negotiating an exceptionally low price with potential suppliers. The ability to select the right specifications at the right time can confer "an increased power of negotiation from the side of the user" (ibid., section 1.2). This benefit failed to materialise from Greece's telecommunications procurement, due to the delays caused by "bad organisation" and poor co-ordination of the procurement process. Contributing to these delays were "an outdated institutional framework, an unstable environment trying to extend the period allowing protectionism, and a delayed deregulation of the telecommunications sector" (ibid.: part 6). A similar set of difficulties also appears in the Italian case (Chapter 8). Conversely, the Swedish X2000 case (Chapter 2) shows that the timing of procurement was dependent upon the procurer's development of adequate bargaining power.

Comparison with other cases indicates that this problem can be addressed by using time as a key consideration in making procurement decisions and plans for implementation. The Austrian study (Chapter 10) points to the example of planners who were working from the outset with a calculus of 'time' and 'benefit' to choose options. For instance, success based on careful planning is here attributed in part to the use of a worst-case scenario indicating the highest possible rate of demand over time for new central controls during the period when the new digital switching system was to be installed (ibid., section 4.2). As the case study of Sweden's procurement of a school computer (Chapter 4) indicates in its conclusions, in technology fields exposed to rapid technical change "any delay, no matter how short" can risk the outcome of an entire project (ibid., part 6). A 'time-conscious' approach to planning is therefore essential, particularly in developmental projects launched in such "high- risk" fields.

A final problem emerging in relation to 'Time' is the need to provide for adjustment to ongoing technological change. This problem is raised primarily in cases of 'adaptive' public technology procurement. The Greek case study (Chapter 9) offers a simple but effective solution of demanding that suppliers match or improve upon the 'state of the art'. In Greece, recurrent contracts were designed that required international providers to supply the 'latest technology' on each occasion (ibid., part 4). This requirement proved to have considerable value in terms of the development of local producer competence in the Greek context, which was clearly one of intentionally adaptive public technology procurement. In the planning of developmental public technology procurement, of course, such a requirement would be, by definition, redundant. However, as the Swedish study of HVDC transmission (Chapter 3) shows, developmental procurements are often preceded and followed by adaptive ones.

3.2 Competence & User-Producer Interaction

'Competence and User-Producer Interaction' is a very strongly pronounced theme of the case studies. One of the primary problems articulated in relation to this theme is a need for balance and complementarity between competencies of 'buyers' and 'sellers'. The struggle to achieve such a balance in order to carry out an important procurement is, for example, a central focus of the X2000 case study (Chapter 2). There, the competence of the procuring agency is shown to have lagged behind that of the leading potential suppliers (ibid., part 4). The cases of telecommunications procurement in Greece and Italy (Chapters 9 and 8, respectively) also provide clear demonstrations of the negative impacts on innovation where a balance of competencies is not present.

The most positive solution to this problem that emerges from the cases is to create 'frameworks for learning' within identifiable 'development blocks'. This can be done by forming 'focal organisations' for emerging networks.[6] These organisations can provide vertical linkages between producers and users, as well as horizontal linkages within each category. Specialised organisations of this kind can function as co-ordinating mechanisms, developing and transferring competencies. In France (Chapter 7), for example, the public laboratory, CNET, was created to develop an autonomous technological capability by 'linking with external competencies' (ibid., section 3.1). As a focal organisation, CNET was able to build technical expertise and transfer it to both the public user and its private suppliers. The interactive 'framework' was SOCOTEL, a national 'pool' of producers whose R&D efforts were led and co-ordinated by CNET (ibid.).

A second problem apparent with respect to 'Competence and Interaction' is that user competence must include both technical and organisational competence. This is shown, for example, in the Compis study (Chapter 4). The case demonstrates that technical competence alone is insufficient, and that organisational capabilities, including administrative competencies geared to the co-ordination of multiple actors on both the supply and demand sides, are necessary to accomplish technology procurement successfully (ibid., parts 3, 4 and 6).

As the Greek study of telecommunications procurement (Chapter 9) suggests, a strategy for solving this problem is to acquire technical and business competencies from relevant external sources and house them within a form of organisation that can direct the process of procurement (ibid., section 5.2). Specialised organisations of this kind function effectively as 'co-ordinating mechanisms' when they provide focus for interactive learning between users and producers and build a 'co-ordination culture'. This was done, for example, in Austria (Chapter 10), through the creation of ÖFEG (ibid., part 3). This joint development company represented not only the procuring agency, ÖPTV, but also the potential suppliers. Mandated to carry out the selection process, ÖFEG pooled technical competence from the 'supply side' and then mobilised this

[6] See the discussion of the concepts of 'development block' and 'focal organisation' in Chapter 1 of this book, sections 3.2 and 4.4.

expertise on the 'demand side'. Key to its organisational competence was co-operation based on reconciling disparate economic interests.

Closely related to the above mentioned problem of organisational competence is the need for competence to be developed within appropriate organisational forms. This concern, as noted above, is foregrounded in both the Italian and Greek case studies of telecommunications procurement (Chapters 8 and 9, respectively). The Italian study (Chapter 8), in particular, draws attention to the negative consequences following from "the lack of co-ordination of the national research activities" over an extended period of time (ibid., section 3.1).

From other case studies, it emerges that 'loose' co-ordination is appropriate during initial periods of search and learning involving formation of new collaborative relations. More formal, focused co-ordination, based on well defined goals, is appropriate to implementing the acquired knowledge. This approach is most clearly demonstrated in the case of Sweden's telecommunications procurement (Chapter 5). The study focuses attention on the changing relations between the procuring agency, STA, and the supplier, LME. This 'Development Pair' progressed from an initial situation of loose and largely informal 'limited collaboration' around electronic switching to a more formal, 'procurement-directed' relation (ibid., part 3). The latter type of relation was realised in ELLEMTEL, a joint development organisation that made both the 'user' and the 'producer' into 'procurers' (ibid., part 4). The evolution of a Development Pair and the relations between its partners is also described in the study of HVDC transmission in Sweden (Chapter 3). There, too, the point is made that the eventual success of the developmental procurement depended on the "maturation" of relations of mutual trust between the user and producer involved (ibid., part 2). At the outset of the actual procurement process, developing "social collaboration" was considered to be "as important" for the producer, ASEA, as "developing the technology" (ibid., section 3.1).

A final problem in 'Competence and User Producer Interaction', evident to some extent in nearly all of the cases, is the need for mechanisms by which competence and specifications for its development and application can be transferred effectively between users and producers. The stability of such mechanisms proved to be an problem, for example, in the case of Compis (Chapter 4). There, "unforeseen institutional changes" occurred which fundamentally altered the terms of procurement: "the educational policy base of Compis stressing uniform technical equipment throughout Swedish schools had been replaced by another stressing a more function-oriented public purchase" (ibid., part 3).

A particularly relevant strategy for resolving the problem of effective transfer mechanisms is to use standards and standard-setting by procuring agencies (in their capacity of 'focal organisations' within emerging networks) to communicate competencies and information about emerging patterns of user demand to producers prior to, or as part of, the procurement process. This is most clearly exemplified in the case of telecommunications procurement in Finland (Chapter 6). The specific example is the PTT's involvement of supplier firms, such as Nokia, in the formulation of technical standards, such as the NMT standard in mobile telecommunications. The PTT monitored technological development and commercial demand through participation in international standards organisations, then transferred the relevant information to industry *via* technical specifications tested by the producers as prototypes (ibid., part 3).[7]

3.3 Political Influence and Levels of Intervention

Some of the most complex problems addressed are those considered under the heading of 'Political Influence and Levels of Intervention'. Illustrating the complexity alluded to, the first of these is the danger that risk-aversion can prevent public agencies from supporting innovation.[8] One of the clearest examples is provided by the Greek study of telecommunications procurement (Chapter 9). This case study outlines an historically based conflict between formal and informal institutions, leading to a situation wherein "the public service is dominated by a reluctance to take on a responsibility that would, fairly or not, risk creating allegations of corruption" (ibid., part 2).

The case study of Austria (Chapter 10) indicates that, as a solution to such problems, informal institutions – such as those of the Austrian 'social partnership' – can provide channels for industry initiatives that counter the conservatism and short-term 'pragmatism' of public agencies and government bureaucracies. In Austria, the decision on procurement was made and implemented within an "economic culture of close personal contacts and joint orientation" that complemented, rather than detracted from, formal institutional arrangements for management of the economy (ibid., section 2.1).

Political decision makers can assist this process by co-ordinating initiatives 'from below', not simply imposing decisions 'from above', as is also shown in the Austrian case. By including all relevant interests, moreover, decision processes are made more transparent and legitimate. These principles were followed in the case of Austria (ibid., part 3). There, activism based within the 'social partnership' of firms and labour organisations within the telecommunications sector led to a revision of the original procurement decision of ÖPTV/ÖFEG. Government responsiveness to this protest led to a decision that increased the societal and industrial benefits, as well as aiding the procurement process. The accommodation of industry interests gave more authority to ÖPTV by reinforcing the principle of 'inclusion'.

A second problem that arises in relation to 'Political Influence and Levels of Intervention' is the risk that the continuity of longer-term initiatives in innovation policy can be interrupted by sudden shifts in 'political influence', particularly when altered political priorities lead to changes in policy. This was clearly a problem, for example, in the case of Italy (Chapter 8). There, aspects of policy that should have complemented each other remained at odds with one another during crucial periods. Thus, "when the national digital system was ready and needed to be sustained by the internal demand, the

---

[7] As noted earlier, international standards were also used strategically in the Greek case.

[8] This problem can be exacerbated by a reactive orientation on part of political decision makers and poor formal organisation of responsible government departments. Conversely, 'political vision' can be stymied by reluctance on the part of a public agency lacking authority based on clear accountability rules and fearing implication in charges of corruption or nepotism from unsuccessful bidders.

government procurement was not able to start a massive introduction of the new technology because of the obstacles and impediments to a prompt and effective decision making" (ibid., part 4). Political opposition leading to a reversal of original expectations on the part of producers also spelled trouble for Compis in Sweden (Chapter 4). This occurred when "one of the buyer organisations originally involved in the project, the Board of Education, ... acted against the idea of a monopsonistic school computer market and allowed several different computers to be sold to Swedish schools " (ibid., part 6).

Other cases, such as that of HVDC transmission in Sweden (Chapter 3), indicate that this problem can be avoided by organisational and institutional means that enable long-term innovation processes to be maintained (Chapter 3). A protective buffer can be formed by organisational repertoires and institutional arrangements creating and preserving 'diversity'. In the case of HVDC, the Swedish development pair, Vattenfall and ASEA, were able to rely on technological alternatives maintained and developed over a long period of time. Thus, when an unexpected shift in market conditions 'de-selected' their original plan to implement HVDC in a new development in Northern Sweden, they were able to use another system instead (ibid., section 3.2). Not long afterwards, they could pursue their development of an HVDC system under nearly ideal conditions through 'the return of an old project' (ibid., section 3.3).

In another variation on this strategy of maintaining options, alternative centres of policy leadership and corresponding frameworks for implementing immediate priorities allow for the continuation of longer-term 'strategic' policies. In France (Chapter 7), for example, there were dual structures in both the public administration and the organisation of industry (ibid., part 2). The development in industry of two 'national champions', each representing a different technological trajectory, was paralleled in government by the existence of two public agencies (CNET and DGT) capable of 'leading' public provision of telecommunication services. This dual structure preserved the continuity of a national technology strategy despite major shifts in political influence and policy (ibid., part 3).

A third problem discerned under this heading is that lack of coherence in policy, administrative structures and regulatory regimes can prevent prompt and effective decision making, while maximising organisational rivalry and political interference in the procurement-related activities and plans of public agencies. Again, the cases of Compis in Sweden (Chapter 4) and telecommunications procurement in Italy (Chapter 8) exemplify this malaise. The Italian study of telecommunications procurement emphasises in its conclusions that "the lack of a coherent political framework for the sector", prior to the 80s, was closely related to the absence of "a comprehensive framework for R&D" (Chapter 8, part 4). For Compis, one of the main lessons drawn is "to try to keep down the number of public organisations involved in a technology procurement project, especially when institutional changes can be expected" (Chapter 4, part 6).

A remedy suggested by other cases is to design administrative structures and regulatory regimes to maximise the degree of responsible autonomy exercised by public agencies. This allows 'informal' initiatives with industrial partners to proceed without direct government involvement during an exploratory phase. This postpones 'formalisation' until it is required to resolve disputes and legitimise incorporation of a specifically focused procurement relation. This approach was practised in telecommunications procurement in Sweden (Chapter 5). The telecommunications service provider, STA, although subordinate to two ministries, could act as a regulator and pursue domestic procurement activities without submitting a procurement order to higher government authorities (ibid., section 2.1). This enabled the agency to form a 'Development Pair' with LME, only 'hardening' the relation when a specific goal could be defined. Direct government intervention was limited to approving the partners' contract governing ELLEMTEL, the joint development company that was then formed (ibid., section 3.5).

A final problem encountered in relation to 'Political Influence and Levels of Intervention' is the threat posed to 'innovation policy' by other aspects of economic policy.[9] To the extent that public agencies consider it their responsibility to maximise economic welfare, there is a danger that they may try to accomplish this at the expense of innovation policies – to the detriment of national industry and economic welfare over the longer term. This problem is encountered, for example, in the case study of telecommunications procurement in Austria (Chapter 10). In that study, it is observed that "There was a clear commitment to strengthen the Austrian telecom sector but no commitment to endow it with technological superiority" through the co-ordination of procurement with other policy measures, such as "the installation of a university institute on digital telecommunication or the harmonised subsidisation of private research activities" (ibid., section 3.3). Similarly, there was a conflict, though not a pronounced one, between protectionist and innovative motivations in the case of the X2000 in Sweden (Chapter 2, part 1).

The appropriate solution suggested by the results of comparative analysis is that regulatory regimes and the mandates governing public agencies can be designed so that they are constrained to pursue economic goals in a manner that gives priority to 'innovation policy'. Public agencies can be required to compete with, or complement, the activities of private ones. Instruments such as competition policy can be used to enhance market conditions supporting high standards of competition. In Finland (Chapter 6), for example, a decentralised system of telecommunications services, in which the PTT was exposed to competitive pressure from local 'telco's', was complemented by market regulation ensuring a highly competitive producer industry. High standards of competition were maintained through formal procurement regulations. Thus, 'infant industry protection' *via* the informal procurement norms of government was minimised, and the PTT was forced to concentrate on technical and cost criteria in procurements of innovative technology (ibid., part 4).

3.4 Interaction Environment

Under the heading of 'Interaction Environment', related problems of structure and dynamics in buyer and seller markets are discussed. The first of these is monopsony – a type of demand structure common in publicly controlled 'natural monopolies', and

[9] See the discussion of the term 'innovation policy' in part 2 of the Introduction to this book.

in nearly all of the cases discussed here. Monopsony, as a demand structure, can provide strong 'demand pull' but tends to encourage rigid supplier-buyer relationships, which may suppress, not encourage, innovation. The Italian case (Chapter 8) provides insight into several aspects of such rigidity. For example, at a time when the 'buyer market' in Italy had been consolidated and could have exercised strong enough demand -pull to encourage the further development of digital switching systems by Italian producers, the telecommunications sector remained 'non-strategic' so far as innovation-oriented public procurement was concerned: "During the 1970s, the telecommunication industry was mainly considered as a way to create and sustain employment and the industrial development of the South of Italy, given the labour intensive character of the production technology during the electromechanical phase" (ibid., section 2.2).

However, other cases indicate that monopsonists can become flexible when they are introduced to economic pressures and unfamiliar technologies that can not be managed with their existing competencies. Such conditions can lead to the formation of innovative relationships with suppliers who have incentives to seize new technical opportunities. Such alliances may be formed, for example, when public agencies share a 'defensive' interest with their domestic suppliers in reducing the domination of 'home markets' by foreign firms. This was the case in France's procurement of telecommunications technology (Chapter 7, part 2). In Austria (Chapter 10), ÖPTV, a powerful monopsonist, was under great pressure to modernise its system and recognised the need to implement digital switching. However, its lack of competence in the new technology required ÖPTV to enlist its suppliers in building up expertise before taking a system decision (ibid., section 3.1). This was done through the creation of new organisational forms and relationships that promoted both interactive learning and technical innovation, not only between user and producers, but also among producers.

Monopsonists can also be made more flexible and receptive to innovation on the basis of competences and competitive interests which they share with producers. In situations where buyers are also suppliers, they may develop an 'offensive' interest in common with other producers. Thus, competitive tension, as well as collaboration, can be introduced into interactive user-producer relationships. This was the case in Swedish telecommunications procurement (Chapter 5). LME shared the producer's market with the public service provider, STA. The latter organisation showed a strong interest in developing digital switching as a source of competitive advantage at an earlier point than did LME. This interest was based on STA's exceptional competence as a producer and its ambition to develop export markets for its switching technology. STA's self-interest motivated it to 'pull' LME into a new trajectory. (Ibid., parts 2 and 3.)

There are also problems related to established producer oligopolies on the 'supply side' of public markets. Strong "domestic" firms within producer oligopolies, particularly when they are associated with segmented markets, may either choose or be forced to concentrate on maintaining their established 'shares' of existing markets. This focus can then prevent them from entering new markets with new technology. This occurred, for instance, in Italy's telecommunications industry (Chapter 8). There, "Given the rigidity of the supply of electromechanical switches, it was very difficult for SIP to change the historical shares in the switching market" (ibid., section 3.1).

As a solution, public agencies can exercise monopsonistic power to encourage innovation on the part of sluggish entrants into new markets. This, in turn, requires the public agencies to have a clearly defined interest in taking such initiatives. Adequate incentives can arise from competitive forces in an 'open' market situation. This is evident in the account of Finland's telecommunication procurements (Chapter 6). Although it had only limited monopsonistic power, the Finnish PTT used it to pressure domestic producers such as Nokia into mobile telecommunications. In the case of the NMT mobile telephony standard, buyer competition at the Nordic level called for supplier competition at the national level. Above all, the PTT wished to avoid a monopoly of supply, where producers would dictate prices (ibid., part 4).

Another problem within producer oligopolies is that the market positions, technical capabilities, and competitive strategies of established firms may discourage them from participating in technological initiatives launched by governments and implemented *via* procurement and other activities of public agencies. Risk aversion and a desire to concentrate on existing 'core' activities may be the basis of such conservatism. LME in Sweden initially displayed such tendencies, for example, with respect to STA's procurements related to the NMT standard (Chapter 5, part 5).

As a remedy, incentive structures can be created that will induce firms to participate in strategies for innovation spearheaded by public agencies. Incentives may be provided directly or indirectly through a variety of instruments and application should be differentiated according to type of firm. In France (Chapter 7), a public strategy for a 'national technology' and a 'national champion' initially faced a producer's market dominated by foreign subsidiaries and national firms with no technological capability of their own (ibid., part 1). Producers were organised into SOCOTEL, within which the foreign subsidiaries were both induced and helped to develop one technology and the national firms another. Industry restructuring and procurement were used to 'nationalise' implementation of the first technology. This provided time for realisation of the second technology and later entry into international competition of the 'national champion', ALCATEL. (Ibid., part 3.)

A final problem, prevalent in smaller countries, is that smaller 'national' or 'local' firms may be unable to command the resources that would enable them to improve their market position vis-à-vis large, international firms. This is observed, for example, in the case study of Austria's procurement of a digital telecommunications switching system, OES (Chapter 10, part 5). The study reports, for instance, that despite domestic success, "There were few exports of telecom equipment produced by Siemens Austria. ... On the contrary, Siemens Germany used its superior market power to compete against Austrian products on the European market" (ibid.).

As a first solution, public agencies, as buyers, can exercise the power of monopsony to negotiate the most favourable terms for the local producers to whom transfers of technology are made by international firms. In the case of telecommunications procurement in Greece (Chapter 9), local producers acting as the contracting subsidiaries to international technology providers gained considerable benefit from procurement agreements ensuring the highest possible local value-added (ibid., parts 4 and 5). This approach was also used in Austria (Chapter 10) with favourable results in the 'home' market for domestic producers, despite their failure to conquer export markets.

A second, more 'offensive' strategy of progressive competency-building and systematic exploitation of established competencies over the longer term, is suggested by the case study of telecommunications procurement in Finland (Chapter 6). The Finnish case study draws attention to the role of Tekes as an organisation bridging the industry-industry and university-industry interface and the vertically and horizontally integrated relationships between the various telecommunications producers and Nokia predecessors (ibid., part 4). The importance to Finland's success of these phenomena, largely absent in the Greek and Austrian cases, suggests that smaller countries interested in strengthening their systems of innovation and the positions of their 'national' firms should institute policies for the establishment of co-operative linkages between public and private research organisations, and among private initiatives in R&D.

4. CONCLUSIONS

As a basis for our development of general conclusions it will be necessary, first of all, to revisit the Overview of cases that introduced Part II of this book. This is in order to consider the insights it offers on the distinction between developmental public technology procurement and adaptive public technology procurement. As indicated in the Overview, some of the difficulties involved in making this distinction are analytical – for example, the period of time over which a procurement process is studied. But there are also empirical factors determining whether or not a given procurement will be primarily 'adaptive' or 'developmental'. These include the strategic orientation (or 'vision') guiding the procurement, the extent of co-ordination with complementary policy instruments, the degree of 'fit' between formal and informal institutional arrangements, the relative maturity of the technology being procured, and the 'state of advance' of knowledge.

A comparison of cases indicates that public technology procurement should proceed with a planning perspective that takes into account not only market opportunities and technological possibilities, but also the historical and political-economic context within which procurement intervenes. Since 'developmental' procurement is often the culmination of longer-term processes of collaboration involving both 'adaptive' and 'regular' procurement, it should not be regarded as an isolated economic event. Rather, "history matters", as demonstrated by the crucial importance of problems related to timing, competence, institutions and interaction.

Our findings related to *The Dimension of Time* indicate that timing is of strategic importance to the procurer. Early and sophisticated demand for new technologies is a main determinant of success, defined in terms of the maximisation of economic benefits. Consequently, user competence, based on systematic search and learning about markets and technologies, is of key importance. At the same time, the high risks involved in the selection of new technological trajectories indicates that, in early stages of development, more than one trajectory should be supported. In the later stages of a technology's development, the identification of new market opportunities becomes particularly important to the procurer's ability to acquire the technology on advantageous terms. Benefit can be increased by making 'time' an explicit consideration in planning, and by anticipating problems that might delay implementation.

Findings related to *Competence and User-Producer Interaction* indicate that competence is essentially relational in character. Accordingly, relations between users and producers have a decisive influence in processes of public technology procurement. Effective public technology procurement requires broad organisational competence as well as more specific technical and economic competencies. The competences needed, moreover, vary greatly from case to case.

Thus, what is most fundamental is the creation of institutional frameworks and organisational forms that will enable users and producers to acquire the necessary competence. 'Focal' organisations capable of developing 'poles of competence' (i.e., organised 'pools' of knowledge and expertise) and co-ordinating the R&D efforts of users and producers are needed. However, to act effectively they also require the markets within which they operate to be organised, through frameworks such as those provided by producer associations, in a manner that supports learning based on recurrent interaction.

Institutional frameworks that facilitate the transfer of relevant knowledge work, over the longer term, to create balance and complementarity between the competencies of 'buyers' and 'sellers'. At the same time, they operate as selection mechanisms that provide stabilising influences by reducing uncertainty. This dual effect is, for example, accomplished through supplier firms' participation in standard-setting by public agencies charged with technology procurement. Like technical standards, moreover, institutional frameworks and organisational forms evolve over time. As technologies mature, the pace of development quickens and competitive forces make implementation and commercialisation imperative. Consequently, procurement relations tend to become more tightly integrated, more specifically focused, and more formalised, taking commercial interests into explicit account. The balance between formal and informal institutions changes.

Our findings related to *Political Influence and Levels of Intervention* indicate that the architecture of formal institutions, including the design of administrative structures and regulatory regimes, best promotes successful public technology procurement when it accomplishes several related purposes. We address these purposes below.

The first purpose is to provide for a central focus on 'innovation policy'. This can be accomplished through market regulation encouraging public agencies to pursue innovation on the basis of self-interest. The second purpose is to ensure that the public agencies charged with carrying out technology procurement have sufficient autonomy and responsibility to conduct technological initiatives informally during earlier stages of development. This means that higher level political decision makers are confined primarily to acting in a supportive capacity at later stages – for example, by resolving conflicts through formal, institutional means.

A third, and closely related purpose, is to ensure continuity. Accomplishing this purpose requires the development of differentiated formal institutional arrangements allowing governments to act on immediate priorities without sacrificing longer-term undertakings. A fourth and final purpose, closely connected with the second and third, is to counter-balance the 'pragmatism' of public agencies and political decision makers. Towards this end, informal institutions should be taken into account as important channels for 'bottom-up' initiatives that can 'correct' those from 'top-down'. This requires the inclusion of all relevant interests in decision making, leading to greater transparency and legitimacy of the decision process among the parties to be involved with its implementation. Such arrangements assure a higher level of co-operation in implementing the decisions reached.

Our findings related to the *Interaction Environment* concern the structure and dynamics of publicly controlled 'natural monopolies'. They indicate that monopsonistic power on the part of buyers need not act as a brake to innovation. Monopsony is often the basis of leverage by which producer firms can be induced to enter new technologies and markets. However, both monopsonistic public agencies and their suppliers, often oligopolists, require adequate incentives to make such transitions. For the former, these incentives can come from pressures to provide improved services at lower cost. For the latter, such incentives can come from competitive pressures on either or both domestic and international markets.

Monopsonistic public agencies can develop relations with producers that are conducive to innovation when they are forced to do so by the challenges posed by their lack of competence in new technologies. These conditions can lead to the formation of alliances with suppliers, and the creation of new organisational forms supporting interactive learning. Alternatively, similar results can be achieved under conditions that give public agencies a competitive interest in forming strategic alliances with key suppliers. Firms can be induced to enter into such relations through market incentives, secured through a variety of policy instruments geared to address the different types of firms and positions in producer markets. Thus, effective public technology procurement will depend critically on effective policy co-ordination.

REFERENCES

Edquist, C. (1997). Systems of Innovation approaches: Their emergence and characteristics. In C. Edquist (Ed.), *Systems of innovation: Technologies, organisations and institutions*. London: Pinter Publishers/ Casell Academic.

Edquist, C., & Johnson, B. (1997). Institutions and organisations in systems of innovation. In C. Edquist (Ed.), *Systems of innovation: Technologies, organisations and institutions* (pp. 41 - 63). London: Pinter Publishers/Casell Academic.

Hersen, M., & Barlow, D. H. (1976). *Single-case experimental designs: Strategies for studying behaviour*. New York: Pergamon.

Johnson, B. (1992). Institutional learning. In B.-Å. Lundvall (Ed.), *National systems of innovation: Towards a theory of innovation and interactive learning* (pp. 23 - 67). London: Pinter Publishers.

Luhmann, N. (1995). *Social systems* (Bednarz, J. with D. Baecker, Trans.). (English ed.). Stanford, California: Stanford University Press.

Yin, R. K. (1994). *Case study research: Design and methods*. (2nd ed.). London: Sage.

12. POLICY IMPLICATIONS

C. Edquist, L. Hommen and L. Tsipouri

1. INTRODUCTION

Our investigation and analysis of public technology procurement has demonstrated that, as argued in Chapter 1 of this book, it is a concept with salient features. These distinguish it from other types of public procurement. With this particular type of procurement, policy makers are faced with new challenges. The reasons pursuing for public technology procurement, and consequently its processes and rules, are different from those of the usual, off-the-shelf, or 'regular', public procurement. Rules and procedures for conducting the latter kind of procurement are by now well tested and internationally established. But public technology procurement, as *procurement under uncertainty*, runs the risk of failing to meet the initial milestones set for it. In particular, the chances of total failure are often much higher than the average decision maker in the public service would be willing to bear. Generally, utility managers and the public service are risk-averse. They therefore tend to disregard the likelihood of potential above-average (or even spectacular) returns, which can be realised in the form of social and economic benefits from the rapid introduction of technological change and its 'spillover' effects.

Our previous analysis (in Chapter 11) demonstrates that public technology procurement can in specific cases be an effective innovation policy tool for enhancing economic growth. Certainly, adopting a positive attitude and special procedures for public technology procurement under all circumstances and at any cost would be a mistake. But what we argue here is that it would be equally a mistake not to take advantage of this instrument. It should not be ignored simply because the behaviour of policy makers implies their lack of will to experiment with such a tool. We have, on the one hand, recognised the basic risk aversion of policy makers. On the other hand, we have identified the parameters and potential benefits of both 'ideal' and 'real' cases of public technology procurement. We now pose some key questions for policy making. These concern whether, how, and under what circumstances, one may introduce formal rules and/or incentives that would facilitate active public technology procurement policies.

2. ALTERNATIVE APPROACHES TO PUBLIC TECHNOLOGY PROCUREMENT POLICIES

Let us, for a moment, postpone making specific public policy prescriptions and, instead, consider the standpoint of policy makers. When it comes to technology procurement, specific demands or proposals reflecting genuine social needs for larger, not yet developed, systems are put forward. In this situation, policy makers seem to have three options: A) postponing procurement, B) borrowing procedures, or C) deciding under uncertainty.

- *A) Postponing procurement*: This first option is consistent with risk aversion. The policy maker *postpones the decision* for technology procurement until the private sector or other countries have undertaken the necessary research and development. Then, innovations come to the market for which the policy maker is responsible as standardised and mature, rather than innovative, products. The benefit of this approach is the absence of risk. At the same time, of course, there is an opportunity cost. However, due to the uncertainty circumstances, nobody can easily quantify this cost and thus no policy maker can readily be held accountable for it. The opportunity cost is that the introduction of technological change is delayed, allowing competitors (other countries, regions or firms) to take the lead in capturing its economic and social benefits.[1] Alternatively, the opportunity cost is that the innovation does not materialise at all.

There are, of course, no statistics on how many cases of public technology procurement have been conceived as ideas and then abandoned. Similarly, there are no figures for how many ideas could be generated if the environment were conducive. But if at world level all policy makers would act according to the 'postponement' principle, it is very likely that technological change on the world scale would be significantly slower. By this we do not mean to suggest that every public service should try to maximise its involvement in public technology procurement. On the contrary, our analysis indicates that in many cases such involvement would have a high likelihood of failure. The same logic applies in the private sector. But we argue here that involvement in public technology procurement must be a deliberate decision, rather than a non-decision based on inertia. There are many cases where the skills of the public agencies and the political environment are not favourable to public technology procurement.

- *B) Borrowing procedures*: The second option for policy makers is to treat technology procurement in the same way as their familiar practice of 'regular', off-the-shelf procurement. In that case, according to current EU legislation, processes and decisions are based on the rules of pure competition : There is an auction for a fixed anticipated price and quality. Policy makers are likely to follow this approach, which consists of borrowing rules from a fundamentally different kind of process and applying it to public technology procurement. Their propensity for doing so arises

[1] This approach is reminiscent of J. M. Keynes's comment that 'it is better to fail conventionally...'

for one or another of two reasons: Either policy makers don't recognise the specificities of public technology procurement, or, even if they do, the borrowed procedure at least protects them from sharing the responsibility for failing to meet milestones. The benefit is that the policy makers cannot be accused of favouritism, lack of transparency, or nepotism and corruption. But, again, there is an opportunity cost: Technological change will be treated as linear and predetermined, flexibility will be minimal and no 'value-added', synergetic benefits, or feedback relationships can be expected. Thus, important opportunities to enhance technological capabilities and enter the market with first mover advantages are lost. In a sense, we can argue that in this case the risk of failure is not reduced. On the contrary, it is increased, because there is no interaction to help eliminate bottlenecks. At the same time, the benefit itself is reduced, since the selection mechanism does not take innovative potential into consideration.

In the past, governments have often decided to award contracts to national companies in order to gain competitiveness benefits. Their actions have had the effect of triggering the protectionism debate without generating really beneficial forms of user-producer interaction. Instead, they have merely offered the benefit of lucrative contracts to 'national' producers. Their hope has been that this would automatically result in economic development that would ultimately benefit the country. In some cases, when the national companies selected were competent and able to learn, the experiment succeeded. In others, it failed.

Unlike the 'postponing' option, which has a sound rationale that in many cases should be followed, the 'borrowing procedures' option is, in our view, an approach to be avoided. As the protection argument is no longer valid in the EU it does not make sense to use a procedure that amplifies risks and reduces profits, unless there are serious suspicions of incapability or corruption. In the latter case the best policy is to remedy such problems.

- *C) Deciding under uncertainty*: Finally, the third option for policy makers is to make decisions based on the recognition that public technology procurement is a specific case that merits its own treatment. The right procedure in this case would be to adapt the basic principle of competition. That is, the policy maker or public agency responsible should launch an open, competitive bid, but then undertake to select the winner based on its innovative capabilities and abilities to interact, rather than on the basis of a pure price competition. The range of potential benefits in such a scenario is indicated by the accounts of successful cases of public technology procurement that are included in this volume (Part II) and analysed above (Chapter 11). However, costs encountered in this context may be considerable. They are associated with the potential inability of the public agency acting as procurer to monitor the process, and also with the discretionary freedom involved. The latter, especially, leaves open many possibilities for inefficiencies, nepotism and corruption in systems where trust and interaction are not deeply rooted.

We base the framework that follows on the above options. The framework tries to suggest how policy makers should act to facilitate the introduction of public technology procurement and to trigger a new mentality when necessary. The baseline is that policy should be decided on the basis of knowledge and decision criteria, rather than inertia or risk aversion.

3. A FRAMEWORK FOR POLICY ORIENTATION

We now turn to the policy questions raised at the outset of this book. Our point of departure for policy orientation is that there are environments – sectors, regions and countries – where the conditions for technology procurement are favourable, whereas in others they are not favourable. These can be termed 'innovative' and 'immature' environments, respectively. In innovative environments, the general conditions for innovation policy have matured and are already present, so that concrete policy instruments for improving efficiency and effectiveness can be discussed. In immature policy environments, which are usually associated with less developed economies, technology procurement is far less likely to occur in the private sector, and there is thus an especially strong justification for public technology procurement. Government has a key role to play in creating an environment more conducive to economic growth based on innovation, but it will be constrained to introduce the concepts of innovation policy incrementally, in relation to specific demonstration projects. In public technology procurement, this means proceeding to select cases with a high likelihood of success. The concepts of timing, competence and interaction remain crucial in both types of environment.

3.1 Immature Environments

Our analysis has demonstrated that in the most successful cases the informal contacts between actors were established first, often in the context of public technology procurement directed primarily towards adaptation. It is out of those contacts that development has sometimes followed. One barrier, thus, in immature environments is that the *informal contacts are considered as inhibiting the impartiality* of the public agency acting as procurer. Therefore, they are avoided. The need to safeguard competition in a strict sense, together with the drive to avoid potential accusations of favouritism, often leads public services and agencies to view with suspicion any interaction that goes beyond formalities. Needless to say, this attitude has not always prevented corruption.

For these reasons, the first act for national and regional authorities or the lobbies of public utilities that are interested in pursuing public technology procurement in such environments is to *raise the issue*. Public technology procurement is a seldom used instrument and is therefore mostly unknown to the public. Raising the issue means, on the one hand, creating a committee to study its characteristics, inform potential actors in procurement, and set the rules. On the other hand, it means promoting an awareness campaign that would make politicians and the general public less suspicious. In other words, the first task of policy makers is to try to change informal rules, slowly but coherently. This is the main practical difference distinguishing 'immature' from more 'innovative', environments. In the latter, the informal rules are more favourable. Established procedures and learning from experience dictate the policy priorities.

Once this double awareness-raising at the two levels of relevant actors and the general public is completed, cases with a high likelihood of success can be selected. To help them meet their targets it is important to strengthen the public agency that is to act as a procurer of new products or systems. Specifically, competencies should be developed in terms of:

- *timing*, i.e., providing agencies with the necessary legal and administrative framework to decide rapidly and flexibly; only then can they be held accountable for delays.
- *creating technical competence* inside and/or outside the procuring organisation. It is very clear from our analysis how important competence is as a basis for interaction. In many cases, depending not only on the overall level of development of the country but also on that of the specific organisation, state-of-the-art knowledge and learning abilities are limited. In those cases the potential benefits of the procurement will not be fully achieved unless the technical competence is created, early and ambitiously. This can be done either through internal measures, such as recruitment, or through the formalisation of external co-operation with competent advisers. It goes without saying that the technical advisers should not necessarily come from within the country or region. Rather, they would have to be selected on the basis of their ability to interact with both the procurer and the supplier, i.e., to play a bridging role.
- *developing organisational competencies*. Finally, technical competence is not the only need, since organisational skills are at least as important as technical ones for public technology procurement. A significant difference is that organisational skills are less specific than technical ones. Thus, if countries decide to adopt public technology procurement policies, organisational skills may be developed centrally and made available to the procurers as needed, on a case by case basis.

It may be argued that in immature environments the case for public technology procurement is not worth debating. Such objections could be based on the unduly pessimistic view that public technology procurement is simply not possible within an economy unused to innovation and a political system unfamiliar with innovation policy. Yet we have evidence from the case studies that if public technology procurement is well and timely managed, the benefits can be considerable in such contexts. Moreover, there is also theoretical argumentation concerning the creation or speeding-up of diffusion and spillovers that explains the rationale for introducing public technology procurement policies. Arguably, immature environments are those in which such policies can be most beneficial, provided that they can be introduced successfully. It is a matter of judging specific opportunities to decide whether, in an immature environment, the first initiatives taken under such policies should be as ambitious as targeting developmental procurement, or might better start with the less demanding objectives of adaptive procurement.

3.2 Innovative Environments

Innovative environments spare policy makers the full effort of awareness-raising that we have called for as an initial stage. But things are not always smooth in such environments either. Our case studies demonstrate overwhelmingly that even the most successful public technology procurements could not respect their original budgets and schedules. However, the returns on both private and social investment for the successful cases were so high that they made the exercise more than worthwhile. It follows that policy makers in innovative environments need to:

- strengthen the mechanisms that may help reduce the danger of going overboard with time and money
- realise the fact that under uncertainty one may decide for procurements which may be unable to respect initial plans. Allowing for this possibility does not ipso facto mean mismanagement or corruption.

In innovative environments it is easier to place innovation and learning processes at the centre of focus. It is also easier to accept the complex relations that support successful innovation as necessary forms of collaboration, rather than as dangers to competition. Laws, regulations and informal rules are more favourable in innovative environments, at least to the extent that they allow for the implementation of a technology procurement perspective. Thus, as indicated by the discussion of 'private technology procurement' in Chapter 1 (part 3), technology procurement based on close relationships and interactive learning among private firms is a common occurrence in innovative environments.

Yet, even in such environments, institutions are different and path dependent. Thus, each system will need to build on its own strengths. We provide no general recommendation on 'one best way' of how to organise competence or increase responsibility and accountability. Similarly, we identify no 'typical' nodes and links (i.e., kinds of intermediaries, numbers and size of players). However, it is an important task for the development and application of public technology procurement as a policy instrument to identify nodes and links that suit the national/regional system. There is, according to our analysis, a need for balance and complementarity between the competencies of 'buyers' and 'sellers'. It is also of particular importance for a given system to identify potential 'development blocks' and to respond to these perceived structures of opportunity by forming 'focal organisations'.[2]

Asking policy makers for such initiatives may appear to entail an element of illusion. But new situations require new policy capabilities and ultimately policy makers always work with incomplete and imperfect knowledge (Edquist et al., 1998: 17). Sometimes, however, they act deliberately. At other times, policy is made by default, i.e., by postponing or altering decisions. There is no doubt that there are also dangers associated with public technology procurement in innovative environments: Timing, from

[2] See the discussion of the concepts of 'development block' and 'focal organisation' in Chapter 1 of this book, sections 3.2 and 4.4. See also Chapter 11, section 3.2.

an historical perspective, assumes more and different dimensions. Life cycles become shorter, and positioning in the global market acquires an explicitly strategic character. And, last but not least, the creation of lock-in situations is also a potential danger for public technology procurement in innovative environments.

4. COMPETITION RULES, LEGAL ISSUES AND THE EU FRAMEWORK

4.1 The Interpretation of Competition

Our suggested policy measures depend on the targets of the policy and may thus differ very widely in application, according to the goals set by policy makers. For example, as we point out in our 'conclusions' to the preceding chapter (Chapter 11, part 4), the broader strategic orientation (or 'vision') guiding a particular initiative in public technology procurement can be one of the main factors determining whether the procurement will be primarily 'adaptive' or 'developmental in character. The assumptions on which our recommendations to policy makers are made are the following:

- policy makers are interested in long-term and not only short-term growth, so that
- innovation plays a central role in their model, and
- competition is not viewed as an end in itself, but as one way to increase efficiency.

The two first assumptions can hardly be disputed theoretically. The extent to which they are not true 'in practice' is related partly to the inefficiencies of public administration. It is also related to the fact that politicians (who are not necessarily policy makers) tend to have a short term perspective, i.e., they often want to see results before the next election. But the third assumption is much more controversial. It expresses views that are not compatible with the current orthodoxy. If price competition is the only determinant of public policy, then public technology procurement is not the correct instrument to adopt. But we argue here that if quality is taken into consideration and the introduction of innovative products becomes a priority, then *too great a stress on 'perfect competition' can undermine competitiveness* (Edquist et al., 1998: 8).

There are many areas where economics identifies price competition as being a less-than-ideal tool, even an inappropriate one. In particular, this is the case when dealing with asymmetric information and adverse selection of risk. As pointed out in the earlier discussion of 'auction theory' the assumptions that standard economic theory applies to such situations do not adequately take into account the 'knowledgeability' of buyers or the need for collaborative interaction between users and producers of new products (Chapter 1, section 1.4). Arguably, rationing, rather than auctioning, works best in such circumstances (Yotopoulos, 1996).

There is often a conflict between institutions designed by policy makers and evolved 'co-operative' institutions concerning relations between buyers and sellers. The rules designed by policy-makers promote anonymous 'market' relations and 'perfect competition'. The evolved institutions reflect close co-operation between procurers and suppliers, which is a necessary element in effective technology procurement and, more generally, in many innovation processes (Edquist et al., 1998: 7). Our analysis indicates a considerable degree of tension between the EU procurement rules and the need to accommodate informal co-operation in the form of user-producer interaction related to technical change. Our point, however, is not to suggest a reversion to protectionist policies. Rather, we warn against allowing the ghost of protectionism to undermine future-oriented policies.

4.2 The European Union's Legal Provisions

The legal system of the European Union's member states in matters of public procurement is adopted via transnational legislation.[3] The essence of the framework is that contracts should be awarded after a formal tender, rather than being negotiated.[4] There are only a few exceptions:

- Directive 93/36/EEC on the award of public supply contracts foresees in art. 6, 3b that contracting authorities may award contracts by negotiated procedure without prior publication of a tender notice when the products involved are manufactured purely for the purpose of research, experiment, study or development.[5] Through art. 8, the directive implies that standards of some kind already exist and does not foresee a procedure for the case where standards do not yet exist.[6]

---

[3] There are some areas in which the EU procurement rules do not apply. There is, for example, no application of the rules to items covered by Article 223.1(b) of the *Treaty of Rome* and maintained on the Commission's list of material "intended specifically for military purposes" — (although the rules do apply to procurement of dual purpose materials) (Business International Ltd., 1991: 19). Under Article 4 of Council Directive 93/36/EEC, which concerns international projects, other areas in which the rules do not apply include the following:

(a) in pursuance of an international agreement concluded in conformity with the Treaty, between a Member State and one or more non-member countries and covering supplies intended for the joint implementation or exploitation of a project by the signatory States; all agreements shall be communicated to the Commission, which may consult the Advisory Committee for Public Contracts set up by Decision 71/306/EEC;

(b) to undertakings in a Member State or a non-member country in pursuance of an international agreement relating to the stationing of troops;

(c) pursuant to the particular procedure of an international organisation. (Westling, 1996: App. 5, 4)

[4] See the discussion in Chapter 1 of this book, section 1.1. It describes three basic procedures recognised under the EU procurement regime: the open, restricted and negotiated procedures. Of these, the 'open' procedure, involving a formal, public tender, is the preferred procedure. The less common 'restricted' procedure, allowable only under certain circumstances, also involves formal tendering. In this case, though, tenders are only invited from a list of 'pre-qualified' bidders. Finally, there is the 'negotiated' procedure, which does not involve tendering but instead relies on closed negotiations. However, it is only allowed under exceptional circumstances.

[5] This is the 'negotiated' or closed procedure referred to in note 4. It can be used only after the procuring agency has established adequate justification — e.g., following irregular tenders, or no tenders at all, in response to an open or restricted procedure, or for reasons of extreme urgency that would render the other procedures unviable. One of the main justifications for the use of the negotiated procedure refers to reasons of technical, artistic, or proprietary 'exclusivity', because of which only certain suppliers can be relied on to deliver the goods.

- Directive 93/37/EEC concerning the co-ordination of procedures for the award of public work contracts equally states that contracting authorities may award contracts by negotiated procedure without prior publication of a tender notice when the products involved are manufactured purely for the purpose of research, experiment, study or development and not to establish commercial viability or to recover research and development costs.
- Directive 93/38/EEC co-ordinating the procurement procedures of entities operating in the water, energy, transport and telecommunications sectors, i.e., the utilities primarily of interest for public technology procurement, does not foresee any similar exception.[7]

These exceptions would not apply to technology procurement as we have defined it, but to a sort of first stage of it, in which there would be insufficient cause or incentive for industry to take the necessary risks. The exceptions thus demonstrate that the legislator has been sensitised to the issue of innovation policy, but has not had either the means or the will or political possibility to go further than that.

In our view, the extent to which the present legislation allows for the introduction of public technology procurement at European Union or member state level is a matter of debate and interpretation. We suggest that if a formal tender were introduced specifying that its content refers to innovative products that have to be jointly developed, the EU level would be particularly appropriate. Suppliers from all member states

---

[6] Development projects, however, constitute an important exception to the rules regarding standards, which like the greater part of the EU procurement regime, have been devised for application in the public purchasing of already existing goods that are, or should be, standardised. As observed in section 5.2, under Council Directive 93/36/EEC projects of a "genuinely *innovative nature*" in which the use of existing European Standards or common technical specifications "would not be appropriate" are exempt from these rules. Nevertheless, just as in cases where a departure is made from normal (open) tendering procedures, it is necessary to record the reasons for not utilising European Standards. Paragraph 4, Article 8 of the same Directive (93/36/EEC) states:

> ... Contracting authorities ... shall record wherever possible the reasons for doing so in the tender notice published in the Official Journal of the European Communities or in the contract documents and in all cases shall record these reasons in their internal documentation and shall supply such information on request to Member States and to the Commission. (Westling, 1996: App. 5, 5)

Provided these conditions are met, the rules regarding standards present no hindrance to innovation. This is also the case for procurement by public utilities, which is governed by special provisions (see note 7, below).

[7] Generally, it is accepted and anticipated in the EU procurement regime that public utilities will make wide use of the 'restricted' procedure and occasional use of the 'negotiated' procedure discussed in note 4. Public utilities operating in the water, energy, transport and telecommunications sectors have greater freedom in source selection and advertising than do public authorities. Article 20 of Council Directive 93/38/ EEC states that "Contracting entities may choose any of the procedures described ... provided that a call for competition has been made" through notice, periodic indicative notice, or notice on the existence of a prequalification system (Westling, 1996: App. 5, 8). With regard to cases of innovation and departures from existing technical standards, the Utilities Directive (93/38/EEC), Article 8, Paragraph 6 states that "contracting entities may derogate" if:

> (d) the relevant European specification is inappropriate for the particular application or does not take account of technical developments which have come about since its adoption ... (Westling, 1996: App 5, 5)

would be eligible to bid. There is no doubt that the prerequisite of joint development could be interpreted as favouring companies speaking the same language or being in geographical proximity. Yet this would not be as influential as innovative capabilities in many cases. Besides, the idea may be extended to include multinational co-operations covering combined needs. In such instances, consortia of European firms could respond, as is often the case in traditional or 'regular' public procurement. For communication and interaction across national borders, the EU is certainly the most appropriate level at which to introduce co-ordination and transnational consortia.

But, going beyond the legal aspects, present EU legislation merely allows interaction to take place – and then only in special cases. Under EU procurement legislation, collaboration between firms and public agencies has been only tolerated, not fostered. Implicitly, the legislation regards interaction as an aberration from 'normal' market relations. Possibilities for interactive learning have thereby been diminished. Yet, as we have shown, this type of relationship between buyers and sellers is crucial to the development of technological innovation. To remedy this problem, we suggest that a committee should be appointed to investigate possible revisions to existing EU procurement rules (Edquist et al., 1998: 8).

The restructuring process and the transition from old to new knowledge-based industries have experienced many difficulties in Europe. Innovative thinking in policy making is now required, if the EU, its member states, and their respective economies are to adapt better than in the past. Public technology procurement is one area in which a strong case can be made for this kind of policy innovation. To trigger the change at EU, rather than national, level would be both logical and appropriate, since innovation policy has been higher on the EU's agenda than in most of the member states.

5. CONCLUDING REMARKS: TOWARDS NEW LEGISLATION OR NEW INFORMAL RULES?

There is no doubt that what we call public technology procurement exists and that its practice will continue. More opportunities of this kind may occur in the near future, as technologies become more complex and systems more integrated. There is also little doubt that this type of procurement has attracted little attention because it is associated with uncertainties and risks. In this volume, we have tried to redress that problem, replacing neglect with critical attention.

Our main recommendation is that policy makers should take public technology procurement into serious reconsideration. This means analysing the ways in which this policy instrument could be used as a leverage for the development of new technological advantages. It also means taking account of the reasons why such options should – or should not – be exercised. This should be done on a case by case basis, depending on the features of the national innovation systems and the specificities of particular projects. Policy makers need, at least, to have an explicit policy based on good knowledge of the issues pertaining to public technology procurement – even though, in many cases, they may end up by not using it. In some respects, it may be necessary to adapt or change the formal legal framework, as suggested above. But the most important consideration is to try to introduce new informal rules. Even in countries where legislation still permits protectionism, there is still widespread suspicion towards interaction and the formation of 'development pairs' between public agencies and private firms. This indicates that a very basic misunderstanding of public technology procurement, expressed as the fear of potential abuse of 'privilege', remains prevalent in public opinion and in the political arena.

REFERENCES

Business International Ltd. (1991). *Public procurement in Europe: How EC rules will open up new cross-border markets*. London: Business International Ltd.

Edquist, C., Hommen, L., Johnson, B., Lemola, T., Malerba, F., Reiss, T., & Smith, K. (1998). *The ISE policy statement: The innovation policy implications of the 'Innovation Systems and European Integration' research project*. Linköping, Sweden: Unitryck (University of Linköping Press).

Westling, H. (1996). *Co-operative procurement: Market acceptance for innovative energy-efficient technologies* (1996 - 3). Stockholm: NUTEK B.

Yotopoulos, P. (1996). *Exchange rate parity for trade and development*. Cambridge: Cambridge University press.

SUBJECT INDEX

A

B

C

D

Q

R

S

Economics of Science, Technology and Innovation

1. A. Phillips, A.P. Phillips and T.R. Phillips:
   Biz Jets. Technology and Market Structure in the Corporate Jet Aircraft Industry. 1994 — ISBN 0-7923-2660-1
2. M.P. Feldman:
   The Geography of Innovation. 1994 — ISBN 0-7923-2698-9
3. C. Antonelli:
   The Economics of Localized Technological Change and Industrial Dynamics. 1995 — ISBN 0-7923-2910-4
4. G. Becher and S. Kuhlmann (eds.):
   Evaluation of Technology Policy Programmes in Germany. 1995 — ISBN 0-7923-3115-X
5. B. Carlsson (ed.): *Technological Systems and Economic Performance: The Case of Factory Automation.* 1995 — ISBN 0-7923-3512-0
6. G.E. Flueckiger: *Control, Information, and Technological Change.* 1995 — ISBN 0-7923-3667-4
7. M. Teubal, D. Foray, M. Justman and E. Zuscovitch (eds.):
   Technological Infrastructure Policy. An International Perspective. 1996 — ISBN 0-7923-3835-9
8. G. Eliasson:
   Firm Objectives, Controls and Organization. The Use of Information and the Transfer of Knowledge within the Firm. 1996 — ISBN 0-7923-3870-7
9. X. Vence-Deza and J.S. Metcalfe (eds.):
   Wealth from Diversity. Innovation, Structural Change and Finance for Regional Development in Europe. 1996 — ISBN 0-7923-4115-5
10. B. Carlsson (ed.):
    Technological Systems and Industrial Dynamics. 1997 — ISBN 0-7923-9940-4
11. N.S. Vonortas:
    Cooperation in Research and Development. 1997 — ISBN 0-7923-8042-8
12. P. Braunerhjelm and K. Ekholm (eds.):
    The Geography of Multinational Firms. 1998 — ISBN 0-7923-8133-5
13. A. Varga:
    University Research and Regional Innovation: A Spatial Econometric Analysis of Academic Technology Transfers. 1998 — ISBN 0-7923-8248-X
14. J. de la Mothe and G. Paquet (eds.):
    Local and Regional Systems of Innovation — ISBN 0-7923-8287-0
15. D. Gerbarg (ed.):
    The Economics, Technology and Content of Digital TV — ISBN 0-7923-8325-7
16. C. Edquist, L. Hommen and L. Tsipouri
    Public Technology Procurement and Innovation — ISBN 0-7923-8685-X

KLUWER ACADEMIC PUBLISHERS — BOSTON / DORDRECHT / LONDON

GPSR Compliance

The European Union's (EU) General Product Safety Regulation (GPSR) is a set of rules that requires consumer products to be safe and our obligations to ensure this.

If you have any concerns about our products, you can contact us on ProductSafety@springernature.com

In case Publisher is established outside the EU, the EU authorized representative is:

Springer Nature Customer Service Center GmbH
Europaplatz 3
69115 Heidelberg, Germany

Batch number: 10739606

Printed by Printforce, the Netherlands